OXFORD MASTER SERIES IN CONDENSED MATTER PHYSICS

The Oxford Master Series in Condensed Matter Physics is designed for final year undergraduate and beginning graduate students in physics and related disciplines. It has been driven by a perceived gap in the literature today. While basic undergraduate condensed matter physics texts often show little or no connection with the huge explosion of research in condensed matter physics over the last two decades, more advanced and specialized texts tend to be rather daunting for students. In this series, all topics and their consequences are treated at a simple level, while pointers to recent developments are provided at various stages. The emphasis in on clear physical principles of symmetry, quantum mechanics, and electromagnetism which underlie the whole field. At the same time, the subjects are related to real measurements and to the experimental techniques and devices currently used by physicists in academe and industry.

Books in this series are written as course books, and include ample tutorial material, examples, illustrations, revision points, and problem sets. They can likewise be used as preparation for students starting a doctorate in condensed matter physics and related fields (e.g. in the fields of semiconductor devices, opto-electronic devices, or magnetic materials), or for recent graduates starting research in one of these fields in industry.

M. T. Dove: *Structure and dynamics*
J. Singleton: *Band theory and electronic properties of solids*
A. M. Fox: *Optical properties of solids*
S. J. Blundell: *Magnetism in condensed matter*
J. F. Annett: *Superconductivity*
R. A. L. Jones: *Soft condensed matter*

# Magnetism in Condensed Matter

STEPHEN BLUNDELL

*Department of Physics*
*University of Oxford*

# OXFORD

UNIVERSITY PRESS

Great Clarendon Street, Oxford, OX2 6DP,
United Kingdom

Oxford University Press is a department of the University of Oxford.
It furthers the University's objective of excellence in research, scholarship,
and education by publishing worldwide. Oxford is a registered trade mark of
Oxford University Press in the UK and in certain other countries

First Edition published in 2001
Reprinted 2003, 2004, 2006 (twice), 2007, 2008, 2009 (twice),
2010, 2011, 2012 (twice), 2014

Published in the United States of America by Oxford University Press
198 Madison Avenue, New York, NY 10016, United States of America

British Library Cataloguing in Publication Data
Data available

Library of Congress Cataloging in Publication Data
Data available

ISBN 978-0-19-850591-4

Printed and bound by CPI Group (UK) Ltd, Croydon, CR0 4YY

# Preface

'... in Him all things hold together.'
*(Colossians 1[17])*

Magnetism is a subject which has been studied for nearly three thousand years. Lodestone, an iron ore, first attracted the attention of Greek scholars and philosophers, and the navigational magnetic compass was the first technological product resulting from this study. Although the compass was certainly known in Western Europe by the twelfth century AD, it was not until around 1600 that anything resembling a modern account of the working of the compass was proposed. Progress in the last two centuries has been more rapid and two major results have emerged which connect magnetism with other physical phenomena. First, magnetism and electricity are inextricably linked and are the two components that make up light, which is called an electromagnetic wave. Second, this link originates from the theory of relativity, and therefore magnetism can be described as a purely relativistic effect, due to the relative motion of an observer and charges moving in a wire, or in the atoms of iron. However it is the magnetism in condensed matter systems including ferromagnets, spin glasses and low-dimensional systems, which is still of great interest today. Macroscopic systems exhibit magnetic properties which are fundamentally different from those of atoms and molecules, despite the fact that they are composed of the same basic constituents. This arises because magnetism is a collective phenomenon, involving the mutual cooperation of enormous numbers of particles, and is in this sense similar to superconductivity, superfluidity and even to the phenomenon of the solid state itself. The interest in answering fundamental questions runs in parallel with the technological drive to find new materials for use as permanent magnets, sensors, or in recording applications.

This book has grown out of a course of lectures given to third and fourth year undergraduates at Oxford University who have chosen a condensed matter physics option. There was an obvious need for a text which treated the fundamentals but also provided background material and additional topics which could not be covered in the lectures. The aim was to produce a book which presented the subject as a coherent whole, provided useful and interesting source material, and might be fun to read. The book also forms part of the Oxford Master Series in Condensed Matter Physics; the other volumes of the series cover electronic properties, optical properties, superconductivity, structure and soft condensed matter.

The prerequisites for this book are a knowledge of basic quantum mechanics and electromagnetism and a familiarity with some results from atomic physics. These are summarized in appendices for easy access for the reader and to present a standardized notation.

Structure of the book:

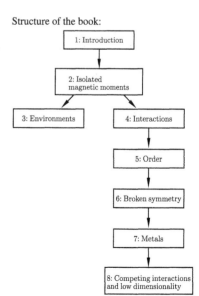

Some possible course structures:

(1) Short course (assuming Chapter 1 is known):

  - Chapter 2 (omit 2.6–2.8)
  - Chapter 3 (omit 3.2)
  - Chapter 4 (omit 4.2.5, 4.2.6)
  - Chapter 5 (omit 5.4–5.7)
  - Chapter 6 (omit 6.4–6.5)
  - Chapter 7 (omit 7.1–7.2)

(2) Longer course:

  - Chapters 1–4
  - Chapter 5 (5.6 and 5.7 as background reading)
  - Chapter 6
  - Chapter 7 (7.5–7.9 as background reading)
  - Chapter 8, selected topics

The interesting magnetic effects found in condensed matter systems have two crucial ingredients: first, that atoms should possess *magnetic moments* and second, that these moments should somehow *interact*. These two subjects are discussed in Chapters 2 and 4 respectively. Chapter 2 answers the question 'why do atoms have magnetic moments?' and shows how they behave and can be studied if they do not interact. Chapter 3 describes how these magnetic moments can be affected by their local environment inside a crystal and the techniques which can be used to study this. Chapter 4 then answers the question 'how do the magnetic moments on different atoms interact with each other?' With these ingredients in place, *magnetic order* can occur, and this is the subject of Chapters 5 and 6. Chapter 5 contains a description of the different types of magnetic order which can be found in the solid state. Chapter 6 considers order again, but starts from basic ideas of broken symmetry and describes phase transitions, excitations and domains. A strong emphasis is the link between magnetic order and other types of broken-symmetry ground states like superconductivity. Chapter 7 is devoted to the magnetic properties of metals, in which magnetism can often be associated with delocalized conduction electrons. Chapter 8 describes some of the subtle and complex effects which can occur when competing magnetic interactions are present and/or the system has a reduced dimensionality. These topics are the subject of intense research activity and there are many outstanding questions which remain to be resolved. Throughout the text, I discuss properties and applications to demonstrate the implications of all these ideas for real materials, including ferrites, permanent magnets and also the physics behind various magneto-optical and magnetoresistance effects which have become of enormous technological importance in recent years. This is a book for physicists and therefore the emphasis is on the clear physical principles of quantum mechanics, symmetry, and electromagnetism which underlie the whole field. However this is not just a 'theory book' but attempts to relate the subject to real measurements and experimental techniques which are currently used by experimental physicists and to bridge the gulf between the principles of elementary undergraduate physics and the topics of current research interest.

Chapters 1–7 conclude with some further reading and problems. The problems are of varying degrees of difficulty but serve to amplify issues addressed in the text. Chapter 8 contains no problems (the subjects described in this chapter are all topics of current research) but has extensive further reading.

It is a great pleasure to thank those who have helped during the course of writing this book. I am grateful for the support of Sönke Adlung and his team at Oxford University Press, and also to the other authors of this Masters series. Mansfield College, Oxford and the Oxford University Department of Physics have provided a stimulating environment in which to work. I wish to record my gratitude to my students who have sometimes made me think very hard about things I thought I understood. In preparing various aspects of this book, I have benefitted greatly from discussions with Hideo Aoki, Arzhang Ardavan, Deepto Chakrabarty, Amalia Coldea, Radu Coldea, Roger Cowley, Steve Cox, Gillian Gehring, Matthias Gester, John Gregg, Martin Greven, Mohamedally Kurmoo, Steve Lee, Wilson Poon, Francis Pratt, John Singleton and Candadi Sukumar. I owe a special debt of thanks to the friends and colleagues who have read the manuscript in various drafts and whose exacting criticisms and insightful questions have immensely improved the final result: Katherine Blundell, Richard Blundell, Andrew Boothroyd, Geoffrey Brooker, Bill Hayes, Brendon Lovett, Lesley Parry-Jones and Peter Riedi. Any errors in this book which I discover after going to press will be posted on the web-site for this book which may be found at:

```
http://users.ox.ac.uk/~sjb/magnetism/
```

Most of all, I want to thank Katherine, dear wife and soulmate, who more than anyone has provided inspiration, counsel, friendship and love. This work is dedicated to her.

*Oxford*                                                        S.J.B.
May 2001

Note added at reprinting:
Thanks are due to the following who pointed out errors in the first printing of this book: Michael Brooks, Jonathan Coe, Ted Davis, Jonathan Fitt, Lucy Helme, Tom Lancaster, Gavin Morley, Oscar Moze, Shoichi Nagata, Toby Perring, Christopher Steer and David Thouless.

Oxford                                                         S.J.B.
January 2003

# Contents

# Introduction

This book is about the manifestation of magnetism in condensed matter. Solids contain magnetic moments which can act together in a cooperative way and lead to behaviour that is quite different from what would be observed if all the magnetic moments were isolated from one another. This, coupled with the diversity of types of magnetic interactions that can be found, leads to a surprisingly rich variety of magnetic properties in real systems. The plan of this book is to build up this picture rather slowly, piece by piece. In this introductory chapter we shall recall some facts about magnetic moments from elementary classical and quantum physics. Then, in the following chapter, we will discuss how magnetic moments behave when large numbers of them are placed in a solid but are isolated from each other and from their surroundings. Chapter 3 considers the effect of their immediate environment, and following this in Chapter 4, the set of possible magnetic interactions *between* magnetic moments is discussed. In Chapter 5 we will be in a position to discuss the occurrence of long range order, and in Chapter 6 how that is connected with the concept of broken symmetry. The final chapters follow through the implications of this concept in a variety of different situations. SI units are used throughout the book (a description of cgs units and a conversion table may be found in Appendix A).

## 1.1 Magnetic moments

The fundamental object in magnetism is the **magnetic moment**. In classical electromagnetism we can equate this with a current loop. If there is a current $I$ around an elementary (i.e. vanishingly small) oriented loop of area $|\mathbf{dS}|$ (see Fig. 1.1(a)) then the magnetic moment $\mathbf{d}\mu$ is given by

$$\mathbf{d}\mu = I\mathbf{dS}, \qquad (1.1)$$

and the magnetic moment has the units of A m$^2$. The length of the vector $\mathbf{dS}$ is equal to the area of the loop. The direction of the vector is normal to the loop and in a sense determined by the direction of the current around the elementary loop.

This object is also equivalent to a **magnetic dipole,** so called because it behaves analogously to an electric dipole (two electric charges, one positive and one negative, separated by a small distance). It is therefore possible to imagine a magnetic dipole as an object which consists of two magnetic monopoles of opposite magnetic charge separated by a small distance in the same direction as the vector $\mathbf{dS}$ (see Appendix B for background information concerning electromagnetism).

(a)                                          (b)

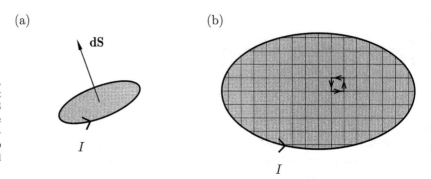

**Fig. 1.1** (a) An elementary magnetic moment, $\mathbf{d\mu} = I\mathbf{dS}$, due to an elementary current loop. (b) A magnetic moment $\boldsymbol{\mu} = I \int \mathbf{dS}$ (now viewed from above the plane of the current loop) associated with a loop of current $I$ can be considered by summing up the magnetic moments of lots of infinitesimal current loops.

The magnetic moment $\mathbf{d\mu}$ points normal to the plane of the loop of current and therefore can be either parallel or antiparallel to the angular momentum vector associated with the charge which is going around the loop. For a loop of finite size, we can calculate the magnetic moment $\boldsymbol{\mu}$ by summing up the magnetic moments of lots of equal infinitesimal current loops distributed throughout the area of the loop (see Fig. 1.1(b)). All the currents from neighbouring infinitesimal loops cancel, leaving only a current running round the perimeter of the loop. Hence,

$$\boldsymbol{\mu} = \int \mathrm{d}\boldsymbol{\mu} = I \int \mathbf{dS}. \tag{1.2}$$

Albert Einstein (1879–1955)

Wander Johannes de Haas (1878–1960)

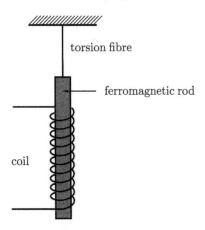

torsion fibre

ferromagnetic rod

coil

**Fig. 1.2** The Einstein–de Haas effect. A ferromagnetic rod is suspended from a thin fibre. A coil is used to provide a magnetic field which magnetizes the ferromagnet and produces a rotation. The experiment can be done resonantly, by periodically reversing the current in the coil, and hence the magnetization in the ferromagnet, and observing the angular response as a function of frequency.

Samuel Jackson Barnett (1873–1956)

### 1.1.1   Magnetic moments and angular momentum

A current loop occurs because of the motion of one or more electrical charges. All the charges which we will be considering are associated with particles that have mass. Therefore there is also orbital motion of mass as well as charge in all the current loops in this book and hence a magnetic moment is always connected with angular momentum.

In atoms the magnetic moment $\boldsymbol{\mu}$ associated with an orbiting electron lies along the same direction as the angular momentum $\mathbf{L}$ of that electron and is proportional to it. Thus we write

$$\boldsymbol{\mu} = \gamma\mathbf{L}, \tag{1.3}$$

where $\gamma$ is a constant known as the **gyromagnetic ratio**. This relation between the magnetic moment and the angular momentum is demonstrated by the **Einstein–de Haas effect**, discovered in 1915, in which a ferromagnetic rod is suspended vertically, along its axis, by a thin fibre (see Fig. 1.2). It is initially at rest and unmagnetized, and is subsequently magnetized along its length by the application of a vertical magnetic field. This vertical magnetization is due to the alignment of the atomic magnetic moments and corresponds to a net angular momentum. To conserve total angular momentum, the rod begins turning about its axis in the opposite sense. If the angular momentum of the rod is measured, the angular momentum associated with the atomic magnetic moments, and hence the gyromagnetic ratio, can be deduced. The Einstein–de Haas effect is a rotation induced by magnetization, but there is also the reverse effect, known as the **Barnett effect** in which magnetization is induced by rotation. Both phenomena demonstrate that magnetic moments are associated with angular momentum.

## 1.1.2  Precession

We now consider a magnetic moment $\mu$ in a magnetic field $\mathbf{B}$ as shown in Fig. 1.3. The energy $E$ of the magnetic moment is given by

$$E = -\mu \cdot \mathbf{B}, \tag{1.4}$$

(see Appendix B) so that the energy is minimized when the magnetic moment lies along the magnetic field. There will be a torque $\mathbf{G}$ on the magnetic moment given by

$$\mathbf{G} = \mu \times \mathbf{B}, \tag{1.5}$$

(see Appendix B) which, if the magnetic moment were not associated with any angular momentum, would tend to turn the magnetic moment towards the magnetic field.[1]

However, since the magnetic moment *is* associated with the angular momentum $\mathbf{L}$ by eqn 1.3, and because torque is equal to rate of change of angular momentum, eqn 1.5 can be rewritten as

$$\frac{d\mu}{dt} = \gamma \mu \times \mathbf{B}. \tag{1.6}$$

This means that the change in $\mu$ is perpendicular to both $\mu$ and to $\mathbf{B}$. Rather than turning $\mu$ towards $\mathbf{B}$, the magnetic field causes the direction of $\mu$ to precess around $\mathbf{B}$. Equation 1.6 also implies that $|\mu|$ is time-independent. Note that this situation is exactly analogous to the spinning of a gyroscope or a spinning top.[2]

In the following example, eqn 1.6 will be solved in detail for a particular case.

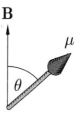

**Fig. 1.3** A magnetic moment $\mu$ in a magnetic field $\mathbf{B}$ has an energy equal to $-\mu \cdot \mathbf{B} = -\mu B \cos\theta$.

[1]For an electric dipole $\mathbf{p}$, in an electric field $\boldsymbol{\mathcal{E}}$, the energy is $E = -\mathbf{p} \cdot \boldsymbol{\mathcal{E}}$ and the torque is $\mathbf{G} = \mathbf{p} \times \boldsymbol{\mathcal{E}}$. A stationary electric dipole moment is just two separated stationary electric charges; it is not associated with any angular momentum, so if $\boldsymbol{\mathcal{E}}$ is not aligned with $\mathbf{p}$, the torque $\mathbf{G}$ will tend to turn $\mathbf{p}$ towards $\boldsymbol{\mathcal{E}}$. A stationary magnetic moment *is* associated with angular momentum and so behaves differently.

[2]Imagine a top spinning with its axis inclined to the vertical. The weight of the top, acting downwards, exerts a (horizontal) torque on the top. If it were not spinning it would just fall over. But because it is spinning, it has angular momentum parallel to its spinning axis, and the torque causes the axis of the spinning top to move parallel to the torque, in a horizontal plane. The spinning top precesses.

---

**Example 1.1**

Consider the case in which $\mathbf{B}$ is along the $z$ direction and $\mu$ is initially at an angle of $\theta$ to $\mathbf{B}$ and in the $xz$ plane (see Fig. 1.4). Then

$$\dot{\mu}_x = \gamma B \mu_y \tag{1.7}$$
$$\dot{\mu}_y = -\gamma B \mu_x \tag{1.8}$$
$$\dot{\mu}_z = 0, \tag{1.9}$$

so that $\mu_z$ is constant with time and $\mu_x$ and $\mu_y$ both oscillate. Solving these differential equations leads to

$$\mu_x(t) = |\mu| \sin\theta \cos(\omega_L t) \tag{1.10}$$
$$\mu_y(t) = |\mu| \sin\theta \sin(\omega_L t) \tag{1.11}$$
$$\mu_z(t) = |\mu| \cos\theta, \tag{1.12}$$

where

$$\omega_L = \gamma B \tag{1.13}$$

is called the **Larmor precession frequency.**

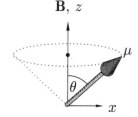

**Fig. 1.4** A magnetic moment $\mu$ in a magnetic field $\mathbf{B}$ precesses around the magnetic field at the Larmor precession frequency, $\gamma B$, where $\gamma$ is the gyromagnetic ratio. The magnetic field $\mathbf{B}$ lies along the $z$-axis and the magnetic moment is initially in the $xz$-plane at an angle $\theta$ to $\mathbf{B}$. The magnetic moment precesses around a cone of semi-angle $\theta$.

Joseph Larmor (1857–1942)

Note that the gyromagnetic ratio $\gamma$ is the constant of proportionality which connects both the angular momentum with the magnetic moment (through eqn 1.3) and the precession frequency with the magnetic field (eqn 1.13). The phenomenon of precession hints at the subtlety of what lies ahead: magnetic fields don't only cause moments to line up, but can induce a variety of dynamical effects.

### 1.1.3   The Bohr magneton

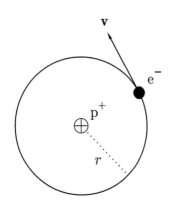

**Fig. 1.5** An electron in a hydrogen atom orbiting with velocity **v** around the nucleus which consists of a single proton.

Before proceeding further, it is worth performing a quick calculation to estimate the size of atomic magnetic moments and thus deduce the size of the gyromagnetic ratio. Consider an electron (charge $-e$, mass $m_e$) performing a circular orbit around the nucleus of a hydrogen atom, as shown in Fig. 1.5. The current $I$ around the atom is $I = -e/\tau$ where $\tau = 2\pi r/v$ is the orbital period, $v = |\mathbf{v}|$ is the speed and $r$ is the radius of the circular orbit. The magnitude of the angular momentum of the electron, $m_e v r$, must equal $\hbar$ in the ground state so that the magnetic moment of the electron is

$$\mu = \pi r^2 I = -\frac{e\hbar}{2m_e} \equiv -\mu_B \tag{1.14}$$

where $\mu_B$ is the Bohr magneton, defined by

$$\mu_B = \frac{e\hbar}{2m_e}. \tag{1.15}$$

Niels Bohr (1885–1962)

This is a convenient unit for describing the size of atomic magnetic moments and takes the value $9.274 \times 10^{-24}$ Am$^2$. Note that sign of the magnetic moment in eqn 1.14 is negative. Because of the negative charge of the electron, its magnetic moment is antiparallel to its angular momentum. The gyromagnetic ratio for the electron is $\gamma = -e/2m_e$. The Larmor frequency is then $\omega_L = |\gamma|B = eB/2m_e$.

### 1.1.4   Magnetization and field

A magnetic solid consists of a large number of atoms with magnetic moments. The magnetization **M** is defined as the magnetic moment per unit volume. Usually this vector quantity is considered in the 'continuum approximation', i.e. on a lengthscale large enough so that one does not see the graininess due to the individual atomic magnetic moments. Hence **M** can be considered to be a smooth vector field, continuous everywhere except at the edges of the magnetic solid.

In free space (vacuum) there is no magnetization. The magnetic field can be described by the vector fields **B** and **H** which are linearly related by

$$\mathbf{B} = \mu_0 \mathbf{H}, \tag{1.16}$$

where $\mu_0 = 4\pi \times 10^{-7}\,\mathrm{Hm^{-1}}$ is the **permeability of free space**. The two magnetic fields **B** and **H** are just scaled versions of each other, the former measured in Tesla (abbreviated to T) and the latter measured in $\mathrm{A\,m^{-1}}$.

In a magnetic solid the relation between **B** and **H** is more complicated and the two vector fields may be very different in magnitude and direction. The general vector relationship is

$$\mathbf{B} = \mu_0(\mathbf{H} + \mathbf{M}). \qquad (1.17)$$

In the special case that the magnetization **M** is linearly related to the magnetic field **H**, the solid is called a **linear material**, and we write

$$\mathbf{M} = \chi\mathbf{H}, \qquad (1.18)$$

where $\chi$ is a dimensionless quantity called the **magnetic susceptibility**. In this special case there is still a linear relationship between **B** and **H**, namely

$$\mathbf{B} = \mu_0(1 + \chi)\mathbf{H} = \mu_0\mu_r\mathbf{H}, \qquad (1.19)$$

where $\mu_r = 1 + \chi$ is the **relative permeability** of the material.

A cautionary tale now follows. This arises because we have to be very careful in defining fields in magnetizable media. Consider a region of free space with an applied magnetic field given by fields $\mathbf{B_a}$ and $\mathbf{H_a}$, connected by $\mathbf{B_a} = \mu_0\mathbf{H_a}$. So far, everything is simple. Now insert a magnetic solid into that region of free space. The internal fields inside the solid, given by $\mathbf{B_i}$ and $\mathbf{H_i}$ can be very different from $\mathbf{B_a}$ and $\mathbf{H_a}$ respectively. This difference is because of the magnetic field produced by all magnetic moments in the solid. In fact $\mathbf{B_i}$ and $\mathbf{H_i}$ can both depend on the position inside the magnetic solid at which you measure them.[3] This is true except in the special case of an ellipsoidal shaped sample (see Fig. 1.6). If the magnetic field is applied along one of the principal axes of the ellipsoid, then throughout the sample

$$\mathbf{H_i} = \mathbf{H_a} - N\mathbf{M}, \qquad (1.20)$$

where $N$ is the appropriate demagnetizing factor (see Appendix D). The 'correction term' $\mathbf{H_d} = -N\mathbf{M}$, which you need to add to $\mathbf{H_a}$ to get $\mathbf{H_i}$, is called the **demagnetizing field**. Similarly

$$\mathbf{B_i} = \mu_0(\mathbf{H_i} + \mathbf{M}) = \mathbf{B_a} + \mu_0(1 - N)\mathbf{M}. \qquad (1.21)$$

For historical reasons, standard convention dictates that **B** is called the **magnetic induction** or **magnetic flux density** and **H** is called the **magnetic field strength**. However, such terms are cumbersome and can be misleading. Following common usage, we refer to both simply as the **magnetic field**. The letters 'B' and 'H' will show which one is meant.

[3] A magnetized sample will also influence the magnetic field outside it, as well as inside it (considered here), as you may know from playing with a bar magnet and iron filings.

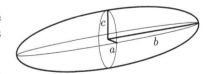

**Fig. 1.6** An ellipsoidal shaped sample of a magnetized solid with principal axes $a$, $b$ and $c$. This includes the special cases of a sphere ($a = b = c$) and a flat plate ($a, b \to \infty$, $c = 0$).

---

**Example 1.2**

For a spherically shaped sample, $N = \frac{1}{3}$ and so the internal fields inside the sphere are

$$\mathbf{H_i} = \mathbf{H_a} - \frac{\mathbf{M}}{3}, \qquad (1.22)$$

$$\mathbf{B_i} = \mathbf{B_a} + \frac{2\mu_0\mathbf{M}}{3}. \qquad (1.23)$$

When the magnetization is large compared to the applied field $|\mathbf{H_a}| = |\mathbf{B_a}|/\mu_0$ (measured before the sample was inserted) these demagnetizing corrections need to be taken seriously. However, it is possible to sweep these complications under the carpet for the special case of weak magnetism. For a linear material with $\chi \ll 1$, we have that $M \ll H$, $H_i \approx H_a$ and $B_i \approx \mu_0 H_i$. We can then get away with imagining that the magnetic field in the material is the same as the magnetic field that we apply. This approximation will be used in Chapters 2 and 3 concerning the relatively weak effects of diamagnetism.[4] In ferromagnets, demagnetizing effects are always significant.

[4]In accurate experimental work on even these materials, demagnetizing fields must still be considered.

---

**Example 1.3**

The intrinsic magnetic susceptibility of a material is

$$\chi_{\text{intrinsic}} = \frac{M}{H_i}. \tag{1.24}$$

This intrinsic material property is not what you measure experimentally. This is because you measure the magnetization $M$ in response to an applied field $H_a$. You therefore measure

$$\chi_{\text{experimental}} = \frac{M}{H_a}. \tag{1.25}$$

The two quantities can be related by

$$\chi_{\text{experimental}} = \frac{M}{H_i + NM} = \frac{M/H_i}{1 + NM/H_i} = \frac{\chi_{\text{intrinsic}}}{1 + N\chi_{\text{intrinsic}}}. \tag{1.26}$$

When $\chi_{\text{intrinsic}} \ll 1$, the distinction between $\chi_{\text{intrinsic}}$ and $\chi_{\text{experimental}}$ is academic. When $\chi_{\text{intrinsic}}$ is closer or above 1, the distinction can be very important. For example, in a ferromagnet approaching the Curie temperature from above (see Chapter 4), $\chi_{\text{intrinsic}} \to \infty$, but $\chi_{\text{experimental}} \to 1/N$.

---

[5]See Section 6.7 for more on magnetic domains.

One last word of warning at this stage: a ferromagnetic material may have no net magnetic moment because it consists of magnetic domains.[5] In each domain there is a uniform magnetization, but the magnetization of each domain points in a different direction from its neighbours. Therefore a sample may appear not to be magnetized, even though on a small enough scale, all the magnetic moments are locally aligned.

In the rest of this chapter we will consider some further aspects of magnetic moments that relate to classical mechanics (in Section 1.2) and quantum mechanics (in Section 1.3).

## 1.2   Classical mechanics and magnetic moments

In this section, we describe the effect of an applied magnetic field on a system of charges using purely classical arguments. First, we consider the effect on a single charge and then use this result to evaluate the magnetization of a system of charges. A summary of some important results in electromagnetism may be found in Appendix B.

## 1.2.1   Canonical momentum

In classical mechanics the force $\mathbf{F}$ on a particle with charge $q$ moving with velocity $\mathbf{v}$ in an electric field $\boldsymbol{\mathcal{E}}$ and magnetic field $\mathbf{B}$ is

$$\mathbf{F} = q(\boldsymbol{\mathcal{E}} + \mathbf{v} \times \mathbf{B}) \tag{1.27}$$

and is called the **Lorentz force**. With this familiar equation, one can show how the momentum of a charged particle in a magnetic field is modified. Using $\mathbf{F} = m\,d\mathbf{v}/dt$, $\mathbf{B} = \nabla \times \mathbf{A}$ and $\boldsymbol{\mathcal{E}} = -\nabla V - \partial\mathbf{A}/\partial t$, where $V$ is the electric potential, $\mathbf{A}$ is the magnetic vector potential and $m$ is the mass of the particle, eqn 1.27 may be rewritten as

$$m\frac{d\mathbf{v}}{dt} = -q\nabla V - q\frac{\partial\mathbf{A}}{\partial t} + q\mathbf{v} \times (\nabla \times \mathbf{A}). \tag{1.28}$$

The vector identity

$$\mathbf{v} \times (\nabla \times \mathbf{A}) = \nabla(\mathbf{v} \cdot \mathbf{A}) - (\mathbf{v} \cdot \nabla)\mathbf{A} \tag{1.29}$$

can be used to simplify eqn 1.28 leading to

$$m\frac{d\mathbf{v}}{dt} + q\left(\frac{\partial\mathbf{A}}{\partial t} + (\mathbf{v} \cdot \nabla)\mathbf{A}\right) = -q\nabla(V - \mathbf{v} \cdot \mathbf{A}). \tag{1.30}$$

Note that $m\,d\mathbf{v}/dt$ is the force on a charged particle measured in a coordinate system that moves with the particle. The partial derivative $\partial\mathbf{A}/\partial t$ measures the rate of change of $\mathbf{A}$ at a fixed point in space. We can rewrite eqn 1.30 as

$$\frac{d}{dt}(m\mathbf{v} + q\mathbf{A}) = -q\nabla(V - \mathbf{v} \cdot \mathbf{A}) \tag{1.31}$$

where $d\mathbf{A}/dt$ is the **convective derivative** of $\mathbf{A}$, written as

$$\frac{d\mathbf{A}}{dt} = \frac{\partial\mathbf{A}}{\partial t} + (\mathbf{v} \cdot \nabla)\mathbf{A}, \tag{1.32}$$

which measures the rate of change of $\mathbf{A}$ at the location of the moving particle. Equation 1.31 takes the form of Newton's second law (i.e. it reads 'the rate of change of a quantity that looks like momentum is equal to the gradient of a quantity that looks like potential energy') and therefore motivates the definition of the **canonical momentum**

$$\mathbf{p} = m\mathbf{v} + q\mathbf{A} \tag{1.33}$$

and an effective potential energy experienced by the charged particle, $q(V - \mathbf{v} \cdot \mathbf{A})$, which is velocity-dependent. The canonical momentum reverts to the familiar momentum $m\mathbf{v}$ in the case of no magnetic field, $\mathbf{A} = 0$. The kinetic energy remains equal to $\frac{1}{2}mv^2$ and this can therefore be written in terms of the canonical momentum as $(\mathbf{p} - q\mathbf{A})^2/2m$. This result will be used below, and also later in the book where the quantum mechanical operator associated with kinetic energy in a magnetic field is written $(-i\hbar\nabla - q\mathbf{A})^2/2m$.

Hendrik Lorentz (1853–1928)

See Appendix G for a list of vector identities. Note also that $\mathbf{v}$ does not vary with position.

Niels Bohr (1885–1962)

Hendreka J. van Leeuwen (1887–1974)

Ludwig Boltzmann (1844–1906)

## 1.2.2  The Bohr–van Leeuwen theorem

The next step is to calculate the net magnetic moment of a system of electrons in a solid. Thus we want to find the **magnetization**, the magnetic moment per unit volume, that is induced by the magnetic field. From eqn 1.4, the magnetization is proportional to the rate of change of energy of the system with applied magnetic field.[6] Now, eqn 1.27 shows that the effect of a magnetic field is always to produce forces on charged particles which are perpendicular to their velocities. Thus no work is done and therefore the energy of a system cannot depend on the applied magnetic field. If the energy of the system does not depend on the applied magnetic field, then there can be no magnetization.

This idea is enshrined in the **Bohr–van Leeuwen theorem** which states that in a classical system there is no thermal equilibrium magnetization. We can prove this in outline as follows: in classical statistical mechanics the partition function $Z$ for $N$ particles, each with charge $q$, is proportional to

$$\int \int \cdots \int \exp(-\beta E(\{\mathbf{r}_i, \mathbf{p}_i\}))\, \mathrm{d}\mathbf{r}_1 \cdots \mathrm{d}\mathbf{r}_N\, \mathrm{d}\mathbf{p}_1 \cdots \mathrm{d}\mathbf{p}_N, \qquad (1.34)$$

where $\beta = 1/k_\mathrm{B}T$, $k_\mathrm{B}$ is the Boltzmann factor, $T$ is the temperature, and $i = 1, \ldots, N$. Here $E(\{\mathbf{r}_i, \mathbf{p}_i\})$ is the energy associated with the $N$ charged particles having positions $\mathbf{r}_1, \mathbf{r}_2, \ldots, \mathbf{r}_N$, and momenta $\mathbf{p}_1, \mathbf{p}_2, \ldots, \mathbf{p}_N$. The integral is therefore over a $6N$–dimensional phase space ($3N$ position coordinates, $3N$ momentum coordinates). The effect of a magnetic field, as shown in the preceding section, is to shift the momentum of each particle by an amount $q\mathbf{A}$. We must therefore replace $\mathbf{p}_i$ by $\mathbf{p}_i - q\mathbf{A}$. The limits of the momentum integrals go from $-\infty$ to $\infty$ so this shift can be absorbed by shifting the origin of the momentum integrations. Hence the partition function is not a function of magnetic field, and so neither is the free energy $F = -k_\mathrm{B}T \log Z$ (see Appendix E). Thus the magnetization must be zero in a classical system.

This result seems rather surprising at first sight. When there is no applied magnetic field, electrons go in straight lines, but with an applied magnetic field their paths are curved (actually helical) and perform cyclotron orbits. One is tempted to argue that the curved cyclotron orbits, which are all curved in the same sense, must contribute to a net magnetic moment and hence there should be an effect on the energy due to an applied magnetic field. But the fallacy of this argument can be understood with reference to Fig. 1.7, which shows the orbits of electrons in a classical system due to the applied magnetic field. Electrons do indeed perform cyclotron orbits which must correspond to a net magnetic moment. Summing up these orbits leads to a net anticlockwise circulation of current around the edge of the system (as in Fig. 1.1). However, electrons near the surface cannot perform complete loops and instead make repeated elastic collisions with the surface, and perform so–called **skipping orbits** around the sample perimeter. The anticlockwise current due to the bulk electrons precisely cancels out with the clockwise current associated with the skipping orbits of electrons that reflect or scatter at the surface.

The Bohr–van Leeuwen theorem therefore appears to be correct, but it is at odds with experiment: lots of real systems containing electrons *do* have a net magnetization. Therefore the assumptions that went into the theorem must be in doubt. The assumptions are classical mechanics! Hence we conclude that classical mechanics is insufficient to explain this most basic property of

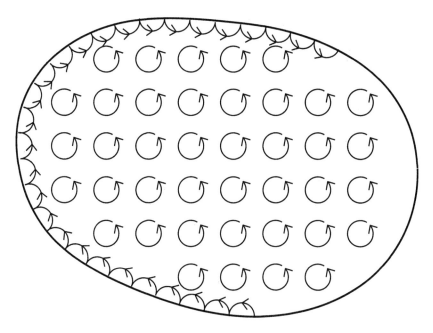

**Fig. 1.7** Electrons in a classical system with an applied magnetic field undergo cyclotron orbits in the bulk of the system. These orbits here precess in an anti-clockwise sense. They contribute a net orbital current in an anti-clockwise sense (see Fig. 1.1). This net current precisely cancels out with the current due to the skipping orbits associated with electrons which scatter at the surface and precess in a clockwise sense around the sample.

magnetic materials, and we cannot avoid using quantum theory to account for the magnetic properties of real materials. In the next section we will consider the quantum mechanics of electrons in some detail.

## 1.3 Quantum mechanics of spin

In this section I will briefly review some results concerning the quantum mechanics of electron spin. A fuller account of the quantum mechanics of angular momentum may be found in the further reading given at the end of the chapter. Some results connected with quantum and atomic physics are also given in Appendix C.

### 1.3.1 Orbital and spin angular momentum

The electronic angular momentum discussed in Section 1.1 is associated with the orbital motion of an electron around the nucleus and is known as the orbital angular momentum. In a real atom it depends on the electronic state occupied by the electron. With quantum numbers $l$ and $m_l$ defined in the usual way (see Appendix C) the component of orbital angular momentum along a fixed axis (in this case the $z$ axis) is $m_l \hbar$ and the magnitude[7] of the orbital angular momentum is $\sqrt{l(l+1)}\hbar$. Hence the component of magnetic moment along the $z$ axis is $-m_l \mu_B$ and the magnitude of the total magnetic dipole moment is $\sqrt{l(l+1)}\mu_B$.

The situation is further complicated by the fact that an electron possesses an intrinsic magnetic moment which is associated with an intrinsic angular momentum. The intrinsic angular momentum of an electron is called spin. It is so termed because electrons were once thought to precess about their own axes, but since an electron is a point particle this is rather hard to imagine.

[7]Strictly, it is the square of the angular momentum and the square of the magnetic dipole moment which are well defined quantities. The operator $\hat{\mathbf{L}}^2$ has eigenvalue $l(l+1)\hbar^2$ and $\hat{L}_z$ has eigenvalue $m_l \hbar$. Similarly, the operator $\hat{\mu}^2$ has eigenvalue $l(l+1)\mu_B^2$ and $\hat{\mu}_z$ has eigenvalue $-m_l \mu_B$.

The concept has changed but the name has stuck. This is not such a bad thing because electron spin behaves so counterintuitively that it would be hard to find any word that could do it full justice!

The spin of an electron is characterized by a **spin quantum number** $s$, which for an electron takes the value of $\frac{1}{2}$. The value of any component of the angular momentum can only take one of $2s + 1$ possible values, namely: $s\hbar, (s - 1)\hbar, \ldots, -s\hbar$. The component of spin angular momentum is written $m_s\hbar$. For an electron, with $s = \frac{1}{2}$, this means only two possible values so that $m_s = \pm\frac{1}{2}$. The component of angular momentum along a particular axis is then $\hbar/2$ or $-\hbar/2$. These alternatives will be referred to as 'up' and 'down' respectively. The magnitude[8] of the spin angular momentum for an electron is $\sqrt{s(s + 1)}\hbar = \sqrt{3}\hbar/2$.

[8] Strictly, the eigenvalue of the operator $\hat{S}^2$ is $s(s + 1)\hbar^2$.

The spin angular momentum is then associated with a magnetic moment which can have a component along a particular axis equal to $-g\mu_B m_s$ and a magnitude equal to $\sqrt{s(s + 1)}g\mu_B = \sqrt{3}g\mu_B/2$. In these expressions, $g$ is a constant known as the **g-factor**. The g-factor takes a value of approximately 2, so that the component of the intrinsic magnetic moment of the electron along the $z$ axis is[9] $\approx \mp\mu_B$, even though the spin is half-integral. The energy of the electron in a magnetic field $B$ is therefore

The g-factor is discussed in more detail in Appendix C.6.

[9] The $\mp$ sign is this way up because the magnetic moment is antiparallel to the angular momentum. This arises because of the negative charge of the electron. When $m_s = +\frac{1}{2}$ the moment is $-\mu_B$. When $m_s = -\frac{1}{2}$ the moment is $+\mu_B$.

Pieter Zeeman (1865–1943)

$$E = g\mu_B m_s B. \tag{1.35}$$

The energy levels of an electron therefore split in a magnetic field by an amount $g\mu_B B$. This is called **Zeeman splitting**.

In general for electrons in atoms there may be both orbital and spin angular momenta which combine. The g-factor can therefore take different values in real atoms depending on the relative contributions of spin and orbital angular momenta. We will return to this point in the next chapter.

The angular momentum of an electron is always an integral or half-integral multiple of $\hbar$. Therefore it is convenient to drop the factor of $\hbar$ in expressions for angular momentum operators, which amounts to saying that these operators measure the angular momentum in units of $\hbar$. In the rest of this book we will define angular momentum operators, like $\hat{L}$, such that the angular momentum is $\hbar\hat{L}$. This simplifies expressions which appear later in the book.

### 1.3.2 Pauli spin matrices and spinors

Wolfgang Pauli (1900–1958)

The behaviour of the electron spin turns out to be connected to a rather strange algebra, based on the three **Pauli spin matrices**, which are defined as

$$\hat{\sigma}_x = \begin{pmatrix} 0 & 1 \\ 1 & 0 \end{pmatrix}, \qquad \hat{\sigma}_y = \begin{pmatrix} 0 & -i \\ i & 0 \end{pmatrix}, \qquad \hat{\sigma}_z = \begin{pmatrix} 1 & 0 \\ 0 & -1 \end{pmatrix}. \tag{1.36}$$

It will be convenient to think of these as a vector of matrices,

$$\boldsymbol{\sigma} = (\hat{\sigma}_x, \hat{\sigma}_y, \hat{\sigma}_z). \tag{1.37}$$

Before proceeding, we recall a few results which can be proved straightforwardly by direct substitution. Let

$$\mathbf{a} = \begin{pmatrix} a_1 \\ a_2 \\ a_3 \end{pmatrix} \tag{1.38}$$

be a three–component vector. Then $\boldsymbol{\sigma} \cdot \mathbf{a}$ is a matrix given by

$$\boldsymbol{\sigma} \cdot \mathbf{a} = \begin{pmatrix} a_3 & a_1 - ia_2 \\ a_1 + ia_2 & -a_3 \end{pmatrix}. \tag{1.39}$$

Such matrices can be multiplied together, leading to results such as

$$(\boldsymbol{\sigma} \cdot \mathbf{a})(\boldsymbol{\sigma} \cdot \mathbf{b}) = \mathbf{a} \cdot \mathbf{b} + i\boldsymbol{\sigma} \cdot (\mathbf{a} \times \mathbf{b}) \tag{1.40}$$

and

$$(\boldsymbol{\sigma} \cdot \mathbf{a})^2 = |\mathbf{a}|^2. \tag{1.41}$$

We now define the spin angular momentum operator by

$$\hat{\mathbf{S}} = \frac{1}{2} \hat{\boldsymbol{\sigma}} \tag{1.42}$$

so that

$$\hat{S}_x = \frac{1}{2} \begin{pmatrix} 0 & 1 \\ 1 & 0 \end{pmatrix}, \quad \hat{S}_y = \frac{1}{2} \begin{pmatrix} 0 & -i \\ i & 0 \end{pmatrix}, \quad \hat{S}_z = \frac{1}{2} \begin{pmatrix} 1 & 0 \\ 0 & -1 \end{pmatrix}. \tag{1.43}$$

Notice again that we are using the convention that angular momentum is measured in units of $\hbar$, so that the angular momentum associated with an electron is actually $\hbar\mathbf{S}$. (Note that some books choose to define $\hat{\mathbf{S}}$ such that $\hat{\mathbf{S}} = \hbar\hat{\boldsymbol{\sigma}}/2$.)

It is only the operator $\hat{S}_z$ which is diagonal and therefore if the electron spin points along the $z$-direction the representation is particularly simple. The eigenvalues of $\hat{S}_z$, which we will give the symbol $m_s$, take values $m_s = \pm\frac{1}{2}$ and the corresponding eigenstates are $|\uparrow_z\rangle$ and $|\downarrow_z\rangle$ where

$$|\uparrow_z\rangle = \begin{pmatrix} 1 \\ 0 \end{pmatrix} \tag{1.44}$$

$$|\downarrow_z\rangle = \begin{pmatrix} 0 \\ 1 \end{pmatrix} \tag{1.45}$$

and correspond to the spin pointing parallel or antiparallel to the $z$ axis respectively. (The 'bra and ket' notation, i.e. writing states in the form $|\psi\rangle$ is reviewed in Appendix C.) Hence

$$\hat{S}_z|\uparrow_z\rangle = \frac{1}{2}|\uparrow_z\rangle, \quad \hat{S}_z|\downarrow_z\rangle = -\frac{1}{2}|\downarrow_z\rangle. \tag{1.46}$$

The eigenstates corresponding to the spin pointing parallel or antiparallel to the $x$- and $y$-axes are

$$|\uparrow_x\rangle = \frac{1}{\sqrt{2}} \begin{pmatrix} 1 \\ 1 \end{pmatrix}, \quad |\downarrow_x\rangle = \frac{1}{\sqrt{2}} \begin{pmatrix} 1 \\ -1 \end{pmatrix} \tag{1.47}$$

$$|\uparrow_y\rangle = \frac{1}{\sqrt{2}} \begin{pmatrix} 1 \\ i \end{pmatrix}, \quad |\downarrow_y\rangle = \frac{1}{\sqrt{2}} \begin{pmatrix} 1 \\ -i \end{pmatrix}. \tag{1.48}$$

This two-component representation of the spin wave functions is known as a **spinor representation** and the states are referred to as **spinors**. A general state can be written

$$|\psi\rangle = \begin{pmatrix} a \\ b \end{pmatrix} = a|\uparrow_z\rangle + b|\downarrow_z\rangle \tag{1.49}$$

where $a$ and $b$ are complex numbers[10] and it is conventional to normalize the state so that

$$|a|^2 + |b|^2 = 1. \tag{1.50}$$

Note that all the terms in eqns 1.40 and 1.41 are matrices. The term $\mathbf{a} \cdot \mathbf{b}$ is shorthand for $\mathbf{a} \cdot \mathbf{b}\,\mathbf{I}$ where $\mathbf{I} = \begin{pmatrix} 1 & 0 \\ 0 & 1 \end{pmatrix}$ is the identity matrix. Similarly $|\mathbf{a}|^2$ is shorthand for $|\mathbf{a}|^2\,\mathbf{1}$.

[10] They could of course be functions of position in a general case.

The total spin angular momentum operator $\hat{\mathbf{S}}$ is defined by

$$\hat{\mathbf{S}} = \begin{pmatrix} \hat{S}_x \\ \hat{S}_y \\ \hat{S}_z \end{pmatrix} = \mathbf{i}\hat{S}_x + \mathbf{j}\hat{S}_y + \mathbf{k}\hat{S}_z, \tag{1.51}$$

where $\mathbf{i}$, $\mathbf{j}$ and $\mathbf{k}$ are the unit cartesian vectors. The operator $\hat{\mathbf{S}}^2$ is then given by

$$\hat{\mathbf{S}}^2 = \hat{S}_x^2 + \hat{S}_y^2 + \hat{S}_z^2. \tag{1.52}$$

Since the eigenvalues of $\hat{S}_x^2$, $\hat{S}_y^2$ or $\hat{S}_z^2$ are always $\frac{1}{4} = (\pm\frac{1}{2})^2$, we have the result that for any spin state $|\psi\rangle$

$$\hat{\mathbf{S}}^2|\psi\rangle = (\hat{S}_x^2 + \hat{S}_y^2 + \hat{S}_z^2)|\psi\rangle = \left(\frac{1}{4} + \frac{1}{4} + \frac{1}{4}\right)|\psi\rangle = \frac{3}{4}|\psi\rangle. \tag{1.53}$$

Many of these results can be generalized to the case of particles with spin quantum number $s > \frac{1}{2}$. The most important result is that the eigenvalue of $\hat{\mathbf{S}}^2$ becomes $s(s+1)$. In the case of $s = \frac{1}{2}$ which we are considering in this chapter, $s(s+1) = \frac{3}{4}$, in agreement with eqn 1.53. The commutation relation between the spin operators is

$$[\hat{S}_x, \hat{S}_y] = i\hat{S}_z \tag{1.54}$$

and cyclic permutations thereof. This can be proved very simply using eqns 1.40 and 1.42. Each of these operators commutes with $\hat{\mathbf{S}}^2$ so that

$$[\hat{\mathbf{S}}^2, \hat{S}_z] = 0. \tag{1.55}$$

Thus it is possible simultaneously to know the total spin and one of its components, but it is not possible to know more than one of the components simultaneously.

A useful geometric construction that can aid thinking about spin is shown in Fig. 1.8. The spin vector $\mathbf{S}$ points in three-dimensional space. Because the quantum states are normalized, $\mathbf{S}$ lies on the unit sphere. Draw a line from the end of the vector $\mathbf{S}$ to the south pole of the sphere and observe the point, $q$, at which this line intersects the horizontal plane (shown shaded in Fig. 1.8). Treat this horizontal plane as an Argand diagram, with the $x$ axis as the real axis and the $y$ axis as the imaginary axis. Hence $q = x + iy$ is a complex number. Then the spinor representation of $\mathbf{S}$ is $\begin{pmatrix} 1 \\ q \end{pmatrix}$, which when normalized is

$$\frac{1}{\sqrt{1 + |q|^2}} \begin{pmatrix} 1 \\ q \end{pmatrix}. \tag{1.56}$$

In this representation the sphere is known as the **Riemann sphere**.

### 1.3.3  Raising and lowering operators

The raising and lowering operators $\hat{S}_+$ and $\hat{S}_-$ are defined by

$$\begin{aligned} \hat{S}_+ &= \hat{S}_x + i\hat{S}_y \\ \hat{S}_- &= \hat{S}_x - i\hat{S}_y. \end{aligned} \tag{1.57}$$

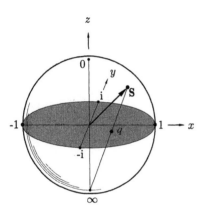

**Fig. 1.8** The Riemann sphere represents the spin states of a spin-$\frac{1}{2}$ particle. The spin vector $\mathbf{S}$ lies on a unit sphere. A measurement of spin along the direction of $\mathbf{S}$ will always give $+\frac{1}{2}$. A line from the south pole of the sphere to $\mathbf{S}$ cuts the horizontal equitorial plane (shaded) at $q = x + iy$ where the horizontal plane is considered as an Argand diagram. The numerical value of the complex number $q$ is shown for six cases, namely $\mathbf{S}$ parallel or antiparallel to the $x$, $y$ and $z$ axes.

George F. B. Riemann (1826–1866)

For an operator $\hat{A}$ to be Hermitian, one must have that $\hat{A}^\dagger = \hat{A}$ where $^\dagger$ implies an adjoint operation (for matrices this means 'take the transpose and then complex conjugate each element'). The raising and lowering operators are not Hermitian (because $\hat{S}_+^\dagger = \hat{S}_-$ and $\hat{S}_-^\dagger = \hat{S}_+$) and therefore they do not correspond to observable quantities. They are nevertheless very useful. Straightforward application of eqns 1.54 and 1.57 yields the following commutation relations :

$$[\hat{S}_+, \hat{S}_-] = 2\hat{S}_z, \tag{1.58}$$

$$[\hat{S}_z, \hat{S}_\pm] = \pm\hat{S}_\pm, \tag{1.59}$$

and

$$[\hat{\mathbf{S}}^2, \hat{S}_\pm] = 0. \tag{1.60}$$

Another useful relation, proven by direct substitution is

$$\hat{S}_+\hat{S}_- + \hat{S}_-\hat{S}_+ = 2(\hat{S}_x^2 + \hat{S}_y^2) \tag{1.61}$$

and this provides a convenient representation for $\hat{\mathbf{S}}^2$, namely

$$\hat{\mathbf{S}}^2 = \hat{S}_x^2 + \hat{S}_y^2 + \hat{S}_z^2 \tag{1.62}$$

$$= \frac{1}{2}\left(\hat{S}_+\hat{S}_- + \hat{S}_-\hat{S}_+\right) + \hat{S}_z^2. \tag{1.63}$$

Expressed as matrices the raising and lowering operators are

$$\hat{S}_+ = \begin{pmatrix} 0 & 1 \\ 0 & 0 \end{pmatrix} \tag{1.64}$$

$$\hat{S}_- = \begin{pmatrix} 0 & 0 \\ 1 & 0 \end{pmatrix}. \tag{1.65}$$

and using eqns 1.43, 1.63, 1.64 and 1.65 this then yields

$$\hat{\mathbf{S}}^2 = \frac{3}{4}\begin{pmatrix} 1 & 0 \\ 0 & 1 \end{pmatrix}, \tag{1.66}$$

in agreement with eqn 1.53.

The raising and lowering operators get their name from their effect on spin states. You can show directly that

$$\hat{S}_+|\uparrow_z\rangle = 0$$
$$\hat{S}_+|\downarrow_z\rangle = |\uparrow_z\rangle$$
$$\hat{S}_-|\uparrow_z\rangle = |\downarrow_z\rangle$$
$$\hat{S}_-|\downarrow_z\rangle = 0.$$

So a raising operator will raise the $z$ component of the spin angular momentum by $\hbar$, a lowering operator will lower the $z$ component of the spin angular momentum by $\hbar$. If the $z$ component of the spin angular momentum is already at its maximum (minimum) level, $\hat{S}_+$ ($\hat{S}_-$) will just annihilate the state.

### 1.3.4   The coupling of two spins

Now consider two spin-$\frac{1}{2}$ particles coupled by an interaction described by a Hamiltonian $\hat{\mathcal{H}}$ given by[11]

$$\hat{\mathcal{H}} = A\hat{\mathbf{S}}^a \cdot \hat{\mathbf{S}}^b, \tag{1.67}$$

where $\hat{\mathbf{S}}^a$ and $\hat{\mathbf{S}}^b$ are the operators for the spins for the two particles. Considered as a joint entity, the total spin can also be represented by an operator:

$$\hat{\mathbf{S}}^{tot} = \hat{\mathbf{S}}^a + \hat{\mathbf{S}}^b \tag{1.68}$$

so that

$$(\hat{\mathbf{S}}^{tot})^2 = (\hat{\mathbf{S}}^a)^2 + (\hat{\mathbf{S}}^b)^2 + 2\hat{\mathbf{S}}^a \cdot \hat{\mathbf{S}}^b. \tag{1.69}$$

[11]The type of interaction in eqn 1.67 will turn out to be very important in this book. The hyperfine interaction (see Chapter 2) and the Heisenberg exchange interaction (see Chapter 4) both take this form.

Combining two spin-$\frac{1}{2}$ particles results in a joint entity with spin quantum number $s = 0$ or 1. The eigenvalue of $(\hat{\mathbf{S}}^{tot})^2$ is $s(s + 1)$ which is therefore either 0 or 2 for the cases of $s = 0$ or 1 respectively. The eigenvalues of both $(\hat{\mathbf{S}}^a)^2$ and $(\hat{\mathbf{S}}^b)^2$ are $\frac{3}{4}$ from eqn 1.53. Hence from eqn 1.69

$$\hat{\mathbf{S}}^a \cdot \hat{\mathbf{S}}^b = \begin{cases} \frac{1}{4} & \text{if} \quad s = 1 \\ -\frac{3}{4} & \text{if} \quad s = 0. \end{cases} \tag{1.70}$$

Because the Hamiltonian is $\hat{\mathcal{H}} = A\hat{\mathbf{S}}^a \cdot \hat{\mathbf{S}}^b$, the system therefore has two energy levels for $s = 0$ and 1 with energies given by

**Table 1.1** The eigenstates of $\hat{\mathbf{S}}^a \cdot \hat{\mathbf{S}}^b$ and the corresponding values of $m_s$, $s$ and the eigenvalue of $\hat{\mathbf{S}}^a \cdot \hat{\mathbf{S}}^b$.

| Eigenstate | $m_s$ | $s$ | $\hat{\mathbf{S}}^a \cdot \hat{\mathbf{S}}^b$ |
|---|---|---|---|
| $\lvert\uparrow\uparrow\rangle$ | 1 | 1 | $\frac{1}{4}$ |
| $\dfrac{\lvert\uparrow\downarrow\rangle + \lvert\downarrow\uparrow\rangle}{\sqrt{2}}$ | 0 | 1 | $\frac{1}{4}$ |
| $\lvert\downarrow\downarrow\rangle$ | $-1$ | 1 | $\frac{1}{4}$ |
| $\dfrac{\lvert\uparrow\downarrow\rangle - \lvert\downarrow\uparrow\rangle}{\sqrt{2}}$ | 0 | 0 | $-\frac{3}{4}$ |

$$E = \begin{cases} \frac{A}{4} & \text{if} \quad s = 1 \\ -\frac{3A}{4} & \text{if} \quad s = 0. \end{cases} \tag{1.71}$$

The degeneracy of each state is given by $2s + 1$, hence the $s = 0$ state is a singlet and the $s = 1$ state is a triplet. The $z$ component of the spin of this state, $m_s$, takes the value 0 for the singlet, and one of the three values $-1, 0, 1$ for the triplet.

Equation 1.70 has listed the eigenvalues of $\hat{\mathbf{S}}^a \cdot \hat{\mathbf{S}}^b$, but it is also useful to describe the eigenstates. Let us first consider the following basis:

$$\lvert\uparrow\uparrow\rangle, \quad \lvert\uparrow\downarrow\rangle, \quad \lvert\downarrow\uparrow\rangle, \quad \lvert\downarrow\downarrow\rangle. \tag{1.72}$$

In this representation the first arrow refers to the $z$ component of the spin labelled $a$ and the second arrow refers to the $z$ component of the spin labelled $b$. The eigenstates of $\hat{\mathbf{S}}^a \cdot \hat{\mathbf{S}}^b$ are linear combinations of these basis states and are listed in Table 1.1. The calculation of these eigenstates is treated in Exercise 1.9. Notice that $m_s$ is equal to the sum of the $z$ components of the individual spins. Also, because the eigenstates are a mixture of states in the original basis, it is not possible in general to know both the $z$ components of the original spins and the total spin of the resultant entity. This is a general feature which will become more important in more complicated situations.

Our basis in eqn 1.72 was unsatisfactory from another point of view: the wave function must be antisymmetric with respect to exchange of the two electrons. Now the wave function is a product of a spatial function $\psi_{\text{space}}(\mathbf{r}_1, \mathbf{r}_2)$ and the spin function $\chi$, where $\chi$ is a linear combination of the states listed in eqn 1.72. The spatial wave function can be either symmetric or antisymmetric with respect to exchange of electrons. For example, the spatial wave function

$$\psi_{\text{space}}(\mathbf{r}_1, \mathbf{r}_2) = \frac{\phi(\mathbf{r}_1)\xi(\mathbf{r}_2) \pm \phi(\mathbf{r}_2)\xi(\mathbf{r}_1)}{\sqrt{2}} \tag{1.73}$$

is symmetric ($+$) or antisymmetric ($-$) with respect to exchange of electrons depending on the $\pm$. This type of symmetry is known as **exchange symmetry**. In eqn 1.73, $\phi(\mathbf{r}_i)$ and $\xi(\mathbf{r}_i)$ are single-particle wave functions for the $i^{\text{th}}$ electron. Whatever the exchange symmetry of the spatial wave function, the spin wave function $\chi$ must have the opposite exchange symmetry. Hence $\chi$ must be antisymmetric when the spatial wave function is symmetric and vice versa. This is in order that the product $\psi_{\text{space}}(\mathbf{r}_1, \mathbf{r}_2) \times \chi$ is antisymmetric overall.

States like $\lvert \uparrow\uparrow\rangle$ and $\lvert \downarrow\downarrow\rangle$ are clearly symmetric under exchange of electrons, but when you exchange the two electrons in $\lvert \uparrow\downarrow\rangle$ you get $\lvert \downarrow\uparrow\rangle$ which is not equal to a multiple of $\lvert \uparrow\downarrow\rangle$. Thus the state $\lvert \uparrow\downarrow\rangle$, and also by an identical argument the state $|\downarrow\uparrow\rangle$, are both neither symmetric nor antisymmetric under exchange of the two electrons. Hence it is not surprising that we will need linear combinations of these two states as our eigenstates. The linear combinations are shown in Table 1.1. $(|\uparrow\downarrow\rangle + |\downarrow\uparrow\rangle)/\sqrt{2}$ is symmetric under exchange of electrons (in common with the other two $s = 1$ states) while $(|\uparrow\downarrow\rangle - |\downarrow\uparrow\rangle)/\sqrt{2}$ is antisymmetric under exchange of electrons.

Another consequence of this asymmetry with respect to exchange is the **Pauli exclusion principle**, which states that two electrons cannot be in the same quantum state. If two electrons *were* in precisely the same spatial and spin quantum state (both in, say, spatial state $\phi(\mathbf{r})$ and both with, say, spin-up), then their spin wave function must be symmetric under the exchange of the electrons. Their spatial wave function must then be antisymmetric under exchange, so

$$\psi_{\text{space}}(\mathbf{r}_1, \mathbf{r}_2) = \frac{\phi(\mathbf{r}_1)\phi(\mathbf{r}_2) - \phi(\mathbf{r}_2)\phi(\mathbf{r}_1)}{\sqrt{2}}$$

$$= 0. \tag{1.74}$$

Hence the state vanishes, demonstrating that two electrons cannot be in the same quantum state.

Very often we will encounter cases in which two spins are coupled via an interaction which gives an energy contribution of the form $A\hat{\mathbf{S}}^a \cdot \hat{\mathbf{S}}^b$, where $A$ is a constant. If $A > 0$, the lower level will be a singlet (with energy $-3A/4$) with a triplet of excited states (with energy $A/4$) at an energy $A$ above the singlet. This situation is illustrated in Fig. 1.9. A magnetic field can split the triplet state into the three different states with different values of $m_s$. If $A < 0$, the triplet state will be the lowest level.

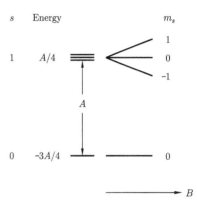

**Fig. 1.9** The coupling of two electrons with an interaction of the form $A\hat{\mathbf{S}}^a \cdot \hat{\mathbf{S}}^b$ gives rise to a triplet ($s = 1$) and a singlet ($s = 0$). If $A > 0$ the singlet is the lower state and the triplet is the upper state. The triplet can be split into three components with a magnetic field $B$.

# Further reading

- B. I. Bleaney and B. Bleaney, *Electricity and Magnetism*, OUP 1989, contains a comprehensive treatment of electromagnetism (see also Appendix B).

- A. I. Rae, *Introduction to Quantum Mechanics*, IOP Publishing 1992 is a clear exposition of Quantum Mechanics at an introductory level.

- A good account of quantum angular momentum can be found in Chapters 1–3 of volume 3 of the *Feynman lectures in Physics*, R. P. Feynman, Addison-Wesley 1975.

- An excellent description of quantum mechanics may be found in J. J. Sakurai, *Modern Quantum Mechanics*, 2nd edition 1994, Addison-Wesley.

# Exercises

(1.1) Calculate the magnetic moment of an electron (with $g = 2$). What is the Larmor precession frequency of this electron in a magnetic field of flux density 0.3 T? What is the difference in energy of the electron if its spin points parallel or antiparallel to the magnetic field? Convert this energy into a frequency.

(1.2) Using the definition of spin operators in eqn 1.43, prove eqn 1.53 and the commutation relations, eqns 1.54 and 1.55.

(1.3) Using the definition of the raising and lowering operators in eqns 1.57, prove eqns 1.58, 1.61.

(1.4) Using the commutation relation for spin, namely that $[\hat{S}_x, \hat{S}_y] = i\hat{S}_z$ (and cyclic permutations), prove that

$$[\hat{\mathbf{S}} \cdot \mathbf{X}, \hat{\mathbf{S}}] = i\hat{\mathbf{S}} \times \mathbf{X}, \tag{1.75}$$

where $\mathbf{X}$ is a vector.

(1.5) Using eqns 1.58 and 1.61, show that

$$\hat{S}_{\pm}|S, S_z\rangle = \sqrt{S(S+1) - S_z(S_z \pm 1)}|S, S_z \pm 1\rangle, \tag{1.76}$$

where $|S, S_z\rangle$ represents a state with total spin angular momentum $S(S+1)\hbar^2$ and $z$ component of spin angular momentum $S_z\hbar$. Hence prove the following special cases of eqn 1.76:

$$\hat{S}_-|S, S\rangle = \sqrt{2S}|S, S-1\rangle \tag{1.77}$$

$$\hat{S}_+|S, S-1\rangle = \sqrt{2S}|S, S\rangle \tag{1.78}$$

(1.6) If the magnetic field $\mathbf{B}$ is uniform in space, show that this is consistent with writing $\mathbf{A} = \frac{1}{2}(\mathbf{B} \times \mathbf{r})$ and show that $\nabla \cdot \mathbf{A} = 0$. Are there other choices of $\mathbf{A}$ that would produce the same $\mathbf{B}$?

(1.7) The kinetic energy operator for an electron is $\hat{\mathbf{p}}^2/2m$. Use eqn 1.41 to show that this can be rewritten

$$\frac{(\boldsymbol{\sigma} \cdot \hat{\mathbf{p}})^2}{2m_e}. \tag{1.79}$$

If a magnetic field is applied one must replace $\hat{\mathbf{p}}$ by $\hat{\mathbf{p}} + e\mathbf{A}$. With the aid of eqn 1.40, show that this replacement substituted into eqn 1.79 leads to kinetic energy of the form

$$\frac{(\hat{\mathbf{p}} + e\mathbf{A})^2}{2me} + g\mu_B\mathbf{B} \cdot \mathbf{S} \tag{1.80}$$

where the g-factor in this case is $g = 2$. (Note that in this problem you have to be careful how you apply eqn 1.40 and 1.41 because $\hat{\mathbf{p}}$ is an operator and will not commute with $\mathbf{A}$.)

(1.8) An atom has zero orbital angular momentum and a spin quantum number $\frac{1}{2}$. It is found to be in the $|\uparrow_z\rangle$ state. A measurement is performed on the value of its angular momentum in a direction at an angle $\theta$ to the $z$ axis. Show that the probability of its angular momentum being parallel to this new axis is $\cos^2(\theta/2)$.

(1.9) Using the basis of eqn 1.72, it is possible to construct matrix representations of operators such as $\hat{S}_z^a \cdot \hat{S}_z^b$ remembering that, for example, an operator such as $\hat{S}_z^a$ only operates on the part of the wave function connected with the first spin. Thus we have

$$\hat{S}_z^a = \frac{1}{2}\begin{pmatrix} 1 & 0 & 0 & 0 \\ 0 & 1 & 0 & 0 \\ 0 & 0 & -1 & 0 \\ 0 & 0 & 0 & -1 \end{pmatrix} \tag{1.81}$$

$$\hat{S}_z^b = \frac{1}{2}\begin{pmatrix} 1 & 0 & 0 & 0 \\ 0 & -1 & 0 & 0 \\ 0 & 0 & 1 & 0 \\ 0 & 0 & 0 & -1 \end{pmatrix} \tag{1.82}$$

Construct similar representations for $\hat{S}_x^a$, $\hat{S}_x^b$, $\hat{S}_y^a$ and $\hat{S}_y^b$ and hence show that

$$\hat{S}^a \cdot \hat{S}^b = \frac{1}{4}\begin{pmatrix} 1 & 0 & 0 & 0 \\ 0 & -1 & 2 & 0 \\ 0 & 2 & -1 & 0 \\ 0 & 0 & 0 & 1 \end{pmatrix}. \tag{1.83}$$

Find the eigenvalues and eigenvectors of this operator and check that your results agree with those in Table 1.1.

(1.10) A magnetic field of 0.5 T is applied to a spherical sample of (a) water and (b) $MnSO_4 \cdot 4H_2O$. In each case, evaluate the fraction the $H$ and $B$ fields inside the sample differ from the free space values. (The magnetic susceptibilities of water and $MnSO_4 \cdot 4H_2O$ are listed in Table 2.1.) You should find that the corrections are very small indeed.

(1.11) Show that the operator

$$\hat{S}_{\theta,\phi} = \sin\theta\cos\phi\,\hat{S}_x + \sin\theta\sin\phi\,\hat{S}_y + \cos\theta\,\hat{S}_z, \tag{1.84}$$

which represents the spin operator for the component of spin along a direction determined by the spherical polar angles $\theta$ and $\phi$, has eigenvalues $\pm\frac{1}{2}$ and eigenstates of the form

$$|\uparrow\rangle = \begin{pmatrix} \cos(\theta/2) \\ \sin(\theta/2)\,e^{i\phi} \end{pmatrix} \tag{1.85}$$

$$|\downarrow\rangle = \begin{pmatrix} \sin(\theta/2) \\ -\cos(\theta/2)\,e^{i\phi} \end{pmatrix}. \tag{1.86}$$

Convince yourself that these results agree with the Riemann sphere representation in Fig. 1.8. Show further that

$$\hat{S}_{\theta,\phi}^2 = \frac{1}{4}\begin{pmatrix} 1 & 0 \\ 0 & 1 \end{pmatrix}. \tag{1.87}$$

(1.12) An electron in a magnetic field aligned along the $z$-direction has a Hamiltonian (energy) operator

$$\hat{\mathcal{H}} = g\mu_B\mathbf{B} \cdot \hat{\mathbf{S}} = g\mu_B B\hat{S}_z. \tag{1.88}$$

The time-dependent Schrödinger equation states that

$$\hat{\mathcal{H}}\psi(t) = i\hbar\frac{d\psi(t)}{dt} \tag{1.89}$$

so that

$$\psi(t) = \exp(-i\hat{\mathcal{H}}t/\hbar)\psi(0). \tag{1.90}$$

Using eqn 1.41, show that

$$\exp(i\alpha\sigma_m) = I\cos\alpha + i\sigma_m\sin\alpha \tag{1.91}$$

where $I$ is the identity matrix, $\sigma_m$ is one of the Pauli spin matrices and $\alpha$ is a real number. Hence show that if $\psi(t)$ is written as a spinor,

$$\psi(t) = \begin{pmatrix} \exp(-ig\mu_B Bt/2\hbar) & 0 \\ 0 & \exp(ig\mu_B Bt/2\hbar) \end{pmatrix} \psi(0) \tag{1.92}$$

and using the results from the previous question, show that this corresponds to the evolution of the spin state in such a way that the expected value of $\theta$ is conserved but $\phi$ rotates with an angular frequency given by $geB/2m$. This demonstrates that the phenomenon of Larmor precession can also be derived from a quantum mechanical treatment.

(1.13) Here is another way to derive spin precession. Start with eqn 1.88 and use eqn C.7 to show that

$$\frac{d}{dt}\langle \hat{\mathbf{S}} \rangle = \frac{1}{i\hbar}\langle [\hat{\mathbf{S}}, \hat{\mathcal{H}}] \rangle \tag{1.93}$$

$$= -\frac{g\mu_B}{\hbar}\langle \hat{\mathbf{S}} \rangle \times \mathbf{B}, \tag{1.94}$$

which is similar to eqn 1.6 with

$$\gamma = -\frac{g\mu_B}{\hbar} = -\frac{ge}{2m_e}. \tag{1.95}$$

The minus sign comes from the negative charge of the electron.

(1.14) This problem is about the corresponding case of an electric dipole. (a) An electric dipole with electric dipole moment $\mathbf{p}$ and moment of inertia $I$ is placed in an electric field $\mathcal{E}$. Show classically that the angle $\theta$, measured between $\mathbf{p}$ and $\mathcal{E}$, obeys the differential equation

$$\ddot{\theta} = -\frac{p\mathcal{E}\sin\theta}{I}. \tag{1.96}$$

Show that this equation leads to simple harmonic motion when $\theta$ is very small.

(b) Now repeat the problem quantum mechanically. Consider the Hamiltonian

$$\hat{\mathcal{H}} = -\frac{\hbar^2}{2I}\frac{\partial^2}{\partial\theta^2} - p\mathcal{E}\cos\theta, \tag{1.97}$$

and justify why this might be an appropriate Hamiltonian to use in this case. Using eqn C.7, show that

$$\frac{d}{dt}\langle \hat{\theta} \rangle = \frac{\langle \hat{L} \rangle}{I} \tag{1.98}$$

where $\hat{L} = -i\hbar\partial/\partial\theta$ and that

$$\frac{d}{dt}\langle \hat{L} \rangle = -p\mathcal{E}\langle\sin\theta\rangle. \tag{1.99}$$

Hence deduce that

$$\frac{d^2}{dt^2}\langle \hat{\theta} \rangle = -\frac{p\mathcal{E}}{I}\langle\sin\theta\rangle, \tag{1.100}$$

which reduces to the classical expression in the appropriate limit. Compare these results to the case of the magnetic dipole. Why are they different? Why does spin precession not result in the electric case?

We have shown that electric dipoles in an electric field oscillate backwards and forwards in the plane of the electric field, while magnetic dipoles precess around a magnetic field. In each case, what is wrong with our familiar idea that if you apply a field (electric or magnetic) then dipoles (electric or magnetic) just line up with the field?

# 2 Isolated magnetic moments

In this chapter the properties of isolated magnetic moments will be examined. At this stage, interactions between magnetic moments on different atoms, or between magnetic moments and their immediate environments, are ignored. All that remains is therefore just the physics of isolated atoms and their interaction with an applied magnetic field. Of course that doesn't stop it being complicated, but the complications arise from the combinations of electrons in a given atom, not from the fact that in condensed matter there is a large number of atoms. Using this simplification, the large number of atoms merely leads to properties like the magnetic susceptibility containing a factor of $n$, the number of atoms per unit volume.

## 2.1 An atom in a magnetic field

In Section 1.1 (see eqn 1.35) it was shown that an electron spin in a magnetic field parallel to the $z$ axis has an energy equal to

$$E = g\mu_B B m_s \tag{2.1}$$

where $g \approx 2$ and $m_s = \pm\frac{1}{2}$. Hence $E \approx \pm\mu_B B$. In addition to spin angular momentum, electrons in an atom also possess orbital angular momentum. If the position of the $i^{\text{th}}$ electron in the atom is $\mathbf{r}_i$, and it has momentum $\mathbf{p}_i$, then the total angular momentum is $\hbar\mathbf{L}$ and is given by

$$\hbar\mathbf{L} = \sum_i \mathbf{r}_i \times \mathbf{p}_i \tag{2.2}$$

where the sum is taken over all electrons in an atom. Let us now consider an atom with a Hamiltonian $\hat{\mathcal{H}}_0$ given by

$$\hat{\mathcal{H}}_0 = \sum_{i=1}^{Z} \left( \frac{p_i^2}{2m} + V_i \right) \tag{2.3}$$

which is a sum (taken over the $Z$ electrons in the atom) of the electronic kinetic energy ($p_i^2/2m_e$ for the $i^{\text{th}}$ electron) and potential energy ($V_i$ for the $i^{\text{th}}$ electron). Let us assume that the Hamiltonian $\hat{\mathcal{H}}_0$ has known eigenstates and known eigenvalues.

We now add a magnetic field $\mathbf{B}$ given by

$$\mathbf{B} = \nabla \times \mathbf{A} \tag{2.4}$$

where $\mathbf{A}$ is the magnetic vector potential. We choose a gauge[1] such that

$$\mathbf{A}(\mathbf{r}) = \frac{\mathbf{B} \times \mathbf{r}}{2}. \tag{2.5}$$

Then the kinetic energy must be altered according to the prescription described in Section 1.2. Since the charge on the electron is $-e$, the kinetic energy is $[\mathbf{p}_i + e\mathbf{A}(\mathbf{r}_i)]^2/2m_e$ and hence the perturbed Hamiltonian must now be written

$$\hat{\mathcal{H}} = \sum_{i=1}^{Z} \left( \frac{[\mathbf{p}_i + e\mathbf{A}(\mathbf{r}_i)]^2}{2m_e} + V_i \right) + g\mu_B \mathbf{B} \cdot \mathbf{S} \tag{2.6}$$

$$= \sum_i \left( \frac{p_i^2}{2m_e} + V_i \right) + \mu_B(\mathbf{L} + g\mathbf{S}) \cdot \mathbf{B} + \frac{e^2}{8m_e} \sum_i (\mathbf{B} \times \mathbf{r}_i)^2 \tag{2.7}$$

$$= \hat{\mathcal{H}}_0 + \mu_B(\mathbf{L} + g\mathbf{S}) \cdot \mathbf{B} + \frac{e^2}{8m_e} \sum_i (\mathbf{B} \times \mathbf{r}_i)^2 \tag{2.8}$$

The dominant perturbation to the original Hamiltonian $\hat{\mathcal{H}}_0$ is usually the term $\mu_B(\mathbf{L} + g\mathbf{S}) \cdot \mathbf{B}$ but, as we shall see, it sometimes vanishes. This is the effect of the atom's own magnetic moment and is known as the **paramagnetic** term. The third term, $(e^2/8m_e) \sum_i (\mathbf{B} \times \mathbf{r}_i)^2$, is due to the **diamagnetic** moment. These contributions will be discussed in greater detail in Section 2.3 (diamagnetism) and Section 2.4 (paramagnetism). In the following section we outline the effects which will need explaining.

## 2.2 Magnetic susceptibility

As shown in Section 1.1.4, for a linear material $\mathbf{M} = \chi\mathbf{H}$ where $\mathbf{M}$ is the magnetic moment per volume (the magnetization) and $\chi$ is the magnetic susceptibility (dimensionless). Note that the definition of $\mathbf{M}$ means that $\chi$ represents the magnetic moment induced by a magnetic field $\mathbf{H}$ *per unit volume*. Magnetic susceptibilities are often tabulated in terms of the **molar magnetic susceptibility**, $\chi_m$, where

$$\chi_m = \chi V_m. \tag{2.9}$$

In this equation $V_m$ is the molar volume, the volume occupied by 1 mole $(6.022 \times 10^{23}$ formula units) of the substance. The molar volume (in m$^3$) is the relative atomic mass[2] of the substance (in kg) divided by the density $\rho$ (in kg m$^{-3}$). The **mass susceptibility** $\chi_g$ is defined by

$$\chi_g = \frac{\chi}{\rho}, \tag{2.10}$$

and has units of m$^3$ kg$^{-1}$. The values of magnetic susceptibility for various substances are listed in Table 2.1. If the susceptibility is negative then the material is dominated by diamagnetism, if it is positive then the material is dominated by paramagnetism.

The magnetic susceptibilities of the first 60 elements in the periodic table are plotted in Fig. 2.1. Some of these are negative, indicative of the dominant rôle of diamagnetism as discussed in Section 2.3. However, some of the values are positive, indicative of paramagnetism and this effect will be discussed in Section 2.4.

[1] Equation 2.4 relates $\mathbf{B}$ and $\mathbf{A}$. However, for a given magnetic field $\mathbf{B}$, the magnetic vector potential $\mathbf{A}$ is not uniquely determined; one can add to $\mathbf{A}$ the gradient of a scalar potential and still end up with the same $\mathbf{B}$. The choice of $\mathbf{A}$ that we make is known as a choice of gauge.

[2] The relative atomic mass is the mass of 1 mole. Note that relative atomic masses are usually tabulated in grams.

**Table 2.1** The magnetic susceptibility $\chi$ and the molar magnetic susceptibility $\chi_m$ for various substances at 298 K. Water, benzene and NaCl are weakly diamagnetic (the susceptibility is negative). CuSO$_4\cdot$5H$_2$O, MnSO$_4\cdot$4H$_2$O, Al and Na are paramagnetic (the susceptibility is positive).

| | $\chi/10^{-6}$ | $\chi_m/10^{-10}$ (m$^3$ mol$^{-1}$) |
|---|---|---|
| water | $-90$ | $-16.0$ |
| benzene | $-7.2$ | $-6.4$ |
| NaCl | $-13.9$ | $-3.75$ |
| graphite ($\parallel$) | $-260$ | $-31$ |
| graphite ($\perp$) | $-3.8$ | $-4.6$ |
| Cu | $-1.1$ | $-0.078$ |
| Ag | $-2.4$ | $-0.25$ |
| CuSO$_4\cdot$5H$_2$O | $176$ | $192$ |
| MnSO$_4\cdot$4H$_2$O | $2640$ | $2.79 \times 10^3$ |
| Al | $22$ | $2.2$ |
| Na | $7.3$ | $1.7$ |

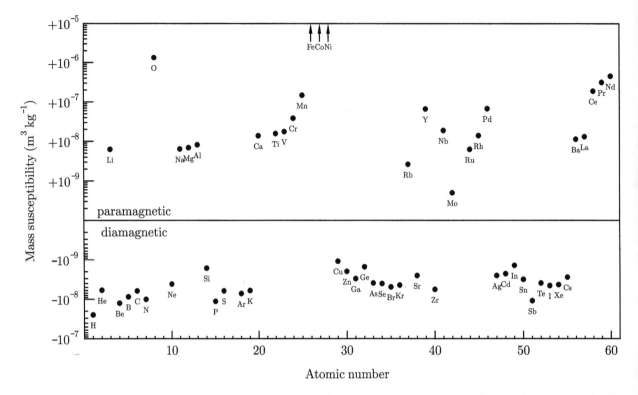

**Fig. 2.1** The mass susceptibility of the first 60 elements in the periodic table at room temperature, plotted as a function of the atomic number. Fe, Co and Ni are ferromagnetic so that they have a spontaneous magnetization with no applied magnetic field.

## 2.3  Diamagnetism

[3]The prefix *dia* means 'against' or 'across' (and leads to words like *diagonal* and *diameter*).

[4]electromotive force

[5] See the further reading.

All materials show some degree of **diamagnetism**,[3] a weak, negative magnetic susceptibility. For a diamagnetic substance, a magnetic field induces a magnetic moment which opposes the applied magnetic field that caused it.

This effect is often discussed from a classical viewpoint: the action of a magnetic field on the orbital motion of an electron causes a back e.m.f.,[4] which by Lenz's law opposes the magnetic field which causes it. However, the Bohr–van Leeuwen theorem described in the previous chapter should make us wary of such approaches which attempt to show that the application of a magnetic field to a classical system can induce a magnetic moment.[5] The phenomenon of diamagnetism is entirely quantum mechanical and should be treated as such.

We can easily illustrate the effect using the quantum mechanical approach. Consider the case of an atom with no unfilled electronic shells, so that the paramagnetic term in eqn 2.8 can be ignored. If $\mathbf{B}$ is parallel to the $z$ axis, then $\mathbf{B} \times \mathbf{r}_i = B(-y_i, x_i, 0)$ and

$$(\mathbf{B} \times \mathbf{r}_i)^2 = B^2(x_i^2 + y_i^2) \tag{2.11}$$

so that the first-order shift in the ground state energy due to the diamagnetic term is

$$\Delta E_0 = \frac{e^2 B^2}{8m_e} \sum_{i=1}^{Z} \langle 0|(x_i^2 + y_i^2)|0\rangle, \tag{2.12}$$

where $|0\rangle$ is the ground state wave function. If we assume a spherically symmetric atom,[6] $\langle x_i^2 \rangle = \langle y_i^2 \rangle = \frac{1}{3}\langle r_i^2 \rangle$ then we have

$$\Delta E_0 = \frac{e^2 B^2}{12m_e} \sum_{i=1}^{Z} \langle 0|r_i^2|0\rangle. \tag{2.13}$$

Consider a solid composed of $N$ ions (each with $Z$ electrons of mass $m$) in volume $V$ with all shells filled. To derive the magnetization (at $T = 0$), one can follow Appendix E, obtaining

$$M = -\frac{\partial F}{\partial B} = -\frac{N}{V}\frac{\partial \Delta E_0}{\partial B} = -\frac{Ne^2 B}{6m_e V}\sum_{i=1}^{Z}\langle r_i^2\rangle, \tag{2.14}$$

where $F$ is the Helmholtz function. Hence we can extract the diamagnetic susceptibility $\chi = M/H \approx \mu_0 M/B$ (assuming that $\chi \ll 1$). Following this procedure, we have the result that

$$\chi = -\frac{N}{V}\frac{e^2\mu_0}{6m_e}\sum_{i=1}^{Z}\langle r_i^2\rangle. \tag{2.15}$$

This expression has assumed first-order perturbation theory. (The second-order term will be considered in Section 2.4.4.) As the temperature is increased above zero, states above the ground state become progressively more important in determining the diamagnetic susceptibility, but this is a marginal effect. Diamagnetic susceptibilities are usually largely temperature independent.

This relation can be rather crudely tested by plotting the experimentally determined diamagnetic molar susceptibilities for various ions against $Z_{\text{eff}}r^2$, where $Z_{\text{eff}}$ is the number of electrons in the outer shell of an ion[7] and $r$ is the measured ionic radius. The assumption is that all the electrons in the outer shell of the ion have roughly the same value of $\langle r_i \rangle^2$ so that

$$\sum_{i=1}^{Z_{\text{eff}}}\langle r_i^2\rangle \approx Z_{\text{eff}}r^2. \tag{2.16}$$

The diamagnetic susceptibility of a number of ions is shown in Fig. 2.2. The experimental values are deduced by comparing the measured diamagnetic susceptibility of a range of ionic salts: NaF, NaCl, NaBr, KCl, KBr, .... The approach is inaccurate since not all the electrons in an ion have the same mean radius squared (so that eqn 2.16 is by no means exact), but the agreement is nevertheless quite impressive. Ions are chosen because, for example, Na and Cl atoms have unpaired electrons but $Na^+$ and $Cl^-$ ions are both closed shell structures, similar to those of Ne and Ar (see the periodic table in Fig. 2.13 below for reference). Thus paramagnetic effects, which would dominate the magnetic response of the atoms, can be ignored in the ions.

Relatively large and anisotropic diamagnetic susceptibilities are observed in molecules with delocalized $\pi$ electrons, such as naphthalene and graphite. Napthalene consists of two benzene molecules joined along one side (Fig. 2.3(a)). The $\pi$ electrons are very mobile and induced currents can run round the edge of the ring, producing a large diamagnetic susceptibility which is largest if the magnetic field is applied perpendicular to the plane of the ring.

[6] This is a good assumption if the total angular momentum $J$ is zero.

H. L. F. von Helmholtz (1821–1894)

[7] For an ion, this value is different from the atomic number $Z$, so we use the symbol $Z_{\text{eff}}$ for an 'effective' atomic number. We are ignoring electrons in inner shells.

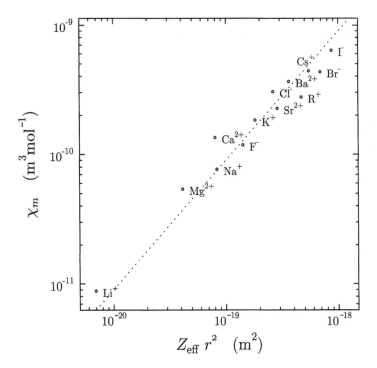

**Fig. 2.2** The measured diamagnetic molar susceptibilities $\chi_m$ of various ions plotted against $Z_{\text{eff}}r^2$, where $Z_{\text{eff}}$ is the number of electrons in the ion and $r$ is a measured ionic radius.

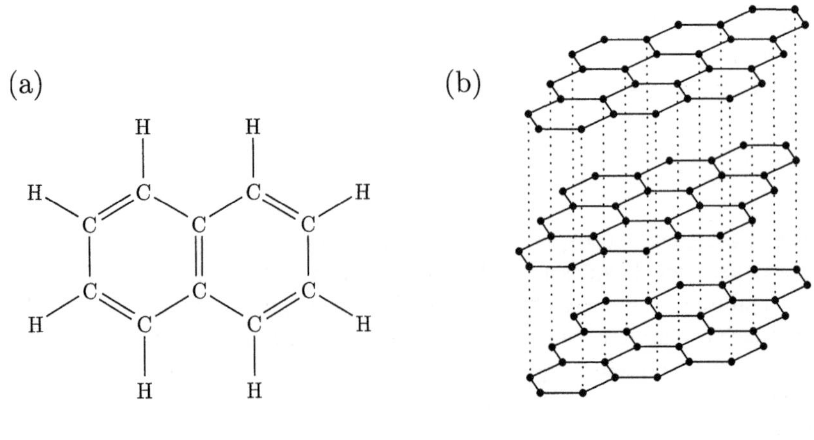

**Fig. 2.3** (a) Naphthalene consists of two fused benzene rings. (b) Graphite consists of sheets of hexagonal layers. The carbon atoms are shown as black blobs. The carbon atoms are in registry in alternate, not adjacent planes (as shown by the vertical dotted lines).

The effective ring diameter is several times larger than an atomic diameter and so the effect is large. This is also true for graphite which consists of loosely bound sheets of hexagonal layers (Fig. 2.3(b)). The diamagnetic susceptibility is much larger if the magnetic field is applied perpendicular to the layers than if it is applied in the parallel direction.

Diamagnetism is present in all materials, but it is a weak effect which can either be ignored or is a small correction to a larger effect.

# 2.4 Paramagnetism

Paramagnetism[8] corresponds to a positive susceptibility so that an applied magnetic field induces a magnetization which aligns parallel *with* the applied magnetic field which caused it. In the previous section we considered materials which contained no unpaired electrons, and thus the atoms or molecules had no magnetic moment unless a field was applied. Here we will be concerned with atoms that *do* have a non-zero magnetic moment because of unpaired electrons. Without an applied magnetic field, these magnetic moments point in random directions because the magnetic moments on neighbouring atoms interact only very weakly with each other and can be assumed to be independent. The application of a magnetic field lines them up, the degree of lining up (and hence the induced magnetization) depending on the strength of the applied magnetic field.

The magnetic moment on an atom is associated with its **total angular momentum J** which is a sum of the orbital angular momentum **L** and the spin angular momentum **S**, so that

$$\mathbf{J} = \mathbf{L} + \mathbf{S}. \tag{2.17}$$

Here, as throughout this book, these quantities are measured in units of $\hbar$. The way in which the spin and orbital parts of the angular momentum combine will be considered in detail in the following sections. In this section we will just assume that each atom has a magnetic moment of magnitude $\mu$.

Although an increase of magnetic field will tend to line up the spins, an increase of temperature will randomize them. We therefore expect that the magnetization of a paramagnetic material will depend on the ratio $B/T$. The paramagnetic effect is in general much stronger than the diamagnetic effect, although the diamagnetism is always present as a weak negative contribution.

## 2.4.1 Semiclassical treatment of paramagnetism

We begin with a semiclassical treatment of paramagnetism (which as we will see below corresponds to $J = \infty$) in which we ignore the fact that magnetic moments can point only along certain directions because of quantization. Consider magnetic moments lying at an angle between $\theta$ and $\theta + \mathrm{d}\theta$ to the applied field **B** which is assumed without loss of generality to be along the $z$ direction. These have an energy $-\mu B \cos\theta$ and have a net magnetic moment along **B** equal to $\mu \cos\theta$. If the magnetic moments could choose any direction to point along at random, the fraction which would have an angle between $\theta$ and $\theta + \mathrm{d}\theta$ would be proportional to the area of the annulus shown in Fig. 2.4 which is $2\pi \sin\theta\, \mathrm{d}\theta$ if the sphere has unit radius. The total surface area of the unit sphere is $4\pi$ so the fraction is $\frac{1}{2}\sin\theta\, \mathrm{d}\theta$. The probability of having angle between $\theta$ and $\theta + \mathrm{d}\theta$ at temperature $T$ is then simply proportional to the product of this statistical factor, $\frac{1}{2}\sin\theta\, \mathrm{d}\theta$, and the Boltzmann factor $\exp(\mu B \cos\theta/k_\mathrm{B}T)$ where $k_\mathrm{B}$ is Boltzmann's constant. The average moment

[8]The prefix *para* means 'with' or 'along' and leads to English words such as *parallel*.

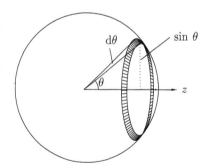

**Fig. 2.4** To calculate the average magnetic moment of a paramagnetic material, consider the probability that the moment lies between angles $\theta$ and $\theta + \mathrm{d}\theta$ to the $z$ axis. This is proportional to the area of the annulus on the unit sphere, shown shaded, which is $2\pi \sin\theta\, \mathrm{d}\theta$.

# 24  *Isolated magnetic moments*

**Fig. 2.5** The magnetization of a classical paramagnet is described by the Langevin function, $L(y) = \coth y - \frac{1}{y}$. For small $y$, $L(y) \approx y/3$, as indicated by the line which is tangential to the curve near the origin. As the magnitude of the magnetic field is increased, or the temperature decreased, the magnitude of the magnetization increases.

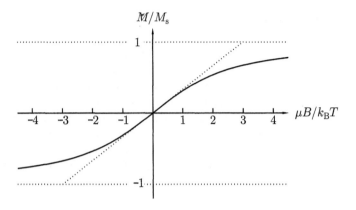

along **B** is then

$$\langle \mu_z \rangle = \frac{\int_0^\pi \mu \cos\theta \exp(\mu B \cos\theta / k_B T)\frac{1}{2}\sin\theta \, d\theta}{\int_0^\pi \exp(\mu B \cos\theta / k_B T)\frac{1}{2}\sin\theta \, d\theta} \tag{2.18}$$

$$= \mu \frac{\int_{-1}^1 x e^{yx} \, dx}{\int_{-1}^1 e^{yx} \, dx}, \tag{2.19}$$

where I have defined $y = \mu B / k_B T$ and $x = \cos\theta$. This leads to

$$\frac{\langle \mu_z \rangle}{\mu} = \coth y - \frac{1}{y} \equiv L(y) \tag{2.20}$$

Paul Langevin (1872–1946)

where $L(y) = \coth y - 1/y$ is the **Langevin function**. It is shown in Fig. 2.5. For small $y$,

$$\coth(y) = \frac{1}{y} + \frac{y}{3} + O(y^3) \tag{2.21}$$

so that

$$L(y) = \frac{y}{3} + O(y^3). \tag{2.22}$$

We will use $n$ to denote the number of magnetic moments per unit volume. The **saturation magnetization**, $M_s$, is the maximum magnetization we could obtain when all the magnetic moments are aligned, so that $M_s = n\mu$. The magnetization that we actually obtain is $M = n\langle \mu_z \rangle$ and the ratio of the magnetization to the saturation magnetization is a useful quantity. Thus we have

$$\frac{M}{M_s} = \frac{\langle \mu_z \rangle}{\mu} \approx \frac{y}{3} = \frac{\mu B}{3k_B T} \tag{2.23}$$

[9] For small fields, $\chi \ll 1$, so $B \approx \mu_0 H$.

and using $\chi = M/H \approx \mu_0 M/B$ which is valid in small fields,[9] we have

$$\chi = \frac{n\mu_0\mu^2}{3k_B T}. \tag{2.24}$$

This demonstrates that the magnetic susceptibility is inversely proportional to the temperature, which is known as Curie's law (after its discoverer, Pierre Curie).[10]

### 2.4.2 Paramagnetism for $J = \frac{1}{2}$

The calculation above will now be repeated, but this time for a quantum mechanical system. The classical moments are replaced by quantum spins with $J = \frac{1}{2}$. There are now only two possible values of the $z$ component of the magnetic moments: $m_J = \pm\frac{1}{2}$. They can either be pointing parallel to $B$ or antiparallel to $B$. Thus the magnetic moments are either $-\mu_B$ or $\mu_B$ (assuming $g = 2$) with corresponding energies $\mu_B B$ or $-\mu_B B$. (These two solutions are sketched in Fig. 2.6.) Thus

$$\langle g\mu_B m_J \rangle = \frac{-\mu_B e^{\mu_B B/k_B T} + \mu_B e^{-\mu_B B/k_B T}}{e^{\mu_B B/k_B T} + e^{-\mu_B B/k_B T}} \quad (2.25)$$

$$= \mu_B \tanh\left(\frac{\mu_B B}{k_B T}\right) \quad (2.26)$$

so writing $y = \mu_B B/k_B T = g\mu_B J B/k_B T$ (where $J = \frac{1}{2}$ and $g = 2$) one has that

$$\frac{M}{M_s} = \frac{\langle m_J \rangle}{J} = \tanh y. \quad (2.27)$$

This function is different from the Langevin function, but actually looks pretty similar (see Fig. 2.7). In small applied fields $\tanh(\mu_B/k_B T) \approx \mu_B/k_B T$ and

$$\chi = \frac{n\mu_0 \mu_B^2}{k_B T}. \quad (2.28)$$

Equation 2.27 can be derived very efficiently using an alternative method. The partition function $Z$ is the sum of the Boltzmann probabilities weighted by any degeneracy. The partition function for one spin is

$$Z = e^{\mu_B B/k_B T} + e^{-\mu_B B/k_B T} = 2\cosh\left(\frac{\mu_B B}{k_B T}\right), \quad (2.29)$$

[10]Often people write

$$\chi = \frac{C_{\text{Curie}}}{T},$$

where $C_{\text{Curie}}$ is the Curie constant.

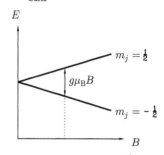

**Fig. 2.6** The energy of a spin-$\frac{1}{2}$ magnetic moment as a function of magnetic field.

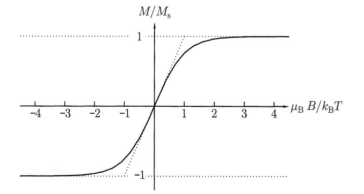

**Fig. 2.7** The magnetization of a spin-$\frac{1}{2}$ paramagnet follows a $\tanh y$ function. For small $y$, $\tanh y \approx y$, as indicated by the line which is tangential to the curve near the origin.

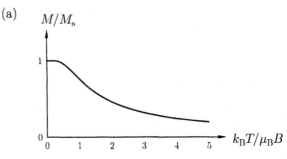

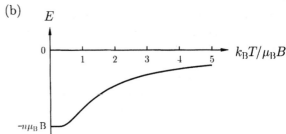

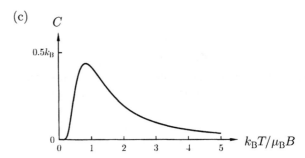

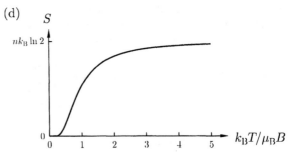

**Fig. 2.8** The (a) magnetization $M$ (normalized by the saturation magnetization), (b) energy $E$, (c) heat capacity $C$ (at constant applied magnetic field) and (d) entropy $S$ of a paramagnetic salt containing $n$ non-interacting spin-$\frac{1}{2}$ ions per unit volume as a function of $k_BT/\mu_BB$. The quantities $E$, $C$ and $S$ are therefore plotted per unit volume of paramagnetic salt.

and the Helmholtz free energy can be evaluated using the expression $F = -k_BT \ln Z$ yielding the Helmholtz free energy for $n$ spins per unit volume as

$$F = -nk_BT \ln\left[2\cosh\left(\frac{\mu_BB}{k_BT}\right)\right]. \qquad (2.30)$$

See Appendix E for more details on $Z$, $F$ and expressions such as $M = -(\partial F/\partial B)_T$.

The magnetization is then given by $M = -(\partial F/\partial B)_T$ which again yields

$$\frac{M}{M_s} = \tanh\left(\frac{\mu_BB}{k_BT}\right), \qquad (2.31)$$

in agreement with eqn 2.27.

This approach can also be used to derive other thermodynamic quantities for this model (see Exercise 2.4), the results of which are plotted in Fig. 2.8 as a function of $k_B T/\mu_B B$. Figure 2.8(a) thus shows the same information as that in Fig. 2.7 but with the horizontal axis inverted. This is because to understand some of the thermal properties of a material we are really interested in the effects of increasing temperature for a fixed magnetic field. As the sample is warmed, the magnetization decreases as the moments randomize but this produces an increase in energy density $E = -M_s B$ (see Fig. 2.8(b)). When $T \to \infty$, the energy is zero since the moments are then completely random with respect to the applied field with the energy gains cancelling the energy losses. Cooling corresponds to an energy decrease (a point we will return to in Section 2.6).

Walter Schottky (1886–1976)

The heat capacity, $C = (\partial E/\partial T)_B$ has a broad maximum close to $k_B T \sim \mu_B B$ which is known as a Schottky anomaly (see Fig. 2.8(c)). This arises because at this temperature, it is possible to thermally excite transitions between the two states of the system. At very low temperature, it is hard to change the energy of the system because there is not enough energy to excite transitions from the ground state and therefore all the spins are 'stuck', all aligned with the magnetic field. At very high temperature, it is hard to change the energy of the system because both states are equally occupied. In between there is a maximum. Peaks in the heat capacity can therefore be a useful indicator that something interesting may be happening. Note however that the Schottky anomaly is not a very sharp peak, cusp or spike, as might be associated with a phase transition, but is a smooth, broad maximum.

The entropy $S = -(\partial F/\partial T)_B$ rises as the temperature increases (see Fig. 2.8(d)), as expected since it reflects the disorder of the spins. Conversely, cooling corresponds to ordering and a reduction in the entropy. This fact is very useful in magnetic cooling techniques, as will be described in Section 2.6.

In the following section we will consider the general case of a paramagnet with total angular momentum quantum number $J$. This includes the two situations, classical and quantum, considered above as special cases.

### 2.4.3 The Brillouin function

The general case, where $J$ can take any integer or half-integer value, will now be derived. Many of the general features of the previous cases ($J = \frac{1}{2}$ and $J = \infty$) are found in this general case, for example an increase in magnetic field will tend to align the moments while an increase in temperature will tend to disorder them.

The partition function is given by

$$Z = \sum_{m_J=-J}^{J} \exp(m_J g_J \mu_B B/k_B T). \tag{2.32}$$

Writing $x = g_J \mu_B B/k_B T$, we have

$$\langle m_J \rangle = \frac{\sum_{m_J=-J}^{J} m_J e^{m_J x}}{\sum_{m_J=-J}^{J} e^{m_J x}} = \frac{1}{Z}\frac{\partial Z}{\partial x}, \tag{2.33}$$

so that

$$M = n g_J \mu_B \langle m_J \rangle = \frac{n g_J \mu_B}{Z} \frac{\partial Z}{\partial B} \frac{\partial B}{\partial x} = n k_B T \frac{\partial \ln Z}{\partial B}. \tag{2.34}$$

Now the partition function $Z$ is a geometric progression with initial term $a = e^{-Jx}$ and multiplying term $r = e^x$. This can therefore be summed using the well-known formula

$$a + ar + ar^2 + \cdots + ar^{M-1} = \sum_{j=1}^{M} ar^{j-1} = \frac{a(1 - r^M)}{1 - r}, \tag{2.35}$$

where $M$ is the number of terms in the series, which in this case is $M = 2J + 1$. After a few manipulations, this leads to

$$Z = \frac{\sinh[(2J + 1)\frac{x}{2}]}{\sinh[\frac{x}{2}]} \tag{2.36}$$

so that with the substitution

$$y = xJ = g_J \mu_B J B / k_B T, \tag{2.37}$$

we find

$$M = M_s B_J(y) \tag{2.38}$$

where the saturation magnetization $M_s$ is

$$M_s = n g_J \mu_B J \tag{2.39}$$

Léon Brillouin (1889–1969)

and where $B_J(y)$ is the Brillouin function given by

$$B_J(y) = \frac{2J + 1}{2J} \coth\left(\frac{2J + 1}{2J} y\right) - \frac{1}{2J} \coth \frac{y}{2J}. \tag{2.40}$$

This function is plotted in Fig. 2.9 for various values of $J$. The Brillouin function has the appropriate limits. For example, when $J = \infty$ it reduces to a Langevin function:

$$B_\infty(y) = L(y), \tag{2.41}$$

and when $J = \frac{1}{2}$ it reduces to a tanh function:

$$B_{1/2}(y) = \tanh(y). \tag{2.42}$$

Hence it reduces to the cases considered in the previous sections.

A typical value of $y$ can be estimated as follows: for $J = \frac{1}{2}$, $g_J = 2$ with $B = 1$ T, $y \sim 2 \times 10^{-3}$ at room temperature. Thus except at very low temperature and/or in extremely large magnetic fields, the experimental situation will correspond to $y \ll 1$ (and hence $\chi \ll 1$). For small $y$ the following result can be derived by using the Maclaurin expansion of $\coth y$:

$$B_J(y) = \frac{(J + 1)y}{3J} + O(y^3). \tag{2.43}$$

Hence for low magnetic fields the susceptibility is given by

$$\chi = \frac{M}{H} \approx \frac{\mu_0 M}{B} = \frac{n \mu_0 \mu_{eff}^2}{3 k_B T} \tag{2.44}$$

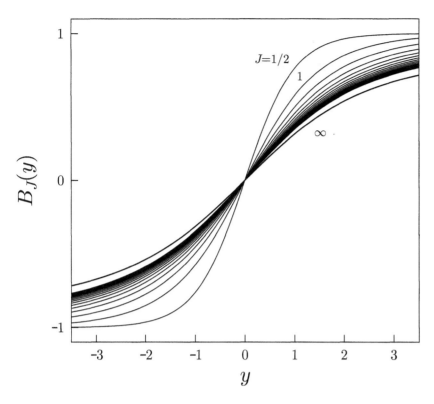

**Fig. 2.9** The magnetization of a paramagnet with magnetic moment quantum number $J$ follows a Brillouin function, $B_J(y)$, which is plotted here for different values of $J$. The values of $J$ are $\frac{1}{2}, 1, \frac{3}{2}, 2, \frac{5}{2}, \ldots$ and $J = \infty$.

which looks like a classical Curie law.[11] A measurement of $\chi$ therefore allows one to deduce $\mu_{\text{eff}}$, the value of the effective moment,

$$\mu_{\text{eff}} = g_J \mu_B \sqrt{J(J+1)} \tag{2.45}$$

where

$$g_J = \frac{3}{2} + \frac{S(S+1) - L(L+1)}{2J(J+1)} \tag{2.46}$$

The constant $g_J$ is known as the **Landé g-value** (see Appendix C).

The Curie's law dependence of the susceptibility leads to $\chi \propto 1/T$ so that a graph of $1/\chi$ against $T$ is a straight line and a graph of $\chi T$ is constant against $T$ (see Fig. 2.10). These points will be useful to keep in mind when in later chapters we consider the rôle of interactions. It is important to note

[11]i.e. $\chi = C_{\text{Curie}}/T$ where $C_{\text{Curie}} = n\mu_0 g_J^2 J(J+1)/3k_B$ is the Curie constant.

Alfred Landé (1888–1975)

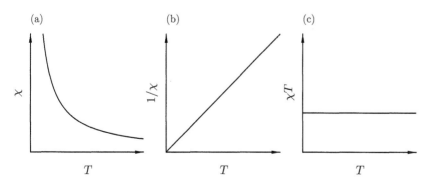

**Fig. 2.10** Curie's law states that $\chi \propto 1/T$ as shown in (a). Thus a straight-line graph is obtained by plotting $1/\chi$ against $T$ as shown in (b). A graph of $\chi T$ against $T$ is constant, as shown in (c).

that the susceptibility is evaluated in the limit of vanishing applied magnetic field and is then given by eqn 2.44 with $\mu_{\text{eff}} = g_J \mu_B \sqrt{J(J+1)}$. However, at high applied magnetic fields, the magnetization saturates to $M_s$ which, using eqn 2.39, is equivalent to a moment of $g_J \mu_B J$ per ion. These two values, $\mu_{\text{eff}} = g_J \mu_B \sqrt{J(J+1)}$ and $M_s/n = g_J \mu_B J$, are unequal except when $J \to \infty$ (the classical limit).

### 2.4.4  Van Vleck paramagnetism

If $J = 0$ in the ground state $|0\rangle$, then there is no paramagnetic effect because

$$\langle 0|\hat{\boldsymbol{\mu}}|0\rangle = g_J \mu_B \langle 0|\hat{\mathbf{J}}|0\rangle = 0. \tag{2.47}$$

This implies that the ground state energy of the system does not change if a magnetic field is applied and so that there is no paramagnetic susceptibility. However, this conclusion is only correct in first-order perturbation theory. Second-order perturbation theory nevertheless predicts a change in the ground state energy $E_0$ because it takes account of excited states with $J \neq 0$ being mixed in. The change of the ground state energy $E_0$ for an ion with $J = 0$ is

$$\Delta E_0 = \sum_n \frac{|\langle 0|(\mathbf{L}+g\mathbf{S})\cdot\mathbf{B}|n\rangle|^2}{E_0 - E_n} + \frac{e^2}{8m_e}\sum_i (\mathbf{B} \times \mathbf{r}_i)^2, \tag{2.48}$$

where the second term is due to the diamagnetism and the sum in the first term is taken over all the excited states of the system. The magnetic susceptibility is then

$$\chi = \frac{N}{V}\left(2\mu_B^2 \sum_n \frac{|\langle 0|(L_z+gS_z)|n\rangle|^2}{E_n - E_0} - \frac{e^2 \mu_0}{6m_e}\sum_{i=1}^{Z}\langle r_i^2\rangle\right), \tag{2.49}$$

where the first term is positive (because $E_n > E_0$) and is called the **van Vleck paramagnetism**. The second term is negative and is the conventional diamagnetic susceptibility that we have already considered (see eqn 2.15). Van Vleck paramagnetism is, like diamagnetism, both small and temperature independent.

## 2.5  The ground state of an ion and Hund's rules

A typical atom does not contain one electron, but many. A lot of these will be in filled shells which have no net angular momentum. However, there may be unfilled shells and the electrons in these unfilled shells can combine to give non-zero spin and orbital angular momentum. In Section 1.3.4 we saw how two spin-$\frac{1}{2}$ electrons could combine into a joint entity with spin equal to either 0 or 1. In an atom, all the spin angular momentum from the electrons in the unfilled shells can combine together and so can all their orbital angular momenta. Thus an atom will have total orbital angular momentum $\hbar\mathbf{L}$ and total spin angular momentum $\hbar\mathbf{S}$. The orbital and spin angular momenta can therefore combine in

$$(2L+1)(2S+1) \tag{2.50}$$

ways. This is the total number of choices of the $z$ component of $\mathbf{L}$ (which is the number of terms in the series $-L, -L+1, \ldots, L-1, L+1$, i.e. $(2L+1)$) multiplied by the total number of choices of the $z$ component of $\mathbf{S}$ (i.e. $(2S+1)$ by a similar argument). These different configurations, obtained by differently combining together the angular momentum (both spin and orbital) from the electrons in the unfilled shells, will cost different amounts of energy. This difference in energy occurs because the choice of spin angular momentum affects the spatial part of the wave function and the orbital angular momentum affects how electrons travel around the nucleus: both therefore affect how well the electrons avoid each other and thus influence the electrostatic repulsion energy. Below we will see how to find the configuration that minimizes the energy.

## 2.5.1 Fine structure

So far we have kept the spin and orbital angular momenta separate since they are independent of one another. However they do weakly couple, via the spin–orbit interaction (see Appendix C), which acts as a perturbation on the states with well defined $\mathbf{L}$ and $\mathbf{S}$. Because of this, $\mathbf{L}$ and $\mathbf{S}$ are not separately conserved but the total angular momentum $\mathbf{J} = \mathbf{L} + \mathbf{S}$ is conserved. If the relativistic effects are considered as a perturbation (which usually can be done) then one can consider $\mathbf{L}^2 = L(L+1)$ and $\mathbf{S}^2 = S(S+1)$ as being conserved. Thus states with $L$ and $S$ are split into a number of levels with differing $J$; this is known as fine structure. $J$ takes the values from $|L-S|$ to $L+S$. From the definition of $\mathbf{J}$ (eqn 2.17),

$$\mathbf{J}^2 = \mathbf{L}^2 + \mathbf{S}^2 + 2\mathbf{L} \cdot \mathbf{S}, \tag{2.51}$$

and since the spin–orbit interaction takes the form $\lambda \mathbf{L} \cdot \mathbf{S}$ (see Appendix C), where $\lambda$ is a constant, the expected value of this energy is

$$\langle \lambda \mathbf{L} \cdot \mathbf{S} \rangle = \frac{\lambda}{2}[J(J+1) - L(L+1) - S(S+1)]. \tag{2.52}$$

The energy of the atom is mainly determined by the values of $S$ and $L$ via electrostatic considerations and so the energy eigenstates can be labelled with values of $S$ and $L$. The precise value that $J$ takes in the range $|L-S|$ to $L+S$ is immaterial in the absence of the spin–orbit interaction. Each level is a multiplet of $(2S+1)(2L+1)$ states. When adding the spin–orbit interaction as a perturbation, the multiplets split up into different fine structure levels labelled by $J$. Each of these levels themselves has a degeneracy of $2J+1$, so that these levels can be split up into their different $m_J$ values by applying a magnetic field. The splitting of the different fine structure levels follows a relationship known as the Landé interval rule which will now be described. The energy separation between adjacent levels $E(J)$ and $E(J-1)$ of a given multiplet is given by

$$\begin{aligned} E(J) - E(J-1) &= \frac{\lambda}{2}[J(J+1) - L(L+1) - S(S+1)] \\ &\quad - \frac{\lambda}{2}[(J-1)J - L(L+1) - S(S+1)] \\ &= \lambda J. \end{aligned} \tag{2.53}$$

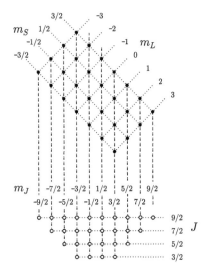

**Fig. 2.11** The combination of angular momentum $L = 3$ and $S = \frac{3}{2}$ leads to values of $J$ from $\frac{3}{2}$ to $\frac{9}{2}$.

Thus the splitting is proportional to $J$ (considering the separation of levels $J - 1$ and $J$).

**Example 2.1**

An example of this combination of angular momentum is shown in Fig. 2.11 for the particular example of $L = 3$ and $S = \frac{3}{2}$. Here $(2S + 1)(2L + 1) = 28$ and the figure demonstrates that if one considers the states for which $J$ takes the values from $|L - S| = \frac{3}{2}$ to $L + S = \frac{9}{2}$, since the degeneracy of each $J$ state is $2J + 1$, there are also 28 possible states. This graphically illustrates the result that

$$\sum_{J=|L-S|}^{L+S} 2J + 1 = (2L + 1)(2S + 1). \tag{2.54}$$

There are clearly a lot of combinations of angular momentum quantum numbers that are possible; which one is the ground state for a particular ion?

### 2.5.2 Hund's rules

The combination of angular momentum quantum numbers which are found to minimize the energy can be estimated using Hund's rules.[12] These three empirical rules are listed in order of decreasing importance, so that one first satisfies the first and then, having done this, attempts to satisfy the second, and so on for the third.

[12] Hund's rules are only applicable to the ground state configuration, and do not imply anything about the ordering of levels above the lowest level. They also assume that there is only one subshell which is incomplete.

Friederich Hund (1896–1997)

(1) Arrange the electronic wave function so as to **maximize** $S$. In this way the Coulomb energy is minimized because of the Pauli exclusion principle, which prevents electrons with parallel spins being in the same place, and this reduces Coulomb repulsion between electrons.

(2) The next step is, given the wave function determined by the first rule, to **maximize** $L$. This also minimizes the energy and can be understood by imagining that electrons in orbits rotating in the same direction can avoid each other more effectively and therefore reduce Coulomb repulsion.

(3) Finally the value of $J$ is found using $J = |L - S|$ if the shell is less than half full and $J = |L + S|$ if it is more than half full. This third rule arises from an attempt to minimize the spin–orbit energy. One should note that the third rule is only applicable in certain circumstances. As will be shown in the following chapter, in many systems, transition metal ions being good examples, the spin–orbit energies are not as significant as some other energy term such as the crystal field so that Hund's third rule is disobeyed. However, as shown below, for rare earth ions Hund's third rule works very well.

Having found values for $S$, $L$ and $J$, this ground state can be summarized using a **term symbol** of the form $^{2S+1}L_J$. Here $L$ is written not as a number, but using a letter according to the sequence

$$
\begin{array}{c|ccccccc}
L & 0 & 1 & 2 & 3 & 4 & 5 & 6 & \dots \\
\hline
& S & P & D & F & G & H & I & \dots
\end{array}
$$

and $2S + 1$ is the spin multiplicity.

## Example 2.2

As an example, consider the rare earth ion $Dy^{3+}$, which has outer shell $4f^9$: $f$ electrons have $l = 3$, so to satisfy Hund's first rule, $2l + 1 = 7$ of them are spin-up, and we then have 2 left for spin-down (see Fig. 2.12). This gives the value of $S$ as $S = 7 \times \frac{1}{2} - 2 \times \frac{1}{2} = \frac{5}{2}$ (which implies that the spin degeneracy is $2S + 1 = 6$). The spin-up electrons give no net orbital angular momentum, so we only get an orbital contribution from the 2 spin-down electrons and it is this which we have to maximize. This then implies that $L = 3 + 2 = 5$ and hence we must use the symbol H. The shell is more than half full, so $J = |5 + \frac{5}{2}| = \frac{15}{2}$. Hence the term symbol is $^6H_{15/2}$.

$$
\begin{array}{c|cc}
& \uparrow & \downarrow \\
\hline
m_l = 3 & \bullet & \bullet \\
2 & \bullet & \bullet \\
1 & \bullet & \\
0 & \bullet & \\
-1 & \bullet & \\
-2 & \bullet & \\
-3 & \bullet &
\end{array}
$$

**Fig. 2.12** The ground state of $Dy^{3+}$.

Hund's rules lead to a prediction of the ground state but tell us nothing about the excited states or how close they are to the ground state. They therefore allow us to estimate the magnetic moment of an ion assuming that only this ground state is populated.

In this and the following chapter we will want to compare these predictions with experimental values and we will concentrate on compounds containing 3d and 4f ions since these are important in many magnetic systems. The 3d elements are the first row of transition metals (Sc–Zn) and the 4f elements are known as the lanthanides or as the rare earths (La–Lu). They are shown in the periodic table in Fig. 2.13.

| 1 H | | | | | | | | | | | | | | | | | 2 He |
|---|---|---|---|---|---|---|---|---|---|---|---|---|---|---|---|---|---|
| 3 Li | 4 Be | | | | | | | | | | | 5 B | 6 C | 7 N | 8 O | 9 F | 10 Ne |
| 11 Na | 12 Mg | | | | | | | | | | | 13 Al | 14 Si | 15 P | 16 S | 17 Cl | 18 Ar |
| 19 K | 20 Ca | 21 Sc | 22 Ti | 23 V | 24 Cr | 25 Mn | 26 Fe | 27 Co | 28 Ni | 29 Cu | 30 Zn | 31 Ga | 32 Ge | 33 As | 34 Se | 35 Br | 36 Kr |
| 37 Rb | 38 Sr | 39 Y | 40 Zr | 41 Nb | 42 Mo | 43 Tc | 44 Ru | 45 Rh | 46 Pd | 47 Ag | 48 Cd | 49 In | 50 Sn | 51 Sb | 52 Te | 53 I | 54 Xe |
| 55 Cs | 56 Ba | * | 72 Hf | 73 Ta | 74 W | 75 Re | 76 Os | 77 Ir | 78 Pt | 79 Au | 80 Hg | 81 Tl | 82 Pb | 83 Bi | 84 Po | 85 At | 86 Rn |
| 87 Fr | 88 Ra | † | 104 Rf | 105 Db | 106 Sg | 107 Bh | 108 Hs | 109 Mt | | | | | | | | | |

| * | 57 La | 58 Ce | 59 Pr | 60 Nd | 61 Pm | 62 Sm | 63 Eu | 64 Gd | 65 Tb | 66 Dy | 67 Ho | 68 Er | 69 Tm | 70 Yb | 71 Lu |
|---|---|---|---|---|---|---|---|---|---|---|---|---|---|---|---|
| † | 89 Ac | 90 Th | 91 Pa | 92 U | 93 Np | 94 Pu | 95 Am | 96 Cm | 97 Bk | 98 Cf | 99 Es | 100 Fm | 101 Md | 102 No | 103 Lr |

**Fig. 2.13** The periodic table. The 3d elements (Sc–Zn) and the 4f elements (La–Lu) are shown shaded. The number by each element is the atomic number $Z$ (the number of protons in the nucleus).

The application of Hund's rules to 3d and 4f ions is shown in Fig. 2.14 (although note that, as discussed in more detail in the following chapter, the Hund's rule predictions for 4f ions agree more with experiment than those for

**Table 2.2** Magnetic ground states for 4f ions using Hund's rules. For each ion, the shell configuration and the predicted values of $S$, $L$ and $J$ for the ground state are listed. Also shown is the calculated value of $p = \mu_{\text{eff}}/\mu_B = g_J[J(J+1)]^{1/2}$ using these Hund's rules predictions. The next column lists the experimental value $p_{\text{exp}}$ and shows very good agreement, except for Sm and Eu. The experimental values are obtained from measurements of the susceptibility of paramagnetic salts at temperatures $k_B T \gg E_{\text{CEF}}$ where $E_{\text{CEF}}$ is a crystal field energy.

| ion | shell | $S$ | $L$ | $J$ | term | $p$ | $p_{\text{exp}}$ |
|---|---|---|---|---|---|---|---|
| $Ce^{3+}$ | $4f^1$ | $\frac{1}{2}$ | 3 | $\frac{5}{2}$ | $^2F_{5/2}$ | 2.54 | 2.51 |
| $Pr^{3+}$ | $4f^2$ | 1 | 5 | 4 | $^3H_4$ | 3.58 | 3.56 |
| $Nd^{3+}$ | $4f^3$ | $\frac{3}{2}$ | 6 | $\frac{9}{2}$ | $^4I_{9/2}$ | 3.62 | 3.3–3.7 |
| $Pm^{3+}$ | $4f^4$ | 2 | 6 | 4 | $^5I_4$ | 2.68 | – |
| $Sm^{3+}$ | $4f^5$ | $\frac{5}{2}$ | 5 | $\frac{5}{2}$ | $^6I_{5/2}$ | 0.85 | 1.74 |
| $Eu^{3+}$ | $4f^6$ | 3 | 3 | 0 | $^7F_0$ | 0.0 | 3.4 |
| $Gd^{3+}$ | $4f^7$ | $\frac{7}{2}$ | 0 | $\frac{7}{2}$ | $^8S_{7/2}$ | 7.94 | 7.98 |
| $Tb^{3+}$ | $4f^8$ | 3 | 3 | 6 | $^7F_6$ | 9.72 | 9.77 |
| $Dy^{3+}$ | $4f^9$ | $\frac{5}{2}$ | 5 | $\frac{15}{2}$ | $^6H_{15/2}$ | 10.63 | 10.63 |
| $Ho^{3+}$ | $4f^{10}$ | 2 | 6 | 8 | $^5I_8$ | 10.60 | 10.4 |
| $Er^{3+}$ | $4f^{11}$ | $\frac{3}{2}$ | 6 | $\frac{15}{2}$ | $^4I_{15/2}$ | 9.59 | 9.5 |
| $Tm^{3+}$ | $4f^{12}$ | 1 | 5 | 6 | $^3H_6$ | 7.57 | 7.61 |
| $Yb^{3+}$ | $4f^{13}$ | $\frac{1}{2}$ | 3 | $\frac{7}{2}$ | $^2F_{7/2}$ | 4.53 | 4.5 |
| $Lu^{3+}$ | $4f^{14}$ | 0 | 0 | 0 | $^1S_0$ | 0 | 0 |

[13]The present chapter deals only with *free* atoms or ions. Things will change when the atoms are put in a crystalline environment. The changes are quite large for 3d ions, as may be seen in chapter 3.

the 3d ions).[13] $S$ rises and becomes a maximum in the middle of each group. $L$ and $J$ have maxima at roughly the quarter and three-quarter positions, although for $J$ there is an asymmetry between these maxima which reflects the differing rules for being in a shell which is less than or more than half full.

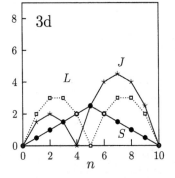 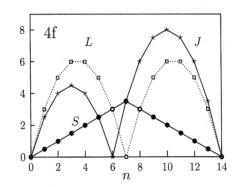

**Fig. 2.14** $S$, $L$ and $J$ for 3d and 4f ions according to Hund's rules. In these graphs $n$ is the number of electrons in the subshell (3d or 4f).

From eqn 2.44 we have found that a measurement of the susceptibility allows one to deduce the effective moment. This effective moment can be expressed in units of the Bohr magneton $\mu_B$ as

$$p = \mu_{\text{eff}}/\mu_B, \tag{2.55}$$

and the Hund's rule predictions would suggest that, using eqn 2.45, we should expect to measure $p = g_J[J(J + 1)]^{\frac{1}{2}}$. Extremely good experimental agreement is usually found between this prediction and the measured values of $p = \mu_{\text{eff}}/\mu_B$ for 4f ions in the solid state, as shown in Table 2.2. A discrepancy does occur for Sm and Eu but this is due to low-lying excited states with different $J$ from the ground states which, because of their close proximity to the ground state in energy, are also significantly populated and cause $p$ to shift from its ground state value. Although the 4f ions generally fit quite well, as will be seen in the following chapter much poorer agreement is found for the 3d ions because of the effect of the local crystal environment. This effect is not so important for 4f ions because the partially filled 4f shells lie deep within the ion, beneath the filled 5s and 5p shells.

In this chapter we have considered the magnetic moment on each atom being associated with the electrons which are localized on that atom. In metals the conduction electrons are not localized on a particular atom but exist in band states and are delocalized over the entire material. Conduction electrons can show both a diamagnetic and paramagnetic effect and this will be considered in chapter 7.

### 2.5.3 L–S and j–j coupling

Henry Norris Russell (1877–1957)

Frederick A. Saunders (1875–1963)

The atomic model that we have considered above has assumed L–S coupling (also known as Russell–Saunders coupling), namely that the spin–orbit interaction is a weak perturbation and the main energy terms are determined by the electrostatic interactions that control the values of $L$ and $S$, i.e. by combining, separately, the orbital and spin angular momenta for the electrons. Only then, when the total orbital and spin angular momenta of the atom as a whole are known, do we consider applying the spin–orbit interaction as a weak perturbation which splits each term into fine structure levels labelled by $J$.

For atoms with high atomic number $Z$, this will not work because the spin–orbit interaction energy is proportional[14] to $Z^4$ (see Appendix C) and therefore cannot be treated as a small perturbation for these atoms. A better scheme is known as j–j coupling: here the spin–orbit interaction is the dominant energy and we couple the spin and orbital angular momentum of each electron separately, and consequently the weaker electrostatic effect may then couple the total angular angular momentum from each electron.

[14]The $Z^4$ dependence works only for hydrogen-like atoms. For neutral atoms it is not so simple, but is close to $Z^2$.

### Example 2.3

As an example of these approaches, take carbon (C) and lead (Pb). They both have two $p$ electrons in their outer shell, but Pb is much further down the periodic table than C and has a value of $Z^4$ which is nearly five orders of magnitude larger. Carbon (configuration: $2p^2$) can be treated with L–S coupling, so using Hund's rules we deduce that $S = 1$ (rule 1 is 'maximize $S$' so combining the two electrons gives $\frac{1}{2} + \frac{1}{2}$), $L = 1$ (rule 2 is 'maximize $L$' and because the electrons are p electrons one of them can have $m_l = 1$, but then the next one must be $m_l = 0$) and $J = 0$ (rule 3, less than half full so $J = |1 - 1|$), so that the term symbol for the ground state of C is $^3P_0$.

In Pb (configuration: $6p^2$) the spin–orbit interaction is the dominant energy and it is more appropriate to use j–j coupling. Each electron has $s = \frac{1}{2}$ and $l = 1$ and so the combination $j$ can be $\frac{1}{2}$ or $\frac{3}{2}$, depending on the sign of the spin–orbit energy. It turns out that this actually leads to $j = \frac{1}{2}$. Both electrons will have $j = \frac{1}{2}$ and these can couple via weaker electrostatic effects to make a resultant $J$ which could be 0 or 1. The ground state is $J = 0$ because of the Pauli exclusion principle (because the two electrons have otherwise identical quantum numbers and therefore must be kept apart in an antisymmetric state), so that the good quantum numbers for the ground state are $j_1 = \frac{1}{2}$, $j_2 = \frac{1}{2}$ and $J = 0$. Note that $L$ and $S$ are not good quantum numbers in this case. (Good quantum numbers are defined in Appendix C.)

## 2.6   Adiabatic demagnetization

In this section we will describe a technique which can be used to cool samples down to very low temperature, and which uses some of the ideas we have developed about paramagnetic materials. This brings out the idea of entropy rather clearly. Consider a sample of a paramagnetic salt, which contains $N$ independent magnetic moments. Without a magnetic field applied, the magnetic moments will point in random directions (because we are assuming that they do not interact with each other) and the system will have no net magnetization. An applied field will however tend to line up the magnetic moments and produce a magnetization. Increasing temperature reduces the magnetization, increasing magnetic field increases the magnetization, emphasized by the fact that the magnetization is a function of $B/T$ (see eqns 2.37 and 2.38).

At very high temperatures, the magnetic moments all point in random directions and the net magnetization is zero. The thermal energy $k_B T$ is so large that all states are equally populated, irrespective of whether or not the state is energetically favourable. If the magnetic moments have $J = \frac{1}{2}$ they can only point parallel or antiparallel to the magnetic field: hence there are $W = 2^N$ ways of arranging up and down magnetic moments. Hence the magnetic contribution to the entropy, $S$, is

$$S = k_B \ln W = N k_B \ln 2. \tag{2.56}$$

In the general case of $J > \frac{1}{2}$, $W = (2J + 1)^N$ and the entropy is

$$S = N k_B \ln(2J + 1). \tag{2.57}$$

To calculate the entropy at a lower temperature, one needs a different approach. The probability $p(m_J)$ that the $z$ axis component of the total angular momentum of an ion takes the value $m_J$ is proportional to the Boltzmann factor $\exp(m_J g \mu_B B / k_B T)$ so that

$$p(m_J) = \frac{1}{Z} \exp(m_J g \mu_B B / k_B T) \tag{2.58}$$

where $Z$ is the partition function (given in eqn 2.32). The amount of order of this system can be described by the entropy $S$ and because we are now treating the system probabilistically, we use the expression:

$$S = -Nk_B \sum_{m_J} p(m_J) \ln p(m_J). \tag{2.59}$$

Alternatively, the equation for the entropy can be generated by computing the Helmholtz free energy, $F$, via $F = -Nk_B T \ln Z$ and then using $S = -(\partial F/\partial T)_B$.

Let us now explore the consequences of eqn 2.59. In the absence of an applied magnetic field, or at high temperatures, the system is completely disordered and all values of $m_J$ are equally likely with probability $p(m_J) = 1/(2J + 1)$ so that the entropy $S$ reduces to

$$S = Nk_B \ln(2J + 1) \tag{2.60}$$

in agreement with eqn 2.57. As the temperature is reduced, states with low energy become increasingly probable; the degree of alignment of the magnetic moments parallel to an applied magnetic field (the magnetization) increases and the entropy falls. At low temperatures, all the magnetic moments will align with the magnetic field to save energy. In this case there is only one way of arranging the system (with all spins aligned) so $W = 1$ and $S = 0$.

The principle of magnetically cooling a sample is as follows. The paramagnet is first cooled to a low starting temperature using liquid helium. The magnetic cooling then proceeds via two steps (see also Fig. 2.15).

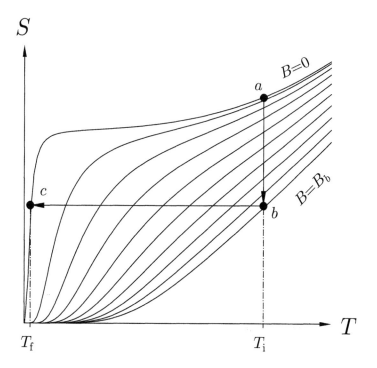

Fig. **2.15** The entropy of a paramagnetic salt as a function of temperature for several different applied magnetic fields between zero and some maximum value which we will call $B_b$. Magnetic cooling of a paramagnetic salt from temperature $T_i$ to $T_f$ is accomplished as indicated in two steps: first, isothermal magnetization from $a$ to $b$ by increasing the magnetic field from 0 to $B_b$ at constant temperature $T_i$; second, adiabatic demagnetization from $b$ to $c$. The $S(T)$ curves have been calculated assuming $J = \frac{1}{2}$ (see eqn 2.76). A term $\propto T^3$ has been added to these curves to simulate the entropy of the lattice vibrations. The curve for $B = 0$ is actually for $B$ small but non-zero to simulate the effect of a small residual field.

The first step is isothermal magnetization. The energy of a paramagnet is reduced by alignment of the moments parallel to a magnetic field. At a given temperature the alignment of the moments may therefore be increased by increasing the strength of an applied magnetic field. This is performed isothermally (see Fig. 2.15, step $a \rightarrow b$) by having the sample thermally connected to a bath of liquid helium (the boiling point of helium at atmospheric pressure is 4.2 K), or perhaps at reduced pressure so that the temperature can be less than 4.2 K. The temperature of the sample does not change and the helium bath absorbs the heat liberated by the sample as its energy and entropy decrease. The thermal connection is usually provided by low pressure helium gas in the sample chamber which conducts heat between the sample and the chamber walls, the chamber itself sitting inside the helium bath. (The gas is often called 'exchange' gas because it allows the sample and the bath to exchange heat.)

The second step is to thermally isolate the sample from the helium bath (by pumping away the exchange gas). The magnetic field is then slowly reduced to zero, slowly so that the process is quasi-static and the entropy is constant. This step is called adiabatic demagnetization (see Fig. 2.15, step $b \rightarrow c$) and it reduces the temperature of the system. During the adiabatic demagnetization the entropy of the sample remains constant; but the entropy of the magnetic moments increases (as the moments randomise as the field is turned down) which is precisely balanced by the entropy of the phonons (the lattice vibrations) which decreases as the sample cools. Entropy is thus exchanged between the phonons and the spins.

Does this method of cooling have a limit? At first sight it looks like the entropy for $B = 0$ would be $S = Nk_B \ln(2J + 1)$ for all temperatures $T > 0$ and therefore would fall to zero only at absolute zero. Thus adiabatic demagnetization looks like it might work as a cooling method all the way to absolute zero. However, in real systems there is always some small residual internal field due to interactions between the moments which ensures that the entropy falls prematurely towards zero when the temperature is a little above absolute zero (see Fig. 2.15). The size of this field puts a limit on the lowest temperature to which the salt can be cooled. In certain salts which have a very small residual internal field, temperatures of a few milliKelvin can be achieved.

## 2.7   Nuclear spins

It is not only the electrons in an atom which have a magnetic moment. The nucleus often has a non-zero spin resulting from the angular momentum of the nucleus. For each nucleus there is a quantum number, $I$, called the nuclear spin quantum number which represents the total angular momentum of the nucleus in units of $\hbar$. However this magnetic moment is very small since its size scales with the inverse of the mass of the particles involved: nuclear moments are thus typically a thousand times smaller than electronic moments (most are between $10^{-3}$ and $10^{-4} \mu_B$). The small size and absence of any strong interaction between nuclei in adjacent atoms precludes the ordering of nuclear spin systems at ordinary laboratory temperatures.

**Table 2.3** Properties of some common nuclear spins. Listed first are the neutron (n), proton (p), deuteron (d) and triton (t). $Z$ is the atomic number (the number of protons), $N$ is the number of neutrons, and nucleus X is listed as $^A$X where $A = Z + N$ is the mass number. $I$ is the nuclear spin quantum number, $\mu$ is the nuclear magnetic moment measured in nuclear magnetons ($\mu_N$) and $g_I$ is the nuclear g-factor. The frequency $\nu$ at which these moments precess in a field of 1 T is shown in the final column.

| Nucleus | $Z$ | $N$ | $I$ | $\mu/\mu_N$ | $g_I$ | $\nu$ (in MHz for $B = 1$ T) |
|---|---|---|---|---|---|---|
| n | 0 | 1 | $\frac{1}{2}$ | −1.913 | −3.826 | 29.17 |
| p=$^1$H | 1 | 0 | $\frac{1}{2}$ | 2.793 | 5.586 | 42.58 |
| d=$^2$H | 1 | 1 | 1 | 0.857 | 0.857 | 6.536 |
| t=$^3$H | 1 | 2 | $\frac{1}{2}$ | −2.128 | 4.255 | 32.43 |
| $^{12}$C | 6 | 6 | 0 | 0 | 0 | 0 |
| $^{13}$C | 6 | 7 | $\frac{1}{2}$ | 0.702 | 1.404 | 10.71 |
| $^{14}$N | 7 | 7 | 1 | 0.404 | 0.404 | 3.076 |
| $^{16}$O | 8 | 8 | 0 | 0 | 0 | 0 |
| $^{17}$O | 8 | 9 | $\frac{5}{2}$ | −1.893 | −0.757 | 5.772 |
| $^{19}$F | 9 | 10 | $\frac{1}{2}$ | 2.628 | 5.257 | 40.05 |
| $^{31}$P | 15 | 16 | $\frac{1}{2}$ | 1.132 | 2.263 | 17.24 |
| $^{33}$S | 16 | 17 | $\frac{3}{2}$ | 0.643 | 0.429 | 3.266 |

The unit of nuclear magnetism is the **nuclear magneton** $\mu_N$ defined by

$$\mu_N = \frac{e\hbar}{2m_p} = 5.0508 \times 10^{-27} \text{ A m}^2, \tag{2.61}$$

where $m_p$ is the mass of the proton. (This is much smaller than the magnetic moment of a 1s electron which is given by the Bohr magneton $\mu_B = e\hbar/2m_e = 9.27 \times 10^{-24}$ A m$^2$, where $m_e$ is the mass of the electron, see eqn 1.15.) The nuclear spin quantum number $I$ takes one of the following values: $0, \frac{1}{2}, 1, \frac{3}{2}, 2, \ldots$ The component of angular momentum in the $z$-direction is given by $m_I$ in units of $\hbar$ where $m_I$ can take one of the following values:

$$-I, -I + 1, \cdots, I - 1, I. \tag{2.62}$$

The magnetic moment of a nucleus takes a value, resolved along the $z$-direction for the spin state with the largest value of $m_I$, of

$$\mu = g_I \mu_N I \tag{2.63}$$

where $g_I$ is the nuclear g-factor, a number of the order of unity which reflects the detailed structure of the nucleus. Values of $I$, $g_I$ and $\mu$ are shown for selected nuclei in Table 2.3. Neither the magnetic moments of the neutron or the proton are simple multiples of $\mu_N$. This is indicative of the complicated internal structure of each particle; both the neutron and the proton are each composed of three quarks. For a proton $I = \frac{1}{2}$ and $g_I = 5.586$. and so its magnetic moment is $\mu_p = 1.410 \times 10^{-26}$ Am$^2$.

Normally we ignore nuclear magnetism: the magnetic moments which exist from the protons in water do not make water stick to the poles of magnets (in fact water is mainly weakly diamagnetic, due to the clouds of electrons in the water molecules). However we can detect the magnetic moments of nuclei by doing experiments which use very sensitive *resonant* techniques. This experimental method, known as nuclear magnetic resonance, is discussed in Section 3.2.1.

**Example 2.4**

Many common nuclei have no magnetic moment at all, for example the commonest forms of carbon and oxygen, $^{12}$C and $^{16}$O. This is because both of these nuclei have an even number of protons and neutrons; pairs of identical particles in a nucleus tend to pair up as singlets. Non-zero magnetic moments are found in nuclei with unpaired nucleons. Thus $^{13}$C has an unpaired neutron and has $I = \frac{1}{2}$, $^{14}$N has an unpaired neutron and an unpaired proton and these combine to give $I = 1$. Though the commonest isotopes of carbon and oxygen have no nuclear moments, the next most common isotopes of these two nuclei do have a magnetic moment (about 1.1% of natural carbon is $^{13}$C and about 0.04% of natural oxygen is $^{17}$O; both of these do have a non-zero moment, see Table 2.3).

## 2.8   Hyperfine structure

Inside an atom the nuclear moment can magnetically interact with the electronic moment but only very weakly. This leads to energy splittings which are even smaller than the fine structure discussed in Section 2.5.1, and so is known as hyperfine structure. The weak magnetic interactions between the nucleus and the electrons are known as hyperfine interactions. Their origin is rather complex, but the essential principle can be understood in the following way. Consider a nuclear magnetic moment $\mu$ which sits in a magnetic field $\mathbf{B}_{\text{electrons}}$ which is produced by the motion and the spin of all the electrons. This produces an energy term $-\mu \cdot \mathbf{B}_{\text{electrons}}$. Now $\mathbf{B}_{\text{electrons}}$ is expected to be proportional to the total angular momentum of all the electrons, $\mathbf{J}$, so that the Hamiltonian for the hyperfine interaction can be written as

$$\hat{\mathcal{H}}_{\text{hf}} = A\mathbf{I} \cdot \mathbf{J}. \tag{2.64}$$

Here $\mathbf{I}$ is the nuclear angular momentum and $A$ is a parameter which can be determined from experiment and measures the strength of the hyperfine interaction.

The precise form of the hyperfine interaction is complicated for a general atom, but can be worked out in detail for a single–electron atom interacting with a point nucleus (see Exercise 2.8). Here we treat the problem in outline only to illuminate the basic mechanism of the interaction.

There are two types of magnetic interaction that are important. The first is the magnetic dipolar interaction which will be discussed in more detail in

Section 4.1. The dipolar interaction between a nuclear moment $\mu_I = g_I \mu_I \mathbf{I}$ and an electronic moment $\mu_e = g_e \mu_B \mathbf{S}$ is given by $\hat{\mathcal{H}}_{\text{dipole}}$ where

$$\hat{\mathcal{H}}_{\text{dipole}} = \frac{\mu_0}{4\pi r^3}\left[\mu_e \cdot \mu_I - \frac{3}{r^2}(\mu_e \cdot \mathbf{r})(\mu_I \cdot \mathbf{r})\right] \tag{2.65}$$

$$= \frac{\mu_0 g_e g_I \mu_B \mu_I}{4\pi r^3}\left[\mathbf{I} \cdot \mathbf{S} - \frac{3}{r^2}(\mathbf{S} \cdot \mathbf{r})(\mathbf{I} \cdot \mathbf{r})\right]. \tag{2.66}$$

If both moments are aligned with the $z$ axis, this reduces to $\hat{\mathcal{H}}_{\text{dipole}} \propto (1 - 3\cos^2\theta)I_z S_z/r^3$ and hence leads to an interaction energy which is proportional to

$$\int d^3\mathbf{r}\,|\psi(\mathbf{r})|^2(1 - 3\cos^2\theta)m_s m_I/r^3, \tag{2.67}$$

an expression which vanishes in the case of an s orbital. This is essentially because a dipole field averages to zero over a spherical surface, as can be understood from Fig. 2.16.

The spherically symmetric s orbital wave function averages this to zero. The integral does not vanish however for an unpaired electron in an orbital with $l > 0$ such as a p orbital. Our treatment has also only considered the spin part of the electronic moment; the orbital moment of the electron also produces a magnetic field at the nucleus which gives a term proportional to $\mathbf{I} \cdot \mathbf{L}$. (All of the terms come out in the wash in the treatment of Exercise 2.8.)

The second mechanism is the **Fermi contact interaction**, which in contrast to the previous mechanism which vanished for an s orbital, vanishes for every orbital *except* an s orbital. Here we consider what happens to eqn 2.66 when $r \to 0$. The nucleus is not of course a point dipole, but has a finite volume, let us call it $V$. We further assume[15] that the nucleus is a sphere of uniform magnetization $\mathbf{M} = \mu_I/V$. The magnetic flux density inside this sphere is then $\mu_0\mathbf{M}$, but we must first subtract the demagnetization field which for a sphere is $\mu_0\mathbf{M}/3$ (see Appendix D) leaving a flux density equal to $2\mu_0\mathbf{M}/3$. Thus an electron which ventures inside the nuclear sphere would experience a magnetic flux density

$$\mathbf{B} = \frac{2\mu_0\mu_I}{3V}, \tag{2.68}$$

so that the energy cost can be obtained by multiplying this by $\mu_e|\psi(0)|^2 V$, the amount of electronic moment which is in the nucleus (note that $|\psi(0)|^2 V$ is the probability of finding the electron inside the nucleus). Thus the energy cost $E_{\text{contact}}$ is then

$$E_{\text{contact}} = \left(\frac{2\mu_0}{3}\right)2\mu_B g_I \mu_N \mathbf{I} \cdot \mathbf{S}, \tag{2.69}$$

where I have used $\mu_e = -2\mu_B\mathbf{S}$ and $\mu_I = g_I\mu_I\mathbf{I}$. The net field experienced by the electron inside the nucleus can also be understood with reference to Fig. 2.16. Only s-electrons have any amplitude at $\mathbf{r} = 0$ and so only they show this effect. The hyperfine splitting caused by the Fermi contact interaction for s-electrons tends to be much larger than the hyperfine splitting due to the magnetic dipolar effect for electrons with $l > 0$.

As shown in Exercise 2.8 the form of the hyperfine interaction which includes both dipolar and Fermi contact terms is quite complicated but essentially consists of a constant times $\mathbf{I} \cdot \mathbf{S}$ for s-electrons ($l = 0$) and $\mathbf{I} \cdot \mathbf{N}$ for

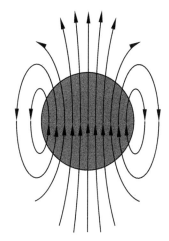

**Fig. 2.16** A schematic diagram of the magnetic field distribution inside and outside of a magnetic nucleus (assumed spherical and shaded in the diagram). The magnetic field distribution is caused by a current which flows around the equator of the sphere shown shaded in the figure. Outside the sphere the magnetic field appears to be due to a point dipole at the centre of the sphere and averaging the field over a spherically symmetric volume gives zero; the field is as often pointing up as it is pointing down. Inside the nuclear sphere, the field is on average pointing up and does not average to zero over a spherically symmetric volume. This is the origin of the Fermi contact interaction.

[15]This assumption is used just to make the calculation simple. It actually gives the right answer, but a better treatment is given in Exercise 2.8.

electrons with $l > 0$, where $\mathbf{N} = \mathbf{L} - \mathbf{S} + 3(\mathbf{S} \cdot \mathbf{r})/r^2$. For s-electrons, $\mathbf{J} = \mathbf{S}$ and so the interaction agrees with eqn 2.64. It turns out that $\mathbf{N}$ can be projected on to $\mathbf{J}$ so that the expected value of $\mathbf{I} \cdot \mathbf{N}$ is proportional to $\mathbf{I} \cdot \mathbf{J}$ (the constant of proportionality is $L(L + 1)/J(J + 1)$ for a single electron atom) so that eqn 2.64 is also appropriate for electrons with $l > 0$.

The total angular momentum $\mathbf{F}$ of the atom, i.e. of the combination of the nucleus and the electron, is given by

$$\mathbf{F} = \mathbf{J} + \mathbf{I}. \tag{2.70}$$

The quantum number $F$ can take values $|J - I|, \ldots, J + I - 1, J + I$. Most of the next few lines closely follow the treatment of the spin–orbit interaction. Using eqn 2.70, we can show that

$$\mathbf{F}^2 = \mathbf{J}^2 + \mathbf{I}^2 + 2\mathbf{I} \cdot \mathbf{J}, \tag{2.71}$$

and since the hyperfine interaction takes the form $A\mathbf{I} \cdot \mathbf{J}$, the expected value of this energy is

$$\langle A\mathbf{I} \cdot \mathbf{J}\rangle = \frac{A}{2}[F(F + 1) - I(I + 1) - J(J + 1)]. \tag{2.72}$$

The hyperfine interaction can be added as a perturbation, and it splits the electronic levels up into different hyperfine structure levels labelled by $F$. Each of these levels has a degeneracy of $2F + 1$, so that these states can be split up into their different $m_F$ values by applying a magnetic field. The splitting of the different hyperfine structure levels also follows a relationship known as the Landé interval rule by direct analogy with eqn 2.53. The energy separation between adjacent levels $E(F)$ and $E(F - 1)$ is therefore given by

$$E(F) - E(F - 1) = AF. \tag{2.73}$$

Thus the splitting is proportional to $F$, the larger of the two quantum numbers of the adjacent levels being considered. This is illustrated in the following example.

(a)

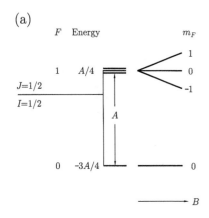

(b)

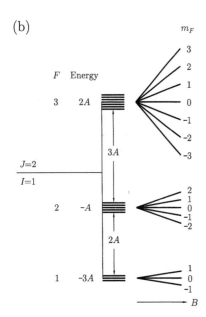

**Fig. 2.17** An illustration of the hypefine splitting for nuclei with (a) $I = \frac{1}{2}$, $J = \frac{1}{2}$ and (b) $I = 1$, $J = 2$. The quantum number $F$ can take values $|J - I|, \ldots, J + I - 1, J + I$. The energy of each state is obtained by $\frac{A}{2}[F(F + 1) - I(I+1) - J(J+1)]$. Each hyperfine state, labelled by $F$, has a degeneracy of $2F + 1$, and these states can be split up into their different $m_F$ values by applying a magnetic field $B$. Note that the Landé interval rule is obeyed. The size of the splitting in a magnetic field depends on the value of $F$, $I$ and $J$ via a g-factor for the quantum number $F$ which is given approximately by $g_F = g_J[F(F + 1) + J(J + 1) - I(I + 1)/2F(F + 1)]$.

### Example 2.5

This effect is demonstrated for two examples in Fig. 2.17. Atomic hydrogen in the ground state contains one proton ($I = \frac{1}{2}$) and one electron ($J = S = \frac{1}{2}$) so that $F = 0$ or 1 (this example is shown in Fig. 2.17(a)). The energy splitting between these states is extremely small, about 6 orders of magnitude smaller than typical electronic energies, and corresponds to an electromagnetic frequency of about 1420 MHz, or a wavelength of about 21 cm. This transition is the basis of the atomic hydrogen maser and is also important in astronomy. Astronomers call this HI emission. If the concentration of atomic hydrogen in space is not so great that $H_2$ molecules can form, emission of radiation at 21 cm can be produced thermally (since the energy required is less than the temperature of the cosmic microwave background, see Fig. 3.8). This is Doppler shifted if the gas is moving with respect to the Earth, so that the rotation of gas within different regions of distant galaxies can be measured from the Earth. It can also be used for our own Galaxy, the Milky Way.

A more complex example is shown in Fig. 2.17(b), for an atom with $I = 1$ and $J = 2$. This shows that the Landé interval rule is obeyed as the size of the splitting between two adjacent levels scales with the value of $F$ for the upper level.

# Further reading

- B. H. Bransden and C. J. Joachain, *Physics of atoms and molecules*, Longman 1983, provides extensive information on isolated atoms.

- Useful background information may also be found in P. W. Atkins, *Molecular quantum mechanics*, OUP 1983.

- Also useful is the comprehensive book by A. Abragam and B. Bleaney, *Electron paramagnetic resonance of transition ions*, Dover 1986.

- D. J. Griffiths, *Introduction to electromagnetism*, Prentice Hall 1989 provides a readable account of magnetostatic fields in matter.

- A discussion of the merits of classical versus quantum mechanical derivations of diamagnetism is given in S. L. O'Dell and R. K. P. Zia, *American Journal of Physics* **54**, 32 (1986).

# Exercises

(2.1) Calculate the diamagnetic orbital susceptibility of a gas of hydrogen atoms (with number density $10^{20}$ m$^{-3}$) in the ground state, and compare this with the paramagnetic spin susceptibility at 100 K.

(2.2) Estimate the diamagnetic susceptibility of a duck (assume it is composed entirely of water). What magnetic field would be necessary to induce the same magnetic moment in the duck as is contained in a magnetized iron filing? Repeat the calculation for a cow.

(2.3) Calculate the paramagnetic moment of a crystal (with dimensions 2 mm × 2 mm × 2 mm) of $CuSO_4 \cdot 5H_2O$ (see Table 2.1, density 2286 kg m$^{-3}$, relative molecular mass 249.7 g) in a field of 1 T at 10 K.

(2.4) Using the expressions for the partition function and the Helmholtz function of a spin-$\frac{1}{2}$ particle in a magnetic field $B$ from eqns 2.29 and 2.30 respectively, show that for $n$ such non-interacting particles per unit volume, the energy $E$ per unit volume is given by

$$E = -n\mu_B B \tanh\left(\frac{\mu_B B}{k_B T}\right), \qquad (2.74)$$

the heat capacity per unit volume is given by

$$C = nk_B \left(\frac{\mu_B B}{k_B T}\right)^2 \operatorname{sech}^2\left(\frac{\mu_B B}{k_B T}\right), \qquad (2.75)$$

and the entropy per unit volume is given by

$$S = nk_B \left[\ln\left(2\cosh\left(\frac{\mu_B B}{k_B T}\right)\right) - \frac{\mu_B B}{k_B T}\tanh\left(\frac{\mu_B B}{k_B T}\right)\right].$$
$$(2.76)$$

These results are plotted in Fig. 2.8.

(2.5) Generalize the results of the previous problem to the case of general $J$. Writing the partition function

$$Z_J(y) = \frac{\sinh[(2J + 1)\frac{y}{2J}]}{\sinh[\frac{y}{2J}]} \qquad (2.77)$$

where $y = xJ = g_J\mu_B JB/k_B T$ (see eqn 2.37), show that the entropy $S(y)$ and the heat capacity at constant field $C(y)$ as a function of $y$ are given by

$$S(y) = nk_B[\ln Z_J(y) - yB_J(y)] \qquad (2.78)$$
$$C(y) = nk_B y^2 \frac{dB_J(y)}{dy}, \qquad (2.79)$$

where $B_J(y)$ is the Brillouin function. Hence show that for $y \ll 1$,

$$Z_J(y) = 2J + 1 \qquad (2.80)$$
$$S(y) = nk_B \left[\ln(2J + 1) - \frac{J + 1}{3J}y^2\right] \qquad (2.81)$$
$$C(y) = nk_B \frac{J + 1}{3J}y^2 \qquad (2.82)$$

and as $y \to \infty$,

$$S(y) \to 0 \qquad (2.83)$$
$$C(y) \to 0. \qquad (2.84)$$

(2.6) Show that Hund's rules for a shell of angular momentum $l$ and containing $n$ electrons can be summarized by

$$S = \frac{2l + 1 - |2l + 1 - n|}{2}$$
$$L = S\,|2l + 1 - n|$$
$$J = S\,|2l - n|.$$

(2.7) Find the term symbols for the ground states of the ions (a) $Ho^{3+}$ ($4f^{10}$), (b) $Er^{3+}$ ($4f^{11}$), (c) $Tm^{3+}$ ($4f^{12}$), and (d) $Lu^{3+}$ ($4f^{14}$).

(2.8) The magnetic vector potential $\mathbf{A}$ at position $\mathbf{r}$ due to a point dipole $\boldsymbol{\mu}$ at the origin is

$$\mathbf{A} = \frac{\mu_0}{4\pi r^3}\boldsymbol{\mu} \times \mathbf{r} = \frac{\mu_0}{4\pi}\nabla \times \left(\frac{\boldsymbol{\mu}}{r}\right). \qquad (2.85)$$

Using this in the Hamiltonian for an electron in an atom

$$\hat{\mathcal{H}} = \frac{1}{2m_e}(\mathbf{p} + e\mathbf{A})^2 + 2\mu_B \mathbf{S}\cdot\mathbf{B} + V(r) \qquad (2.86)$$

and neglecting terms in $A^2$, show that the perturbation, $\hat{\mathcal{H}}' = \hat{\mathcal{H}} - \hat{\mathcal{H}}_0$, to the Hamiltonian $\hat{\mathcal{H}}_0 = p^2/2m_e + V(r)$ due to the presence of a nuclear dipole $\boldsymbol{\mu} = g_I \mu_N \mathbf{I}$ is

$$\hat{\mathcal{H}}' = \frac{\mu_0}{4\pi}2g_I \mu_B \mu_N \left[(\mathbf{S}\cdot\nabla)(\mathbf{I}\cdot\nabla)\frac{1}{r}\right.$$
$$\left. -(\mathbf{S}\cdot\mathbf{I})\nabla^2\frac{1}{r} + \frac{1}{r^3}\mathbf{L}\cdot\mathbf{I}\right]. \qquad (2.87)$$

To prove this you will need the vector identity

$$\nabla \times (\nabla \times \mathbf{Q}) = \nabla(\nabla\cdot\mathbf{Q}) - \nabla^2\mathbf{Q} \qquad (2.88)$$

where $\mathbf{Q}$ is a vector field. Using $\nabla^2(1/r) = -4\pi\delta(\mathbf{r})$, where $\delta(\mathbf{r})$ is a delta function, show that

$$(\mathbf{S}\cdot\nabla)(\mathbf{I}\cdot\nabla)\frac{1}{r}$$
$$= -\frac{1}{r^3}\left[\mathbf{S}\cdot\mathbf{I} - \frac{3(\mathbf{S}\cdot\mathbf{r})(\mathbf{I}\cdot\mathbf{r})}{r^2}\right] - \frac{4\pi}{3}\mathbf{S}\cdot\mathbf{I}\delta(\mathbf{r})$$
$$\qquad (2.89)$$

and using these results deduce that

$$\hat{\mathcal{H}}' = \frac{\mu_0}{4\pi}2g_I \mu_B \mu_N \mathbf{I}$$
$$\cdot\left[\frac{\mathbf{L}}{r^3} - \frac{\mathbf{S}}{r^3} + \frac{3\mathbf{r}(\mathbf{S}\cdot\mathbf{r})}{r^5} + \frac{8\pi}{3}\mathbf{S}\delta(\mathbf{r})\right]. \qquad (2.90)$$

The first term in the square brackets is the coupling of the orbital moment of the electron with the nuclear moment. The second and third terms are the dipolar coupling of the spin of the electron with the nuclear moment. The final term in this equation is equal to the Fermi contact energy (eqn 2.69).

(2.9) The magnetic susceptibility of platinum is $2.61\times10^{-4}$. The density of platinum is 21450 kg m$^{-3}$ and its relative atomic mass is given by 195.09 g mol$^{-1}$. Calculate its molar susceptibility (in m$^3$ mol$^{-1}$) and its mass susceptibility (in m$^3$ kg$^{-1}$). Using Appendix A, translate these results into cgs units to find the molar susceptibility in emu mol$^{-1}$ and the mass susceptibility in emu g$^{-1}$. To understand why the magnetic susceptibility of platinum, a metal, is temperature independent (in contrast with Curie's law), see chapter 7.

(2.10) Consider a set of spins described by the partition function in eqn 2.32. The Helmholtz free energy density is $F = U - TS = -nk_B T \log Z$ where $U = -\mathbf{M}\cdot\mathbf{B}$ is the potential energy density and $S$ is the entropy density. This can be rearranged to show that

$$TS = -\mathbf{M}\cdot\mathbf{B} + nk_B T \log Z. \qquad (2.91)$$

Show that if the magnetic field $\mathbf{B}$ is changed by $\delta\mathbf{B}$, with the temperature held fixed, then

$$T\,\delta S = -\mathbf{B}\cdot\delta\mathbf{M} \qquad (2.92)$$

so that

$$\delta F = \delta U - T\,\delta S = -\mathbf{M}\cdot\delta\mathbf{B}. \qquad (2.93)$$

# Environments

We have seen in the previous chapter that the magnetic properties of many crystals containing rare earths can be deduced by considering the rare earth ions to be behaving as completely free ions without interacting with each other or their surroundings. However, for magnetic ions in certain crystals one cannot usually ignore such interactions and for many materials they are large and significant. In this chapter we will consider the interactions between an atom and its immediate surroundings. In the following chapter we shall consider the direct magnetic interactions between a magnetic atom in a crystal and its neighbouring magnetic atoms.

## 3.1   Crystal fields

To understand the effect of the local environment due to the crystal on the energy levels of an atom, it is necessary to first review the shapes of the atomic orbitals. The angular dependences of the electron density of the s, p and d orbitals are shown in Fig. 3.1. This figure only shows the angular part of the wave functions for each orbital; there is also a radial part (for details, see Appendix C). Only s orbitals are spherically symmetric; the others have a pronounced angular dependence. This is crucial because the local environments are often not spherically symmetric so that different orbitals will behave in different ways.

### 3.1.1   Origin of crystal fields

The crystal field is an electric field derived from neighbouring atoms in the crystal. In crystal field theory the neighbouring orbitals are modelled as negative point charges; an improvement on this approximation is ligand field theory which is essentially an extension of molecular orbital theory that focusses on the rôle of the d orbitals on the central ion and their overlap with orbitals on surrounding ions (ligands). The size and nature of crystal field effects depend crucially on the symmetry of the local environment. A common case to consider is the octahedral environment. This is because in many transition metal compounds a transition metal ion sits at the centre of an octahedron with an ion such as oxygen on each corner. The crystal field in this case arises mainly from electrostatic repulsion from the negatively charged electrons in the oxygen orbitals. Diagrams of both an octahedral and a tetrahedral environment are shown in Fig. 3.2.

The d orbitals fall into two classes, the $t_{2g}$ orbitals which point *between* the $x$, $y$ and $z$ axes (these are the $d_{xy}, d_{xz}$ and $d_{yz}$ orbitals) and the $e_g$ orbitals which

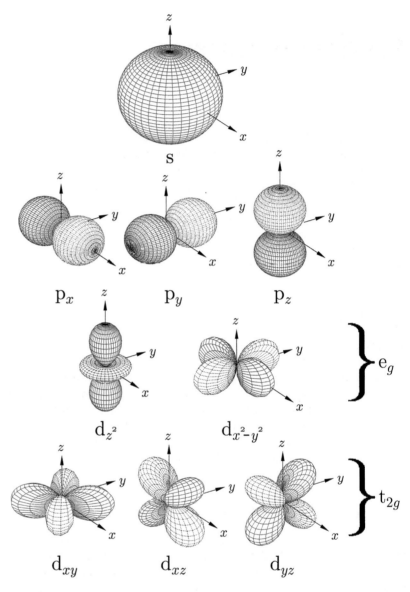

**Fig. 3.1** The angular distribution of the s, p and d orbitals. The $d_{z^2}$ and $d_{x^2-y^2}$ levels are grouped together and called the $e_g$ levels. The $d_{xy}$, $d_{xz}$ and $d_{yz}$ levels are grouped together and called the $t_{2g}$ levels. The $d_{z^2}$ orbital is sometimes referred to as $d_{3z^2-r^2}$.

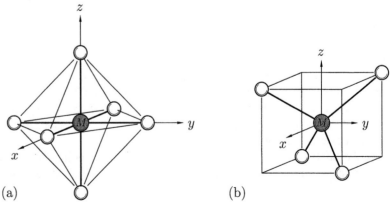

**Fig. 3.2** A metal atom $M$ in an (a) octahedral and (b) tetrahedral environment. The octahedral environment is found in many transition metal oxides where oxygen anions sit on the corners of an octahedron with the metal atom at the centre. The tetrahedral environment can be conveniently described by considering alternate corners of a cube as shown.

point *along* these axes (the $d_{z^2}$ orbital, which has lobes which point along the $z$ axis and the $d_{x^2-y^2}$ orbital, which has lobes which point along both the $x$- and $y$-axes). Suppose a cation containing ten d electrons is placed in the centre of a sphere of radius $r$ which is uniformly negatively charged. The d orbitals will all be degenerate in this spherically symmetric environment, though the presence of the charge will raise the energy of the entire system. Now imagine the charge on the sphere to collect into six discrete point charges, each lying at the vertex of an octahedron, but still on the surface of the sphere. The total electronic energy of all the d orbitals will not change, but the d orbitals will no longer be degenerate. What has now been created is an octahedral environment.

To demonstrate that the environment affects the orbitals in different ways, consider Fig. 3.3 which shows, in plan view, two different d orbitals in an octahedral environment (this is the projection of Fig. 3.2(a) on to the $xy$ plane). The crystal field is largely produced by p orbitals on neighbouring atoms. It is clear that the $d_{xy}$ orbital (Fig. 3.3(a)) has a lower overlap with these neighbouring p orbitals than the $d_{x^2-y^2}$ orbital (Fig. 3.3(b)) and hence will have a lower electrostatic energy.

In an octahedral environment, neighbouring positive charges congregate at the points $(\pm r, 0, 0)$, $(0, \pm r, 0)$ and $(0, 0, \pm r)$, the three orbitals $d_{xy}$, $d_{yz}$, $d_{zx}$ which point between the $x$, $y$ and $z$ axes will be lowered in energy, but the $d_{z^2}$ and $d_{x^2-y^2}$ which point along the $x$, $y$ and $z$ axes will be raised in energy. The five levels therefore split as shown in Fig. 3.4(a), with the threefold $t_{2g}$ levels lowered in energy and the twofold $e_g$ levels raised in energy.

If the local environment is something other than octahedrally symmetric, the crystal field may even work in the opposite sense. For example, in a tetrahedral environment, the orbitals which point along the axes now maximally avoid the charge density associated with the atoms situated on four of the corners of the cube which describe a tetrahedron (see Fig. 3.2(b)). Thus in this tetrahedral case the two-fold $e_g$ levels are lower in energy (see Fig. 3.4(b)).

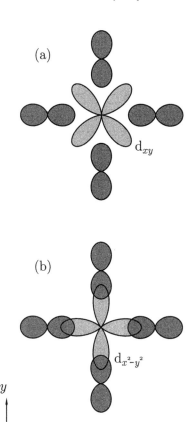

Fig. 3.3 The crystal field originates from an electrostatic interaction. (a) The $d_{xy}$ orbital is lowered in energy with respect to (b) the $d_{x^2-y^2}$ orbital in an octahedral environment.

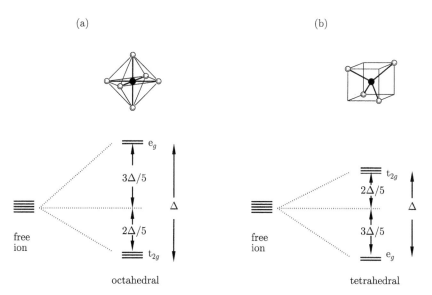

Fig. 3.4 The crystal field in an (a) octahedral and (b) tetrahedral environment.

(a)   $S = 2$

$e_g$

$t_{2g}$

(b)   $S = 0$

$e_g$

$t_{2g}$

**Fig. 3.5** Electronic configurations for the (a) high-spin (weak-field) and (b) low-spin (strong-field) cases for a $3d^6$ ion, e.g. $Fe^{2+}$.

If one is dealing with a transition metal ion in which not all 3d electrons are present, those that are present will fill the lowest (in this case the $t_{2g}$) levels before filling the $e_g$ levels. However, the precise order in which the orbitals fill depends on the competition between the crystal field energy and the Coulomb energy cost of putting two electrons in the same orbital, which is known as the **pairing energy** and is usually positive. If the crystal field energy is lower than the pairing energy (known as the **weak-field** case), then as the electrons are added to the system, they will first singly occupy each orbital before any orbital becomes doubly occupied. If on the other hand the crystal field energy is larger than the pairing energy (the **strong-field** case), electrons will doubly occupy the lower energy orbitals before they consider scaling the dizzy energetic heights of the higher energy orbitals. These cases are illustrated in Fig. 3.5.

If one is dealing with octahedral environments, there is never any doubt about how to add the electrons if you are dealing with adding 1, 2, 3, 8, 9 or 10 electrons. The interesting cases occur with 4, 5, 6 or 7. This will now be illustrated for the case of adding six electrons, as would be appropriate for an $Fe^{2+}$ ion (see also Fig. 3.5).

**Example 3.1**

The $Fe^{2+}$ ion has a $3d^6$ shell. The pairing energy is positive. In the weak–field case, one would fill each orbital once which would leave one electron to spare; this could then be reluctantly paired up with one of the $t_{2g}$ electrons. This would leave four unpaired electrons and a $S = 2$ state (see Fig. 3.5(a)). This is known as a **high-spin** configuration. In the strong field case, all six electrons are shoe-horned into the three $t_{2g}$ orbitals leaving the $e_g$ orbitals unfilled. Since there are no unpaired electrons, the system has $S = 0$ (see Fig. 3.5(b)). This is known as a **low-spin** configuration.

In some materials with $Fe^{2+}$ it is possible to initiate a **spin transition** between the low-spin and high-spin configurations using temperature or pressure or even light irradiation.

### 3.1.2   Orbital quenching

The expected magnetic ground states for 3d ions are shown in Table 3.1. It is a simple matter to calculate the values of $S$, $L$ and $J$ by following Hund's rules as outlined in Section 2.5. However, as mentioned there, the values of the predicted moment given by $g_J[J(J+1)]^{1/2}$ do not always appear to agree with experiment. The exception to this is the case of $3d^5$ and $3d^{10}$ for which there is a half or completely full shell of electrons so that $L = 0$.

The reason for this discrepancy is that for 3d ions the crystal field interaction is much stronger than the spin–orbit interaction. Hence Hund's third rule as stated in Section 2.5, which is based upon the fact that spin–orbit interaction is the next most significant energy term after Coulombic effects, is actually wrong. The data appear to suggest that these systems instead choose a ground

**Table 3.1** Magnetic ground states for 3d ions using Hund's rules. For each ion, the shell configuration and the predicted values of $S$, $L$ and $J$ for the ground state are listed. Also shown is the calculated value of $p = \mu_{eff}/\mu_B$ for each ion using Hund's rules predictions. This is given the symbol $p_1 = g_J[J(J+1)]^{1/2}$ and the next column lists the experimental values $p_{exp}$ which are derived from measurements on paramagnetic salts containing the relevant ions. This agrees much better with $p_2 = 2[S(S + 1)]^{1/2}$, which assumes orbital quenching, so that $L = 0$, $J = S$ and $g_J = 2$.

| ion | shell | $S$ | $L$ | $J$ | term | $p_1$ | $p_{exp}$ | $p_2$ |
|---|---|---|---|---|---|---|---|---|
| $Ti^{3+}$, $V^{4+}$ | $3d^1$ | $\frac{1}{2}$ | 2 | $\frac{3}{2}$ | $^2D_{3/2}$ | 1.55 | 1.70 | 1.73 |
| $V^{3+}$ | $3d^2$ | 1 | 3 | 2 | $^3F_2$ | 1.63 | 2.61 | 2.83 |
| $Cr^{3+}$, $V^{2+}$ | $3d^3$ | $\frac{3}{2}$ | 3 | $\frac{3}{2}$ | $^4F_{3/2}$ | 0.77 | 3.85 | 3.87 |
| $Mn^{3+}$, $Cr^{2+}$ | $3d^4$ | 2 | 2 | 0 | $^5D_0$ | 0 | 4.82 | 4.90 |
| $Fe^{3+}$, $Mn^{2+}$ | $3d^5$ | $\frac{5}{2}$ | 0 | $\frac{5}{2}$ | $^6S_{5/2}$ | 5.92 | 5.82 | 5.92 |
| $Fe^{2+}$ | $3d^6$ | 2 | 2 | 4 | $^5D_4$ | 6.70 | 5.36 | 4.90 |
| $Co^{2+}$ | $3d^7$ | $\frac{3}{2}$ | 3 | $\frac{9}{2}$ | $^4F_{9/2}$ | 6.63 | 4.90 | 3.87 |
| $Ni^{2+}$ | $3d^8$ | 1 | 3 | 4 | $^3F_4$ | 5.59 | 3.12 | 2.83 |
| $Cu^{2+}$ | $3d^9$ | $\frac{1}{2}$ | 2 | $\frac{5}{2}$ | $^2D_{5/2}$ | 3.55 | 1.83 | 1.73 |
| $Zn^{2+}$ | $3d^{10}$ | 0 | 0 | 0 | $^1S_0$ | 0 | 0 | 0 |

state such that $L = 0$ (so that $J = S$, $g_J = 2$) and hence

$$\mu_{eff} = 2\mu_B\sqrt{S(S + 1)}. \tag{3.1}$$

As shown in Table 3.1, this produces a much better degree of agreement with experiment. This effect is known as **orbital quenching** and the orbital moment is said to be **quenched**. For 4f ions, the orbitals are much less extended away from the nucleus and lie beneath the 5s and 5p shells so that the crystal field terms are much less important and Hund's third rule *is* obeyed. The situation in the higher transition metal ions (the 4d and 5d series) is less clear-cut because the heavier ions have a larger spin–orbit splitting and the effects of the crystal field and the spin–orbit interaction can be comparable.

For the case of the 3d ions where spin–orbit interactions can be effectively ignored, let us consider the orbital quenching effect in more detail. The crystal field in an octahedral environment is given by a constant plus a term proportional to $x^4 + y^4 + z^4 - \frac{3}{5}r^4 + O(r^6/a^6)$ (see Exercise 3.2), and is therefore a real function. This is typical of crystal-field Hamiltonians which can be expressed by real functions (without differential operators) and therefore the eigenfunctions of the Hamiltonian are all real. Now the total angular momentum operator $\hat{\mathbf{L}}$ is Hermitian and so has real eigenvalues but the operator itself is purely imaginary.[1] Thus if we have a non-degenerate ground state $|0\rangle$ (non-degenerate so that we can't play tricks by making linear combinations with it and another ground state) which is realized by the crystal-field splitting it must be a real function. Therefore $\langle 0|\hat{\mathbf{L}}|0\rangle$ must be purely imaginary because $\hat{\mathbf{L}}$ is purely imaginary. But since $\hat{\mathbf{L}}$ is Hermitian, $\langle 0|\hat{\mathbf{L}}|0\rangle$ must be purely real. It can only be purely real and purely imaginary if

$$\langle 0|\hat{\mathbf{L}}|0\rangle = 0 \tag{3.2}$$

[1] Equation C.23 shows that the angular momentum operator is $\hat{\mathbf{L}} = -i\hat{\mathbf{r}} \times \nabla$ which is purely imaginary.

so that all components of the orbital angular momentum of a non-degenerate state are quenched.

This result can be seen in another way. Orbital states which are eigenstates of $\hat{L}_z$ with eigenvalue $m_l$ have an azimuthal dependence of the form $e^{im_l\phi}$. The requirement for real eigenfunctions implies that linear combinations of states with $\pm m_l$ must be formed,[2] so the resulting state has $\langle L_z \rangle = 0$. A semiclassical interpretation of orbital quenching is that the orbital angular momentum precesses in the crystal field, so that its magnitude is unchanged but its components all average to zero.

In fact the orbital angular momentum may not be completely quenched because the spin–orbit interaction is not completely ignorable, even in the 3d ions. The spin–orbit interaction can be included as a perturbation and in this case it can mix in states with non-zero angular momentum. This results in a quenched ground state but with a g-factor which is not quite equal to the spin-only value of 2, the difference from 2 reflecting the mixed-in $L > 0$ states. It can also result in the g-factor being slightly anisotropic so that its value slightly depends on which direction you apply the magnetic field with respect to the crystal axes.

[2] This is to make states with azimuthal dependence such as $(\cos m_l \phi)$ or $\sin(m_l\phi)$. Remember that

$$\cos(m_l\phi) = \frac{1}{2}(e^{im_l\phi} + e^{-im_l\phi})$$

$$\sin(m_l\phi) = \frac{1}{2i}(e^{im_l\phi} - e^{-im_l\phi}).$$

### 3.1.3 The Jahn–Teller effect

We have assumed so far that all we need to do is to work out what kind of symmetry the local environment has, then deduce the electronic structure

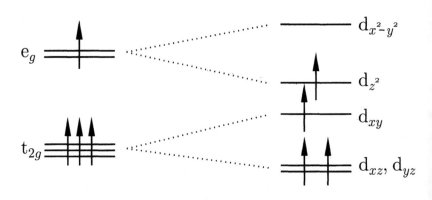

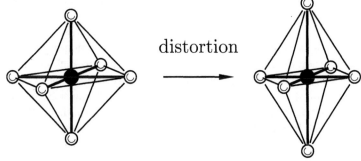

distortion

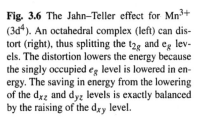

**Fig. 3.6** The Jahn–Teller effect for $Mn^{3+}$ ($3d^4$). An octahedral complex (left) can distort (right), thus splitting the $t_{2g}$ and $e_g$ levels. The distortion lowers the energy because the singly occupied $e_g$ level is lowered in energy. The saving in energy from the lowering of the $d_{xz}$ and $d_{yz}$ levels is exactly balanced by the raising of the $d_{xy}$ level.

and hence figure out the magnetic properties based on how many electrons there are to fill up the energy levels. Sometimes, however, the magnetic properties themselves can influence the symmetry of the local environment! This comes about because it can sometimes be energetically favourable for, say, an octahedron to spontaneously distort as shown in Fig. 3.6 because the energy cost of increased elastic energy is balanced by a resultant electronic energy saving due to the distortion. This phenomenon is known as the **Jahn–Teller effect**. For example, $Mn^{3+}$ ions (which have a configuration $3d^4$) in an octahedral environment show this kind of behaviour (see Fig. 3.6). In contrast, $Mn^{4+}$ ions ($3d^3$) would not show this effect because there is no net lowering of the electronic energy by a distortion.

To describe the effect, at least at the phenomenological level, we will assume that the distortion of the system can be quantified by a parameter $Q$, which denotes the distance of distortion along an appropriate normal mode coordinate. This gives rise to an energy cost which is quadratic in $Q$ and can be written as

$$E(Q) = \frac{1}{2}M\omega^2 Q^2, \tag{3.3}$$

where $M$ and $\omega$ are respectively the mass of the anion and the angular frequency corresponding to the particular normal mode. This relation is plotted in Fig. 3.7(a). Clearly the minimum distortion energy is zero and is obtained when $Q = 0$ (no distortion).

The distortion also raises the energy of certain orbitals while lowering the energy of others. If all orbitals are either completely full or completely empty, this does not matter since the overall energy is simply given by eqn 3.3. However, in the cases of partially filled orbitals this effect can be highly significant since the system can have a net reduction in total energy. The electronic energy dependence on $Q$ could be rather complicated, but one can write it as a Taylor series in $Q$ and provided the distortion is small it is legitimate to keep only the term linear in $Q$. Let us therefore suppose that the energy of a given orbital has a term either $AQ$ or $-AQ$ corresponding to a raising or a lowering of the electronic energy, where $A$ is a suitable constant, assumed to be positive. Then the total energy $E(Q)$ is given by the sum of the electronic energy and the elastic energy

$$E(Q) = \pm AQ + \frac{1}{2}M\omega^2 Q^2, \tag{3.4}$$

where the two possible choices of the sign of the $AQ$ term give rise to two separate curves which are plotted in Fig. 3.7(b). If we consider only one of them we can find the minimum energy for that orbital using $\partial E/\partial Q = 0$ which yields a value of $Q$ given by

$$Q_0 = \frac{A}{M\omega^2} \tag{3.5}$$

and a minimum energy which is given by $E_{min} = -A^2/2M\omega^2$ which is less than zero. If only that orbital is occupied, then the system can make a net energy saving by spontaneously distorting.

What we have been considering so far is essentially a static Jahn–Teller effect because the distortion which can spontaneously occur is fixed on a particular axis of an octahedron. However, the distortion can switch from one

(a)

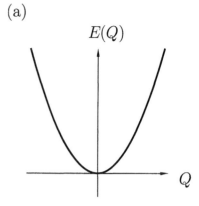

(b)

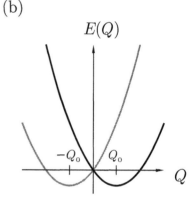

**Fig. 3.7** (a) The energy of an octahedral complex as a function of the distortion $Q$ according to eqn 3.3. (b) The energy of an octahedral complex as a function of the distortion $Q$ according to eqn 3.4.

axis to another at higher temperatures, giving rise to a **dynamic Jahn–Teller effect**. Another type of dynamic effect involves rapid hopping of the distortion from site to site. This is important in, for example, materials which contain a mixture of $Mn^{3+}$ and $Mn^{4+}$ ions. Such phenomena can be detected by their effect on magnetic resonance data.

A further effect is that in certain materials (e.g. $DyVO_4$), below a certain critical temperature, a Jahn–Teller distortion of each complex can occur *cooperatively* throughout the crystal. This is known as a **cooperative Jahn–Teller transition**.

This concludes our discussion of the effect of the crystal field. In the rest of this chapter there follows a description of various experimental techniques which can be used to study the interaction of a magnetic moment with its environment.

## 3.2    Magnetic resonance techniques

The local environment of a magnetic moment is determined by the crystal fields, but spin–orbit couplings and hyperfine interactions with the nuclei also play an important rôle in controlling the electronic structure. These effects can be studied by using a variety of experimental techniques which involve magnetic resonance. These experimental techniques will be described in the remainder of this chapter.

We have already seen in Section 1.1.2 how the application of a magnetic field to a magnetic moment can induce precession of that magnetic moment at an angular frequency given by $|\gamma B|$ where $\gamma$ is the gyromagnetic ratio. A system of magnetic moments in a magnetic field can absorb energy at this frequency and thus one may observe a resonant absorption of energy from an electromagnetic wave tuned to the correct frequency. This is **magnetic resonance** and it can take a number of different experimental forms, depending on what type of magnetic moment is resonating. The appropriate frequency of the electromagnetic wave depends on the sizes of both the magnetic field and the magnetic moment. Figure 3.8 summarizes the range of electromagnetic radiation as a function of frequency for a range of different units. It can be used for reference and referred to throughout the following discussion as necessary.

### 3.2.1    Nuclear magnetic resonance

The most commonly used form of magnetic resonance is **nuclear magnetic resonance** (NMR). This technique is greatly employed in medical imaging, where it goes by the name of **magnetic resonance imaging** (MRI) to avoid the use of the dreaded word "nuclear". Some people may be devastated to discover that the average human body contains over $10^{27}$ nuclei; by mass we are $\sim 99.98\%$ nuclear! MRI measures NMR in the protons in a patient's body and provides a safe, non-invasive technique of yielding detailed cross-sectional images.

To perform any NMR experiment, one needs a nucleus with a non-zero spin. Nuclei which are commonly studied include $^1H$ (proton), $^2H$ (deuteron) and $^{13}C$. In a simple NMR experiment (see Fig. 3.9), a sample is placed inside

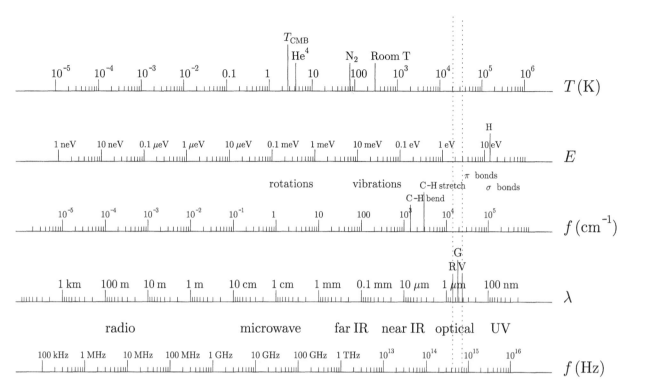

**Fig. 3.8** The electromagnetic spectrum. The energy of a photon is shown as a temperature $T = E/k_B$ in Kelvin and as an energy $E$ in eV. The corresponding frequency $f$ is shown in Hz and, because the unit is often quoted in spectroscopy, in $cm^{-1}$. The $cm^{-1}$ scale is marked with some common molecular transitions and excitations (the typical range for molecular rotations and vibrations are shown, together with the C–H bending and stretching modes). The energy of typical $\pi$ and $\sigma$ bonds are also shown. The wavelength $\lambda = c/f$ of the photon is shown (where $c$ is the speed of light). The particular temperatures marked on the temperature scale are $T_{CMB}$ (the temperature of the cosmic microwave background), the boiling points of liquid Helium ($He^4$) and nitrogen ($N_2$), both at atmospheric pressure, and also the value of room temperature. Other abbreviations on this diagram are IR = infrared, UV = ultraviolet, R = red, G = green, V = violet. The letter H marks 13.6 eV, the magnitude of the energy of the 1s electron in hydrogen. The frequency axis also contains descriptions of the main regions of the electromagnetic spectrum: radio, microwave, infrared (both 'near' and 'far'), optical and UV.

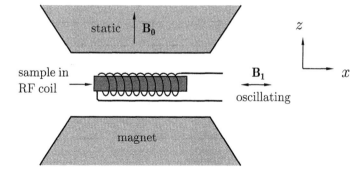

**Fig. 3.9** Schematic diagram of an NMR experiment. The sample sits inside a radio frequency (RF) coil which produces an oscillating RF field. A highly homogenous static magnetic field is provided by a magnet. The static field $\mathbf{B_0}$ and the oscillating field $\mathbf{B_1}$ are perpendicular to each other. In a real experiment, the sample would be much smaller than is shown here so that it experiences a uniform field from the RF coil.

a coil which is mounted between the pole pieces of a magnet. The magnet produces a magnetic field $\mathbf{B_0}$ along a particular direction, say the $z$-direction. We have already seen that the quantity $m_I$, the $z$ component of the angular momentum of the nucleus, can only take integral values between $-I$ and $I$.

[3]The positive charge of the nuclei and the negative charge of the electron is responsible for the differing signs in the expressions:

$$\mu = +g_N\mu_N\mathbf{I}$$
$$\mu = -g_J\mu_B\mathbf{J},$$

for the nuclear and electronic moments respectively.

[4]For the time being, we set $J = 0$ and thus ignore the electron spin.

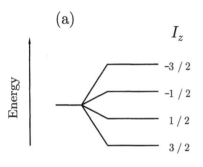

(a)

$I_z$

−3 / 2

−1 / 2

1 / 2

3 / 2

Energy

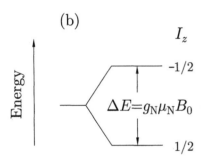

(b)

$I_z$

−1/2

$\Delta E = g_N\mu_N B_0$

1/2

Energy

**Fig. 3.10** (a) The four levels from an $I = \frac{3}{2}$ nucleus. (b) In the simpler case of an $I = \frac{1}{2}$ nucleus, there are just two levels. These two nuclear levels are separated by an energy $\Delta E = g_N\mu_N B_0$; the lower (upper) level corresponds to the nuclear magnetic moment lying parallel (antiparallel) to the magnetic field $B_0$.

[5]Actually you do not have to get it exactly right. You can be plus or minus the linewidth which can be determined by various factors, including variations in the magnetic field at the nucleus due to the dipolar field produced by neighbouring nuclei.

The energy $E$ of the nucleus is the energy of a magnetic moment[3] $\mu$ in the magnetic field $\mathbf{B}_0$, and thus $E = -\mu \cdot \mathbf{B}_0$ so that

$$E = -g_N\mu_N m_I B_0. \tag{3.6}$$

This corresponds to a ladder of $2I + 1$ equally spaced levels with mutual separation $g_N\mu_N B_0$, as illustrated in Fig. 3.10(a) for the case $I = \frac{3}{2}$, as found in Na or Cu.[4] Exciting transitions between adjacent pairs of levels with a radiofrequency (RF) field is the basis of nuclear magnetic resonance. The RF field $\mathbf{B}_1$ is applied in the $x$-direction, and leads to a perturbation of the system which is proportional to $B_1 \hat{I}_x$. The matrix element of the perturbation is proportional to $\langle m'_I | \hat{I}_x | m_I \rangle$ and is zero unless $m'_I = m_I \pm 1$. The allowed transitions are therefore described by the selection rule,

$$\Delta m_I = \pm 1, \tag{3.7}$$

which implies that only transitions between adjacent states may occur. In the most common arrangement, the RF coil not only produces the excitation, but is itself part of a tuned circuit with a large $Q$ factor. As transitions are excited in the nuclei by the RF field, energy is transferred between the RF circuit and the sample, and this results in small changes of the $Q$ factor of the circuit.

There are two ways to perform the experiment: you could keep the frequency of the RF field constant and vary the magnetic field or you could keep the magnetic field constant and vary the frequency of the RF field. It is usually the former which is performed. A crucial factor is to have a highly homogeneous magnet to produce the constant field $B_0$. If this is not the case, different parts of the sample will sit in slightly different magnetic fields and will come into resonance at slightly different points of the magnetic field sweep, causing the measured resonance to be extremely broad, maybe so much that it is washed out altogether.

### Example 3.2

For a proton, $I = \frac{1}{2}$ and $m_I$ can take the values $\pm\frac{1}{2}$ only, as illustrated in Fig. 3.10(b). The two states of the system are separated by an energy $\Delta E = g_N\mu_N B$ which is tiny: for a proton in a typical laboratory magnetic field $B_0 \sim$ 1 T this corresponds to an energy splitting of $\sim 10^{-7}$ eV; when this energy is expressed as $k_B T$ it is equivalent to a temperature of only $\sim 1$ mK.

Therefore at room temperature, at this magnetic field, the nuclei will show only a minute tendency to line up with the applied magnetic field on average because the thermal randomizing energy will *vastly* dominate over the weak alignment energy. Any effect due to the magnetism of the nuclei would be practically impossible to detect if a resonant technique were not employed. Because in NMR one is perturbing the system with a RF signal of *exactly*[5] the right frequency $\omega$ given by

$$\hbar\omega = \Delta E \tag{3.8}$$

we can excite transitions across this energy splitting. Why do we need radiofrequencies? The next example shows why.

**Example 3.3**

For a magnetic field of 1 T, the frequency $\nu$ is given by

$$\nu = \frac{\omega}{2\pi} = \frac{g_N \mu_N B_0}{2\pi\hbar} = 42.58 \text{ MHz} \tag{3.9}$$

which is in the radiofrequency region of the electromagnetic spectrum.

This means one can use oscillators and coils (in comparison with electron spin resonance experiments which have to use Gunn diodes and microwave waveguides, see Section 3.2.2). Therefore at exactly the right frequency of RF excitation the system will absorb energy. This is the nuclear magnetic resonance and is a resonance between the frequency of the RF excitation and the energy separation of the nuclear levels. Modern NMR spectrometers use 12–15 T magnets, so the frequency of the RF is in the $\sim$500–650 MHz range, still in the radio frequency region of the electromagnetic spectrum.

The NMR resonance frequency at a given magnetic field can be slightly shifted up and down from the value $\gamma B/2\pi$ depending on the chemical environment of the nucleus. These **chemical shifts** (typically a few parts per million) are due to the fact that the electrons orbiting the nucleus slightly shield the nucleus from the applied field. The amount by which a nucleus is shielded in a given chemical environment is well known, allowing a given molecule to be 'fingerprinted'. NMR lines can split due to magnetic coupling to neighbouring nuclei. This is known as **spin–spin coupling** and occurs via an indirect contact hyperfine interaction which is mediated by the electrons. This also gives information about the environment of the nucleus.

In order to understand the transitions which we are inducing in a little more detail, let us consider how a two-level spin system ($I = \frac{1}{2}$) will absorb energy from the RF coil. We will label the lower level $-$ and the upper level $+$. The probability of inducing a transition between the two levels is independent of whether the transition is from the lower to the higher level or *vice versa*. This is because the perturbing Hamiltonian $\hat{\mathcal{H}} \propto \hat{I}_x$, appropriate for an oscillating RF field, is Hermitian, so that

$$|\langle\psi_-|\hat{\mathcal{H}}|\psi_+\rangle|^2 = |\langle\psi_+|\hat{\mathcal{H}}|\psi_-\rangle|^2. \tag{3.10}$$

Hence the probability per unit time of these so-called **stimulated transitions** between levels $+$ and $-$ is independent of the direction of the transition and occurs at a rate $W$ which is proportional to the size of the RF power used to excite transitions. At time $t$, if there are $N_-(t)$ spins in the lower level, $WN_-(t)$ will be excited per unit time into the upper level. If there are $N_+(t)$ spins in the upper level, $WN_+(t)$ will be excited per unit time into the lower level. Thus

$$\frac{dN_+(t)}{dt} = WN_-(t) - WN_+(t) \tag{3.11}$$

$$\frac{dN_-(t)}{dt} = WN_+(t) - WN_-(t) \tag{3.12}$$

which when subtracted gives

$$\frac{d}{dt}(N_+(t) - N_-(t)) = -2W(N_+(t) - N_-(t)), \tag{3.13}$$

which can be solved to give

$$N_+(t) - N_-(t) = (N_+(0) - N_-(0))e^{-2Wt}. \tag{3.14}$$

An initial difference in population tends exponentially to zero when driven by a stimulated electromagnetic transition. Let us now define $n(t)$ by

$$n(t) = N_+(t) - N_-(t), \tag{3.15}$$

so that eqn 3.14 becomes

$$n(t) = n(0)e^{-2Wt}. \tag{3.16}$$

The rate of absorption or emission of electromagnetic energy by the spin system is easily calculated since each transition involves an exchange of energy equal to $\hbar\omega$. The energy of the system at time $t$ is

$$E(t) = N_-E_- + N_+(E_- + \hbar\omega), \tag{3.17}$$

where $E_-$ is the energy of the lower level. This expression for $E(t)$ can be rearranged to give a constant term plus $\frac{1}{2}\hbar\omega n(t)$. Thus the rate of absorption of energy is given by

$$\frac{dE}{dt} = -W\hbar\omega n(t) \tag{3.18}$$

and, because it is proportional to $n(t)$, will tend to zero with a time constant of $1/2W$ as the populations of the upper and lower levels become progressively equalized. This demonstrates also that to absorb energy you need $n(t) \neq 0$, i.e. a population difference. This is produced by interaction with the thermal modes of the system, so that the polarization of the system is brought back towards a Boltzmann probability expression:

$$\left(\frac{N_+}{N_-}\right)_0 = e^{-\hbar\omega/k_B T}. \tag{3.19}$$

Here, the subscript 0 denotes the equilibrium value of the system. After absorption of energy the polarization may change, but if the nuclear spin system interacts with the thermal motion and excitations of the sample, the polarization of the spin system can return towards its equilibrium value. If the polarization of the spin system has been reduced to zero by rapid electromagnetic transitions, the spin system will take some time to 'recover' once the electromagnetic transitions are switched off; this time is called $T_1$ and measures the time constant of the interaction between the spin system and its surroundings. $T_1$ is the spin–lattice relaxation time (sometimes called the longitudinal relaxation time). Thus we would expect that the polarization would be restored as

$$n(t) = n_0(1 - e^{-t/T_1}). \tag{3.20}$$

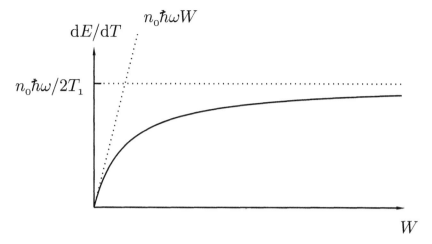

**Fig. 3.11** The rate of absorption of electromagnetic energy given by eqn 3.25 as a function of the transition rate $W$. For low $W$, the expression is proportional to $W$, but at high $W$ it saturates at a value governed by $1/T_1$.

Now let us put both processes in together, the stimulated transitions and the relaxation processes. Thus we may combine our equations to arrive at

$$\frac{\mathrm{d}}{\mathrm{d}t}n(t) = -2Wn(t) + \frac{n_0 - n(t)}{T_1},$$ (3.21)

which has a steady state solution when

$$\frac{\mathrm{d}}{\mathrm{d}t}n(t) = 0.$$ (3.22)

Thus

$$n(t) = \frac{n_0}{1 + 2WT_1},$$ (3.23)

and the steady state rate of absorption of electromagnetic energy is given by

$$\frac{\mathrm{d}E}{\mathrm{d}t} = n(t)\hbar\omega W$$ (3.24)

$$= n_0\hbar\omega\frac{W}{1 + 2WT_1}.$$ (3.25)

This relation is plotted in Fig. 3.11. For low-amplitude RF perturbing fields, the rate of absorption of electromagnetic energy is proportional to $W$. But at large RF perturbing fields, it settles at a level proportional to $T_1$, and independent of the precise magnitude of $W$. This is known as saturation.

The rate of absorption of electromagnetic energy is proportional to the population difference between the upper and lower levels. The rate of absorption may therefore be used to monitor the population difference and to observe its time dependence, subject of course to the constraint that such observation will necessarily perturb the population difference to some extent. Equation 3.25 and Fig. 3.11 demonstrate that you can only increase the power of the RF signal so much to improve the signal. This also demonstrates that $\mathrm{d}E/\mathrm{d}t$ is proportional to $\hbar\omega$; more energetic (i.e. higher frequency) photons lead to a larger signal. But of course this means larger magnetic fields to match $g\mu_N B$ to $\hbar\omega$. This will prove to be an advantage for ESR (see Section 3.2.2 below) since very beefy microwave photons are needed to excite transitions across the larger electronic gaps, and this leads to a higher sensitivity. Because electronic moments are larger than nuclear moments, you can get away with smaller magnetic fields than NMR and still use higher frequency photons.

Let us return to NMR and consider relaxation processes in a little more detail. In a magnetic field a very weak polarization of the spins will exist (weak because the energy separation $\Delta E \ll k_B T$.) Let us say it takes an equilibrium value $M_0$ with the static magnetic field $\mathbf{B}_0$ switched on (recall that $\mathbf{B}_0$ is along the $z$ axis). The effect of the radio frequency excitations will be to progressively destroy this magnetization. Also, as we have seen, if the magnetic field is switched off, the magnetization will relax back to the equilibrium value by the weak interactions with the surroundings with a time constant $T_1$. Thus we expect that

$$\frac{dM_z}{dt} = \frac{M_0 - M_z}{T_1}. \tag{3.26}$$

This $T_1$ relaxation must involve interactions with the lattice because energy must be exchanged with it. This is because changing $M_z$ has energetic consequences since relaxing a spin in the $z$-direction changes its orientation with respect to the applied field, and hence changes its energy.

The $M_x$ and $M_y$ components should be zero, but if they are not they will relax back to zero in a time $T_2$ such that

$$\frac{dM_x}{dt} = -\frac{M_x}{T_2}, \qquad \frac{dM_y}{dt} = -\frac{M_y}{T_2}. \tag{3.27}$$

$T_2$ is the **spin–spin relaxation time** and is the characteristic time for dephasing because it corresponds to the interaction between different parts of the spin system. It can also be due to inhomogeneities in the magnetic field $B_0$. It thus leads to differences in precession frequency due to the interactions between the observed spin and the spins of its neighbours. Changing $M_x$ or $M_y$ has no energetic consequences because the applied field is along $M_z$.

The effect of the applied magnetic field is also to cause spin precession, so that the equations for $M_x$, $M_y$ and $M_z$ are

$$\frac{dM_x}{dt} = \gamma (\mathbf{M} \times \mathbf{B})_y - \frac{M_x}{T_2}, \tag{3.28}$$

$$\frac{dM_y}{dt} = \gamma (\mathbf{M} \times \mathbf{B})_y - \frac{M_y}{T_2}, \tag{3.29}$$

$$\frac{dM_z}{dt} = \gamma (\mathbf{M} \times \mathbf{B})_z + \frac{M_0 - M_z}{T_1}, \tag{3.30}$$

Felix Bloch (1905–1983)

which are known as the **Bloch equations**. They were worked out in 1946 and can be solved for a number of cases of interest, very often by using a rotating reference frame method in which the coordinates are changed to ones which rotate in the $xy$ plane at the resonance frequency.

The spin–spin relaxation time $T_2$ can be measured using the following technique: in thermal equilibrium with $B_1$ switched off, there is a weak magnetization parallel to $z$. A short pulse of RF signal of duration $t_p$, with the frequency of the RF signal set close to the resonance frequency of the spins, causes the spins to rotate by an angle $\gamma B_1 t_p$ where $B_1$ is the amplitude of the RF signal. This angle can be adjusted by changing $t_p$ and if $t_p = \pi / 2\gamma B_1$, it produces what is known as a **90° pulse**. The amplitude and time of the pulse are chosen

signal

$t$

**Fig. 3.12** A simulated free induction decay signal. The oscillations die away with time due to spin–spin relaxation ($T_2$ processes). The data show an interference pattern which results from three inequivalent nuclei, each of which experiences a slightly different magnetic field because of their different sites within a crystal.

so that the magnetization just precesses for a short period and rotates into the $xy$ plane. With $B_1$ switched off, the magnetization now precesses in the static field $\mathbf{B_0}$, rotating steadily in the $xy$ plane at a frequency $\gamma B_0$, producing an induced voltage in the coil. The oscillation would carry on for ever if it were not for $T_2$ relaxation processes (spin–spin relaxation), which cause the oscillations to relax, thereby giving a measurement of $T_2$. The corresponding decay of voltage in the coil is known as **free induction decay** (see Fig. 3.12).

One problem that occurs is that this method also measures relaxation due to the inhomogeneity of the magnet because this will cause different parts of the sample to give slightly different precession frequencies. What is needed is a method to separate the real spin–spin relaxation from the relaxation due to the inhomogeneity of the magnet. One way this can be achieved is to use a **spin echo technique**. The principle of this technique is illustrated in Fig. 3.13 and involves both a 90° pulse (which tips the spins by 90° into the $xy$ plane as before) and the subsequent application of a 180° pulse (which rotates the spins by 180°). The effect can be likened to a marathon in which all athletes run steadily but some athletes are faster than others. After the starter's pistol (the 90° pulse) the athletes begin to run as a pack, but after a while the fitter begin to lead while others begin to trail behind. Then suddenly, at time $\tau$, the rules of the game are changed (the 180° pulse) and the runners are told to turn around and head for the start of the race. Now the slower runners have an advantage since they are closer to the start. But by symmetry it is easy to see that the runners will all arrive at the start of the race simultaneously, at time $2\tau$, regardless of their individual speeds. In the same way, all the spins realign after $2\tau$, regardless of their individual dephasing. This is the spin echo.

Thus in the experiment we can use the spin echo to remove the problem that the magnetic field is inhomogeneous causing spins to precess at different rates. Also the spread in fields due to chemical shifts is removed. But we will not have taken away the effect of the spin-spin interaction which is due to time-dependent fluctuating random magnetic fields due to neighbouring nuclei. This relaxation mechanism cannot be refocussed and so the echo signal will be reduced in amplitude by an amount which depends on $\tau$. The NMR intensity of the echo signal should follow $I(2\tau) = I(0)\mathrm{e}^{-2\tau/T_2}$ and so a measurement of the true $T_2$, due to the spin-spin coupling, is possible.

A measurement of $T_1$ can be performed using a similar method, though this time a 180° pulse is applied first. This causes the magnetization to rotate from

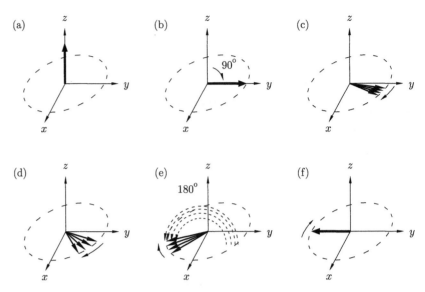

**Fig. 3.13** The spin echo effect. (a) The equilibrium magnetization initially lies along the $z$ direction, parallel to $B_0$. (b) At a time which we will call $t = 0$, a 90° pulse along the $x$ axis is used to rotate the spins into the $xy$ plane. (c) Because of the steady field $B_0$ along the $z$ axis, the spins now precess in the $xy$ plane, though at slightly different rates because of inhomogeneities in the magnetic field. (d) This means they progressively dephase with respect to each other, becoming more separated in angle as they precess round. (e) At time $t = \tau$, a 180° pulse along the $x$ axis rotates the spins by 180° around the $x$ axis, and as they subsequently precess in the field $B_0$, their order has been reversed. (f) They then will come back together at time $2\tau$, producing a spin echo signal, provided no other relaxation process has occurred in the meantime. Although the figure has been drawn for the case of a short delay between the 90° pulse and the 180°, the effect will work if the delay is long and the spins make a number of rotations before the 180° pulse is applied.

along $\hat{\mathbf{z}}$ to along $-\hat{\mathbf{z}}$. It then relaxes back to $\hat{\mathbf{z}}$ with a time constant $T_1$ but does not precess. Thus as a function of time $\tau$ after the 180° pulse

$$\mathbf{M}(\tau) = \hat{\mathbf{z}}M_z(\tau) = \hat{\mathbf{z}}M_0(1 - 2e^{-\tau/T_1}). \tag{3.31}$$

At a time $\tau$ after the 180° pulse, a 90° pulse is used to rotate the magnetization into the $xy$ plane where it now begins to do a free induction decay with initial amplitude $M_z(\tau)$. Thus by measuring the initial amplitude of the free induction decay as a function of the time delay between the 180° pulse and the 90° pulse, $T_1$ can be deduced. A variety of other pulse sequences can also be used to extract additional types of information.

### 3.2.2 Electron spin resonance

One can perform the analogous experiment to NMR with electrons. In this case the effect is known as **electron spin resonance** (ESR), or sometimes as **electron paramagnetic resonance** (EPR). The electron magnetic moment is much larger than that of nuclei so the precession frequencies are also much higher. For typical laboratory magnetic fields, the electromagnetic radiation is now in the microwave regime. A schematic of the experimental setup is shown in Fig. 3.14. The sample sits in a resonant cavity and microwave radiation enters via a waveguide. It is usually much more convenient to keep the frequency of

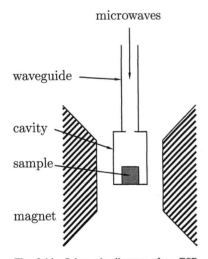

**Fig. 3.14** Schematic diagram of an ESR experiment. Microwaves enter a cavity via a waveguide and the absorption of microwaves induced by a resonance is measured by monitoring the $Q$-factor of the cavity. The sample must be placed in the centre of the magnet, where the field is most clearly uniform.

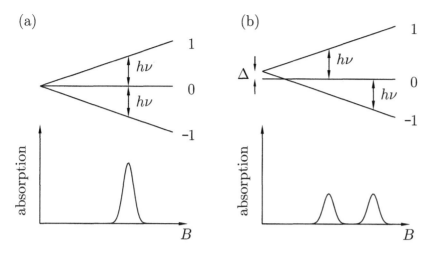

**Fig. 3.15** ESR in an ion with $J = 1$ such as $Ni^{2+}$ with (a) no crystal field splitting (leading to a single ESR line) and (b) crystal field splitting $\Delta$ (leading to two ESR lines).

the microwave radiation fixed and to sweep the magnetic field. The cavity has a very high quality factor ($Q$-factor) and is designed to enhance the sensitivity of the detection of the weak ESR signal. The microwave absorption is modified as a function of the applied magnetic field. Very often a set of modulation coils is added to provide a small oscillatory magnetic field at the sample in addition to the static magnetic field provided by the magnet. The measurement can then be further enhanced by a phase sensitive detection technique.[6]

In an ESR transition the selection rule is

$$\Delta m_J = \pm 1, \tag{3.32}$$

the familiar dipole selection rule because one is inducing magnetic dipole transitions. Figure 3.15 shows these transitions in an ion with $J = 1$ such as $Ni^{2+}$ in an octahedral environment. The experiment is performed at fixed frequency $h\nu$ with the magnetic field swept. Every time two adjacent levels are separated by $h\nu$, there is an ESR transition. Figure 3.15 demonstrates how the position and the number of lines can give information about the crystal field splitting.

Figure 3.16 shows the energy of a free $Mn^{2+}$ ion (spin-$\frac{5}{2}$) as a function of magnetic field. At a magnetic field $B$ this gives a number of possible transitions between the different $m_J$ levels, all of which have size $g\mu_B B$. Thus a single line might be expected to be observed in an ESR experiment at a frequency $\nu$ given by $h\nu = g\mu_B B$. However, in a real experiment on a solid containing $Mn^{2+}$ ions, the crystal field significantly complicates this picture, even in a cubic environment.

However, the hyperfine coupling between the nucleus and the electron gives a term in the Hamiltonian equal to $A\mathbf{I} \cdot \mathbf{J}$ and each $m_J$ level is split into $2I + 1$ hyperfine levels depending on the value of $m_I$. In an ESR transition the selection rules are therefore given by eqn 3.32 and by

$$\Delta m_I = 0. \tag{3.33}$$

[6]A lock-in amplifier is used to produce a signal which drives the oscillatory magnetic field. The microwave signal is then fed into the lock-in amplifier, which rejects any component of this signal which is not at the frequency of the oscillatory magnetic field. In this way the noise is drastically reduced.

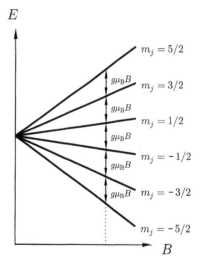

**Fig. 3.16** The energy of a spin-$\frac{5}{2}$ magnetic moment, such as a free $Mn^{2+}$ ion, as a function of magnetic field. The possible transitions with the selection rule $\Delta m_J = \pm 1$ are as shown. This picture ignores the crystal field.

The latter selection rule occurs because the microwave frequency induces only dipole transitions between electronic levels, not between nuclear levels.[7] Since $A$ is very small in comparison with $g\mu_B B$ for typical fields in an ESR experiment, one need only compute the components of $\mathbf{J}$ parallel to $\mathbf{B}$ so that the energy levels in a magnetic field are

$$E = g\mu_B B m_J + A m_I m_J \tag{3.34}$$

where the energy of the nuclear spin in the applied magnetic field has been ignored (since it is much smaller even than the hyperfine energy). Using the selection rule, one can now show that the resonances should occur at

$$hv = [g\mu_B B(m_J + 1) + A m_I(m_J + 1)] - [g\mu_B B m_J + A m_I m_J]$$
$$= g\mu_B B + A m_I \tag{3.35}$$

and each ESR line will split up into $2I + 1$ hyperfine lines.

### Example 3.4

For the case of a free $Mn^{2+}$ ion, $I = \frac{5}{2}$, so this means six lines in the ESR spectrum. This situation is depicted in Fig. 3.17 (although for a $Mn^{2+}$ ion in a crystal, the crystal field causes the splittings to be more complex and to depend on the angle between the magnetic field and the crystal axes in a non-trivial way).

ESR experiments probe electronic spins in a useful range because crystal field splittings are very often at GHz frequencies. ESR studies in paramagnetic salts have provided a great deal of detailed information concerning the crystal and ligand fields. Very often the effect is anisotropic so that lines move as the magnetic field is rotated with respect to the crystal axis.

### Example 3.5

As an example of how this information can be obtained, consider $Ce^{3+}$ ($4f^1$) in an environment with axial symmetry (see Fig. 3.18). This ion has $L = 3$ and $S = \frac{1}{2}$ so the spin–orbit interaction splits this into an eightfold degenerate $J = \frac{7}{2}$ level and a lower sixfold degenerate $J = \frac{5}{2}$. The crystal field is axial and contains only even powers of $J_z$ by symmetry, so the sixfold $J = \frac{5}{2}$ manifold (as a group of lines are called) splits up into three doublets, and the eightfold $J = \frac{7}{2}$ manifold into four. These splittings are much smaller than the spin–orbit interaction in rare earth systems. The doublets can be split by applying a magnetic field which is almost always the smallest energy term for typical laboratory magnetic fields, usually around $10^{-4}$ eV. At low temperature, only the ground state will be occupied and it is this transition which will be observed.

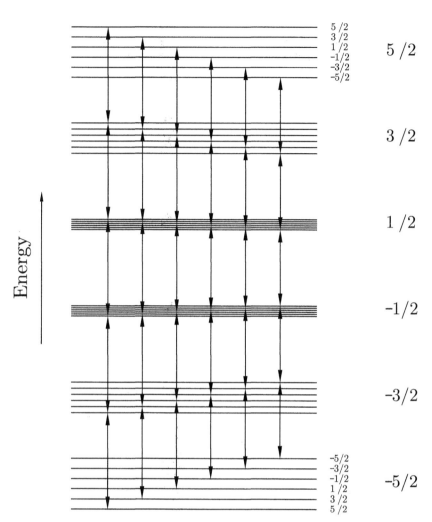

Energy

$5/2$

$3/2$

$1/2$

$-1/2$

$-3/2$

$-5/2$

**Fig. 3.17** The hyperfine splitting in a free $Mn^{2+}$ ion at a fixed magnetic field. $J = \frac{5}{2}$ and $I = \frac{5}{2}$ so each of the six $m_J$ levels is split into six hyperfine levels. The possible transitions with $\Delta m_J = \pm 1$ and $\Delta m_I = 0$ are shown, giving rise to six distinct lines in the ESR spectrum. This picture ignores the effect of the crystal field which leads to these splittings becoming strongly dependent on the direction of the applied magnetic field with respect to the crystalline axes.

The ground state of many systems is, like $Ce^{3+}$, a doublet. This is a consequence of **Kramers theorem**, which states that in a system containing an odd number of electrons, at least two-fold degeneracy must remain in the absence of a magnetic field. The pairs of states involved, **Kramers doublets**, are time conjugate[8] and therefore can be split by a magnetic field but not by an electric field. $Ce^{3+}$ ($4f^1$) has an odd number of electrons in the 4f shells and therefore qualifies as a **Kramers ion**.

The ground state of $Ce^{3+}$ is a Kramers doublet and, viewed in isolation, looks like a spin-$\frac{1}{2}$ state, although its quantum numbers are very different. It can in fact be assigned an **effective spin** $\tilde{S} = \frac{1}{2}$. This is a fictitious angular momentum which is chosen so that the degeneracy of the group of levels under consideration is set equal to $(2\tilde{S} + 1)$. The motivation for this substitution is to find an effective spin Hamiltonian that approximately describes the experimental situation in the language of something much simpler, namely a free atom or ion with spin-$\frac{1}{2}$. This Hamiltonian can be written

$$\tilde{H} = g\mu_B \mathbf{B} \cdot \tilde{\mathbf{S}} \qquad (3.36)$$

Hendrik A. Kramers (1894–1952)

[8] that is, they are complex conjugates of each other, and are thus time reversed versions of each other.

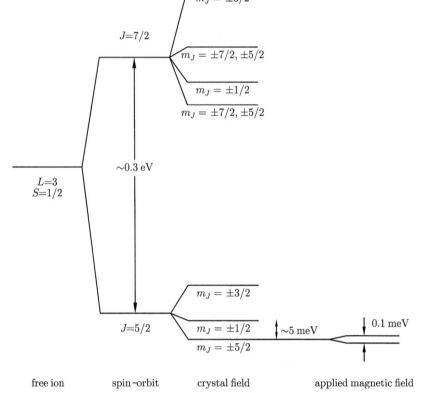

**Fig. 3.18** The energy levels for a $Ce^{3+}$ ion. The dominant energy splitting is the spin–orbit interaction (this energy corresponds to about 3000 K so the upper manifold is not appreciably thermally excited). The two manifolds are split into sets of doublets by an assumed axial crystal field. The crystal field splittings have been exaggerated in this diagram so that they are easily observable. The lower splitting is ∼5 meV which corresponds to about 50 K. Finally an applied magnetic field splits all the doublets (only the lower one is shown split here).

where $g$ is now an effective g-factor (sometimes called a spectroscopic splitting factor) which would be equal to 2 for a real isolated spin-$\frac{1}{2}$ electron but includes the orbital contribution in this case. Very often one needs to use an effective g-tensor, $\mathsf{g}$, to express the fact that the interaction of the spin with the field depends on the orientation of the magnetic field, so that eqn 3.36 must be replaced with

$$\tilde{H} = \mu_B(\mathbf{B} \cdot \mathsf{g} \cdot \tilde{\mathbf{S}}) = \sum_{ij} g_{ij} B_i \tilde{S}_j. \tag{3.37}$$

The effect of the crystal field can then be included by adding in a term to the Hamiltonian to represent the energetic preference for the spin to lie along particular crystalline directions because of the crystal field. This is known as **single ion anisotropy** and the term in the Hamiltonian, $\tilde{H}_{SI}$, is

$$\tilde{H}_{SI} = -DS_z^2, \tag{3.38}$$

for a **uniaxial crystal** (i.e. one with a particular axis such that the energy just depends on the angle of the spin with that axis) and for a cubic crystal

$$\tilde{H}_{SI} = -D(S_x^4 + S_y^4 + S_z^4) \tag{3.39}$$

where $D$ is an anisotropy constant.

ESR is useful not only in studying the physics of salts with paramagnetic ions but is used extensively in chemistry. The resonances can also be extremely sensitive to atomic position in a molecule and the technique is greatly used in the study of chemical reactions and particularly of **free radicals**, which are atoms or fragments of molecules with an unpaired electron and are often highly reactive. Free radicals are of great environmental and biological importance. ESR is not as widely used as NMR because it is applicable only to materials with an unpaired spin.

---

**Example 3.6**

One example of a field in which ESR is very useful is in studies of various biological molecules which often contain transition metal ions. An important example is haem, an iron porphyrin group, which is found in haemoglobin and myoglobin. These molecules are involved in transfer of oxygen and other small molecules such as carbon monoxide. The Fe(II) ion in the haem converts from $S = 2$ to $S = 0$ after binding to the small molecule (this is an example of a spin transition, see Section 3.1.1).

---

The spin-lattice relaxation times in ESR are often very short in comparison with those of NMR; this is because the electronic moment is much more strongly coupled to lattice vibrations than are the nuclei. Lattice vibrations cause a modulation in the crystal field which couples to the electronic moment via the spin–orbit interaction, often making $T_1$ so short that the resulting ESR line is too broad to observe. (The width of the line scales as $T_1^{-1}$). A favourable case is $Mn^{2+}$ which is an S-state ion ($3d^5$, $^6S_{5/2}$) and therefore has no orbital moment to which the spins can couple and ESR can be observed easily in this ion at room temperature. For other transition metal ions, the size of $T_1$ depends on the degree of orbital quenching.[9] The resonance can be narrowed ($T_1$ increased) by cooling the sample, since this reduces the lattice vibrations. This also has the benefit of increasing the relative thermal population of the levels which increases the intensity of the resonance.

A variant of ESR is a technique called **ENDOR** (an acronymn for Electron Nuclear DOuble Resonance) which can be used to measure hyperfine interaction constants and the nuclear Zeeman interaction with great precision. It is often impossible to measure directly NMR transitions between nuclear levels in paramagnetic salts because they are very weak. In ENDOR one uses both an RF signal *and* a microwave signal and capitalizes on the high sensitivity of ESR. The microwave power is tuned to a $\Delta m_I = 0$ transition and the ESR signal then depends on the relative population of those levels. The populations can be changed by using RF power tuned to the NMR transition. The NMR transition can then be detected via its effect on the ESR signal.

[9]The more completely the orbital moment is quenched towards zero, the smaller the coupling of the electronic moment to the lattice vibrations and this leads to a longer $T_1$ and hence a narrower ESR line.

### 3.2.3   Mössbauer spectroscopy

Another form of spectroscopy is based upon the Mössbauer effect which was discovered in 1957. The principle of the experiment is illustrated in Fig. 3.19. A source containing $^{57}Co$ nuclei provides a ready supply of excited $^{57}Fe$ nuclei; these decay to the ground state via a gamma ray cascade which includes a

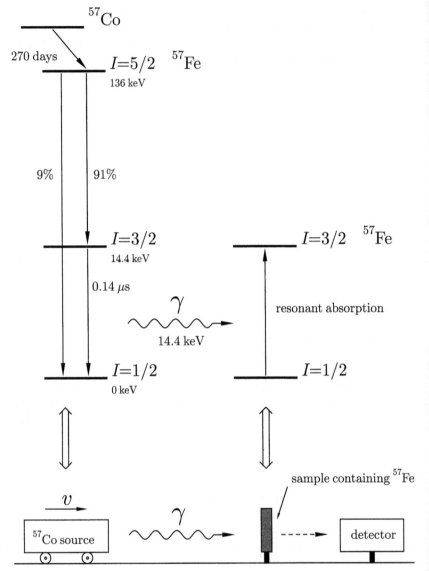

**Fig. 3.19** Schematic illustration of the principles of the Mössbauer technique. $^{57}$Co decays slowly into an excited state of the $^{57}$Fe nucleus. The majority (91%) of the subsequent decay is rapidly to the $I = \frac{3}{2}$ state which then decays to the $I = \frac{1}{2}$ ground state. This decay releases a 14.4 keV photon. The experiment is performed by moving the $^{57}$Co source at a speed $v$ relative to the sample which must contain some $^{57}$Fe nuclei. The detector measures the transmission of $\gamma$-rays through the sample which can be used to deduce the absorption.

14.4 keV gamma ray (corresponding to a frequency $\nu = 3.5 \times 10^{18}$ Hz). This gamma ray can excite a transition in the sample being studied if it is absorbed resonantly. To do this, its energy must match the energy gap in the sample. By moving the source at speed $v$ one can very slightly adjust the frequency of the gamma ray because of the Doppler effect. Because of the high frequency of the photon, the Doppler shifts can be quite significant: a velocity $v = 1$ mm s$^{-1}$ leads to a shift of $\nu v/c \approx 12$ MHz. Thus one can probe any splittings in the ground state in the source or absorber nucleus which might result from magnetic or other interactions.

The technique would be useless if the $\gamma$ ray photon emitted by the excited $^{57}$Fe nucleus did not have a well defined frequency. This is why the relatively slow decay of the $^{57}$Fe nucleus is vital: the rather plodding 0.2 $\mu$s half-life corresponds to an uncertainty of only about 2 MHz in the frequency. A second vital feature is that the $^{57}$Fe atoms are in the solid state. In order to conserve momentum, a free Fe atom would be subject to a recoil velocity of the order of $h\nu/m_{Fe}c$ ∼80 m s$^{-1}$ which would ruin the experiment (we have seen that even a relative velocity of 1 mm s$^{-1}$ can produce a measurable effect). However a $^{57}$Fe atom which is held rigidly in a solid transmits that momentum to the entire crystal. If the recoil energy is lower than a certain quantity, the probability of phonon emission becomes vanishingly small and the gamma ray can be emitted without any loss of recoil energy. This recoil-free emission and resonant absorption of $\gamma$-rays is the essence of the **Mössbauer effect**.

Rudolf L. Mössbauer (1929–)

The conditions described above therefore imply that the effect will be optimized for low-energy $\gamma$-rays associated with nuclei strongly bound in a crystal lattice at low temperatures. $^{57}$Fe is not the only isotope which is ideal for this purpose (other isotopes which can be used include $^{119}$Sn, $^{127}$I, $^{151}$Eu, and $^{197}$Au) but it is the most commonly used. The raw result which one obtains from a Mössbauer measurement is a plot of $\gamma$-ray counts (or relative absorption) against the velocity of the source with respect to the absorber.

Why should this effect tell us anything about the sample? For a start, the resonant absorption may not occur exactly where you would expect it to (i.e. when the source is stationary), but may be slightly shifted (i.e. to when the source is moving at a particular velocity). This **isomer shift** is due to the slight change in the Coulomb interaction between the nuclear and electronic charge distributions over the nuclear volume which is associated with the slight increase of size of the $^{57}$Fe nucleus in the $I = \frac{3}{2}$ state.

Furthermore, one may not necessarily just observe one resonant absorption line as a function of source velocity, but perhaps a number of lines. This can be due to **quadrupole splitting** or **magnetic splitting**. The first effect is due to the electric quadrupole moment of the excited $^{57}$Fe nucleus (although the ground state of $^{57}$Fe has $I = \frac{1}{2}$ and thus has no electric quadrupole moment, the excited state of interest has $I = \frac{3}{2}$ and nuclei with $I > \frac{1}{2}$ can have a non-zero quadrupole moment). If the nucleus is subjected to an electric field gradient, as may be found in certain crystal environments, the interaction between the nuclear quadrupole moment and the electric field gradient splits the excited $I = \frac{3}{2}$ state into a doublet, so that two lines are produced in the Mössbauer spectrum. The second effect is caused by the interaction between the nucleus and the local magnetic field. This can split the $I = \frac{1}{2}$ ground state into a doublet, and the excited $I = \frac{3}{2}$ into a quadruplet, leading to six possible lines in the Mössbauer spectrum (the selection rule is $\Delta m_I = 0, \pm 1$). This effect can be used to detect magnetic exchange interactions and local magnetic fields. All these shifts and splittings are illustrated schematically in Fig. 3.20. Note that the figure is not (and could not be) drawn to scale. The observed splittings are typically in the range $10^7$–$10^8$ Hz and are therefore 11 or 12 orders of magnitude smaller than the energy gap between the ground ($I = \frac{1}{2}$) and excited state ($I = \frac{3}{2}$) of the $^{57}$Fe nucleus. Putting this another way, one is measuring energy splittings of less than 1 $\mu$eV with a 14.4 keV photon! Were it not for the Mössbauer effect, the remarkable recoil-free resonance absorption and emission of these $\gamma$ rays, such experiments would be completely impossible.

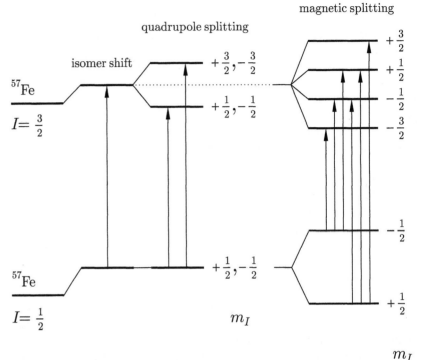

**Fig. 3.20** The effects of chemical shift, quadrupole splitting and magnetic splitting on the nuclear energy levels of $^{57}$Fe. The arrows show the Mössbauer absorption transitions. The difference in the size of the transitions is greatly exaggerated; in reality they typically differ from each other by less than 1 part in $10^{11}$.

### 3.2.4  Muon-spin rotation

The **muon** is a spin-$\frac{1}{2}$ particle (charge $\pm e$, mass $250m_e$) which has a lifetime of 2.2 $\mu$s. The muon is found in nature as the dominant constituent of cosmic rays arriving at sea-level. Muons can be used to study the magnetic properties of samples, but cosmic rays do not provide a sufficiently intense source. For this purpose it is therefore necessary to use the more intense beams of muons available from synchrotrons and cyclotrons.

A technique called **muon-spin rotation** (often abbreviated to $\mu$SR) will now be described in some detail. It is important to realize that in sharp contrast to the neutron and X-ray techniques that will be discussed later in the book, scattering is not involved; muons are implanted into a sample of interest and reside there for the rest of their short lives, never to emerge again. It is the positrons into which they decay that are released from the sample and yield information about the muons from which they came.

The mass of the muon is intermediate between that of the electron and the proton, and thus so are its magnetic moment and gyromagnetic ratio. The muon comes in either charge state, although it is the positive muon, $\mu^+$, which is of particular use in experiments in magnetism. As a small, positively charged particle, it is attracted by areas of large electron density and stops in interstitial sites in inorganic materials or bonds directly on to organic molecules. By contrast the negative muon, $\mu^-$, implants close to an atomic nucleus and is generally much less sensitive to magnetic properties.

Muons can be prepared by colliding a high energy proton beam with a suitable target which produces pions. The pions ($\pi^+$) decay very quickly (26 ns) into muons

$$\pi^+ \to \mu^+ + \nu_\mu,$$

so that if one selects the muons arising from pions which have stopped in the target, the muon beam emerges completely spin-polarized. This is because the neutrino ($\nu_\mu$) which is also produced has its spin antiparallel to its momentum; the pion has no spin, so the muon must have its spin antiparallel to its momentum and hence emerges spin-polarized. These muons can then be implanted into a sample but their energy is large, at least 4 MeV. Following implantation they lose energy very quickly (in 0.1–1 ns) to a few keV by ionization of atoms and scattering with electrons. A muon then begins to undergo a series of successive electron capture and loss reactions which reduce the energy to a few hundred eV in about a picosecond. If muonium (see Fig. 3.21), a hydrogen-like state consisting of $\mu^+$ and $e^-$, is ultimately formed then electron capture ultimately wins and the last few eV are lost by inelastic collisions between the muonium atom and the host atoms. All of these effects are very fast so that the muon (or muonium) is thermalized very rapidly. Moreover the effects are all Coulombic in origin and do not interact with the muon-spin so that the muon is thermalized in matter without appreciable depolarization. This is a crucial feature for muon-spin rotation experiments. One may be concerned that the muon may only measure a region of sample which has been subjected to radiation damage by the energetic incoming muon. This does not appear to be a problem since there is a threshold energy for vacancy production, which means that only the initial part of the muon path suffers much damage. Beyond this point of damage the muon still has sufficient energy to propagate through the sample a further distance, thought to be about 1 $\mu$m, leaving it well away from any induced vacancies and interstitials.

In a $\mu$SR experiment the muons are stopped in the specimen of interest and decay after a time $t$ with probability proportional to $e^{-t/\tau_\mu}$ where $\tau_\mu = 2.2$ $\mu$s is the lifetime of the muon. The muon decay is a three–body process

$$\mu^+ \to e^+ + \nu_e + \bar{\nu}_\mu.$$

The decay involves the weak interaction and thus has the unusual feature of not conserving parity. This phenomenon (which also lies behind the negative helicity of the muon neutrino) leads to a propensity for the emitted positron ($e^+$) to emerge predominantly along the direction that the muon-spin was aligned along when the muon decayed.

The angular distribution of emitted positrons is shown in Fig. 3.22 for the case of the most energetic emitted positrons. In fact positrons over a range of energies are emitted so that the net effect is something not quite as pronounced. In an experiment a magnetic field can be applied perpendicular to the initial muon-spin direction. This can cause the muon-spin to precess. By repeating the experiment for many muons, one can therefore follow the polarization of an ensemble of precessing muons with arbitrary accuracy, provided one is willing to take data for long enough. A schematic diagram of the experiment is shown in Fig. 3.23(a). Consider a muon, with its polarization aligned antiparallel to its momentum, which is implanted in a sample. (It is antiparallel because of

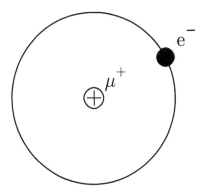

**Fig. 3.21** Muonium, a hydrogen-like state consisting of $\mu^+$ and $e^-$.

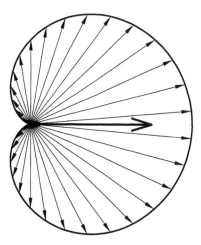

**Fig. 3.22** The angular distribution of emitted positrons with respect to the initial muon-spin direction. The expected distribution for the most energetically emitted positrons is shown.

(a)

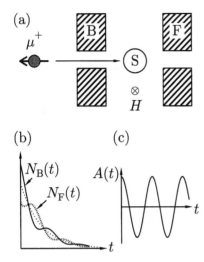

(b)         (c)

**Fig. 3.23** (a) Schematic illustration of a $\mu$SR experiment. A spin-polarized beam of muons is brought to rest in a sample S. Following decay, positrons are detected in either a forward detector F or a backward detector B. If a transverse magnetic field $H$ is applied to the sample as shown then the muons will precess. (b) The number of positrons detected in the forward (broken-line) and backward (solid-line) detectors. The dotted line shows the average of the two signals. (c) The asymmetry function.

the way that it was formed, see above, so the muon enters the sample with its spin pointing along the direction from which it came.) If the muon is unlucky enough to decay immediately, then it will not have time to precess and a positron will be emitted preferentially into the backward detector. If it lives a little longer it will have time to precess so that if it lives for half a revolution the resultant positron will be preferentially emitted into the forward detector. Thus the positron beam from an ensemble of precessing muons can be likened to the beam of light from a lighthouse.

The time evolution of the numbers of positrons detected in the forward and backward detector are described by the functions $N_F(t)$ and $N_B(t)$ respectively and these are shown in Fig. 3.23(b). Because the muon decay is a radioactive process these two terms sum to an exponential decay. Thus the time evolution of the muon polarization can be obtained by examining the normalized difference of these two functions via the **asymmetry function** $A(t)$, given by

$$A(t) = \frac{N_B(t) - N_F(t)}{N_B(t) + N_F(t)}, \tag{3.40}$$

and is shown in Fig. 3.23(c). The muon is produced with 100% spin polarization so that, in contrast with NMR, one does not need to use tricks with pulses to observe a free induction decay. The muon will undergo Larmor precession in a magnetic field, internal or external. This precession can be followed by measuring the asymmetry. The frequency of precession is directly related to the magnetic field via $\omega = \gamma_\mu B$ where $\gamma_\mu = ge/2m_\mu = 2\pi \times 135.5$ MHz T$^{-1}$ is the gyromagnetic ratio for the muon and $m_\mu$ is its mass (here $g \approx 2$). Implanted muons in magnetically ordered materials therefore precess in the internal magnetic field and directly yield oscillating signals whose frequency is proportional to the internal magnetic field. In this respect the muon behaves as a microscopic magnetometer. The Larmor precession frequencies for the proton (for NMR), electron (for ESR) and muon (for $\mu$SR) are shown in Fig. 3.24.

The large magnetic moment of the muon makes it very sensitive to extremely small magnetic fields (down to $\sim 10^{-5}$ T) and thus is useful in studying

**Fig. 3.24** The Larmor precession frequency $f$ in MHz (and the corresponding period $\tau = 1/f$) for the electron, muon and proton as a function of applied magnetic field $B$.

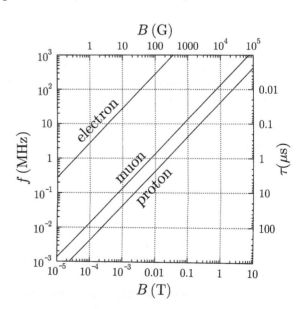

small moment magnetism. It is also valuable in studying materials in which the magnetic order is random or of very short range. Since muons stop uniformly throughout a sample, each signal appears in the experimental spectrum with a strength proportional to its volume fraction, and thus the technique is helpful in cases where samples may be multiphase or incompletely ordered. Because no spatial information is obtained (in contrast to diffraction techniques considered later) single–crystal samples are not essential (though they can be useful in certain cases) and experiments can often provide information on the magnetic order of certain materials where conventional magnetic neutron diffraction cannot be simply performed. To extract quantitative information from $\mu$SR experiments it is necessary to know the muon-site and this can in some cases hinder the search for a straightforward interpretation of the data. (In NMR of course the nucleus localizes the protons rather well so you exactly know the location of your probe.) Usually there is a small set of possible interstitial sites which the muon can occupy and in favourable circumstances only one will be consistent with the observed data. The technique has been widely applied to magnetic materials.

# Further reading

- A good introduction to NMR may be found in P. J. Hore, *Nuclear magnetic resonance* OUP 1995. Also extremely useful is B. Cowan, *Nuclear magnetic resonance and relaxation* CUP 1997. The classic text on NMR is A. Abragam, *Principles of nuclear magnetism* OUP 1961.

- A. Abragam and B. Bleaney, *Electron Paramagnetic Resonance of Transition Ions*, Dover 1986, provides extensive information about crystal fields and ESR experiments in paramagnetic salts.

- The crystal field can be considered using the so-called Stevens operators, see K. W. H. Stevens, *Proc. Phys. Soc. A* **65**, 209 (1952) and M. T. Hutchings, *Solid State Physics* **16**, 227 (1966).

- For further information on the Mössbauer effect see *Mössbauer spectroscopy*, edited by D. P. E. Dickson and F. J. Berry, CUP 1986.

- For further information on $\mu$SR see S. J. Blundell, *Contemp. Phys.* **40**, 175 (1999).

# Exercises

(3.1) A $Sc^{++}$ ion has one electron in the 3d shell. It is in an anisotropic crystal and the crystal field can be written as a potential acting on the 3d electron as $A\hat{l}_z^2$. What are the lowest orbital states of the Sc ion if $A > 0$ and if $A < 0$? The spin–orbit coupling $\lambda \hat{\mathbf{l}} \cdot \hat{\mathbf{s}}$ is much smaller than the crystal field. When this is included, what are the approximate ground states of the ion, for $A < 0$ and $A > 0$? Discuss the effect on these states of applying a small magnetic field along the $z$ axis and perpendicular to the $z$ axis, and sketch the temperature dependence of the susceptibility.

(3.2) Equal point positive charges are placed on each of the six corners of an octahedron. Taking the origin of a set of cartesian coordinates to be at the centre of the octahedron, show that the potential close to the centre is given by

$$V = \frac{q}{4\pi\epsilon_0 a} \left[ 6 + \frac{35}{4a^4} \left( x^4 + y^4 + z^4 - \frac{3}{5}r^4 \right) + O\left( \frac{r^6}{a^6} \right) \right],$$

(3.41)

where $q$ is the magnitude of each charge and $a$ is the distance between the origin and each charge.

(3.3) A compound has $n$ transition-metal ions per unit volume, and each ion sits in a crystal field of the form

$$V = D \left( x^4 + y^4 + z^4 - \frac{3}{5}r^4 \right)$$

(3.42)

which acts as a perturbation on the degenerate d levels. We can choose the unperturbed wave functions in a variety of ways, so let us begin by picking eigenfunctions of $\hat{L}_z$, labelling them by their eigenvalue $m_l$. Thus

$$|\pm 2\rangle = R(r)\sin^2\theta e^{\pm 2i\phi}$$
$$|\pm 1\rangle = \mp 2R(r)\sin\theta\cos\theta e^{\pm i\phi}$$
$$|0\rangle = \sqrt{\frac{2}{3}}R(r)(3\cos^2\theta - 1), \qquad (3.43)$$

where $R(r)$ is the radial part of the wave function which includes a normalization constant. A general state of the system can be written as $|\psi\rangle = \sum_{m_j=-2}^{2} a_j|j\rangle$ and then specified as a vector:

$$|\psi\rangle = \begin{pmatrix} a_2 \\ a_1 \\ a_0 \\ a_1 \\ a_{-2} \end{pmatrix}. \qquad (3.44)$$

In this basis show that

$$V = \begin{pmatrix} A & 0 & 0 & 0 & 5A \\ 0 & -4A & 0 & 0 & 0 \\ 0 & 0 & 6A & 0 & 0 \\ 0 & 0 & 0 & -4A & 0 \\ 5A & 0 & 0 & 0 & A \end{pmatrix}, \qquad (3.45)$$

where $A$ is a constant and show that the eigenvalues and eigenfunctions are:

eigenvalues :  $6A$  $6A$  $-4A$  $-4A$  $-4A$

eigenfunctions :  $|0\rangle$  $\dfrac{|2\rangle+|-2\rangle}{\sqrt{2}}$  $|1\rangle$  $|-1\rangle$  $\dfrac{|2\rangle-|-2\rangle}{\sqrt{2}}$

$$(3.46)$$

A magnetic field $B$ induces a splitting $\mu_B B m_l$ so that $V$ becomes

$$V = \begin{pmatrix} A+2\mu_B B & 0 & 0 & 0 & 5A \\ 0 & -4A+\mu_B B & 0 & 0 & 0 \\ 0 & 0 & 6A & 0 & 0 \\ 0 & 0 & 0 & -4A-\mu_B B & 0 \\ 5A & 0 & 0 & 0 & A-2\mu_B B \end{pmatrix}.$$

$$(3.47)$$

Calculate the eigenvalues and eigenvectors in this case. A plot of the eigenvalues is shown in Fig. 3.25.

Also show that at low temperatures, $k_B T \ll A$, the magnetic susceptibility in vanishing magnetic field is equal to

$$\chi = \begin{cases} \dfrac{2\mu_0\mu_B^2 n}{3k_B T} & \text{for} \quad A > 0 \\[2ex] \dfrac{2\mu_0\mu_B^2 n}{5A} & \text{for} \quad A < 0, \end{cases} \qquad (3.48)$$

the first case being temperature independent and small, while the second case is equivalent to that expected for a free ion with $l = 1$.

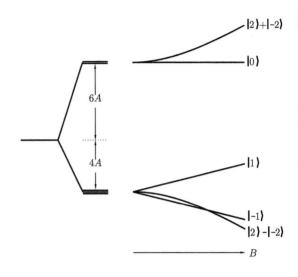

**Fig. 3.25** A crystal field splits an ion with $l = 2$ into a triplet and an excited singlet. A magnetic field $B$ splits these states further. The eigenstates as listed refer to vanishing magnetic field (as the field increases, the higher state acquires progressively more $|2\rangle$ character and the lower state acquires progressively more $|-2\rangle$ character).

(3.4) An ion, whose nucleus has zero nuclear spin, has a ground state comprising two degenerate levels corresponding to an effective spin $S = 1/2$. The application of a magnetic field of flux density $B$ produces a separation of the levels which is linear in $B$. A single electron paramagnetic resonance line is observed for the ion at a frequency of 30 GHz and a magnetic flux density of 0.6 T. An isotope with non-zero nuclear spin $I$ gives rise to a hyperfine structure (described in the spin-Hamiltonian by a term $A\mathbf{I}\cdot\mathbf{S}$) comprising four approximately equally spaced resonance lines, with separation $10^{-2}$ T, symmetrically disposed about the line due to the isotope with zero nuclear spin. What information does this give about (a) the nuclear spin and (b) the nuclear magnetic moment of the isotope? Calculate the value of the parameter $A$.

Why is it sometimes necessary to perform these measurements at low temperature?

(3.5) Figure 3.26 shows the magnetic moment per ion plotted versus $B/T$ for three paramagnetic salts. Show that eqn 2.9 has the same form as the data at both low and high $B/T$. The 3d shells of free $Cr^{3+}$ and $Fe^{3+}$ ions contain three and five electrons respectively, and the 4f shell of $Gd^{3+}$ contains seven. Using Hund's rules, calculate the values of $S$, $L$, $J$ and $g_J$ for these ions, and show that the values you obtain account well for the data in the figure for $Gd^{3+}$ and $Fe^{3+}$, but not for $Cr^{3+}$. Infer the actual g-factor and ground-state quantum numbers for $Cr^{3+}$ from the experimental data, and describe briefly the reason for the discrepancy between the data and your initial calculation.

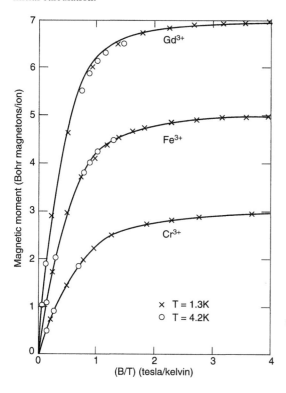

**Fig. 3.26** The magnetic moment per ion for three paramagnetic salts: KCr(SO$_4$)$_2$.12H$_2$O (Cr$^{3+}$), NH$_4$Fe(SO$_4$)$_2$.12H$_2$O (Fe$^{3+}$) and Gd$_2$(SO$_4$)$_3$.8H$_2$O (Gd$^{3+}$). After W. Henry, *Phys. Rev.* **88**, 559 (1952).

(3.6) An NMR spectrometer operates at a frequency of 60 MHz. At what applied magnetic field would you expect to observe the resonance of $^1$H, $^2$H, $^{13}$C and $^{19}$F nuclei? (Use the data in Table 2.3.)

(3.7) An ESR spectrometer operates at a frequency of 9 GHz (known as X-band). What magnetic field would be required to observe a signal from the unpaired electron in DPPH (an organic molecule used to calibrate ESR spectrometers which gives a sharp signal at $g = 2$).

(3.8) A spin Hamiltonian for a system is given by

$$\hat{\mathcal{H}} = g_{\parallel}\mu_B B_z S_z + g_{\perp}\mu_B (B_x S_x + B_y S_y) + D S_z^2. \quad (3.49)$$

Show that for two electrons in a triplet state ($S = 1$) in this system, the energy eigenstates $E$ can be obtained from

$$E(E - D)^2 - E(g_{\parallel}\mu_B B \cos\theta)^2$$
$$- (E - D)(g_{\perp}\mu_B B \sin\theta)^2 = 0. \quad (3.50)$$

Find and sketch the solutions for $\theta = 0$ and $\theta = \pi/2$. Illustrate the possible photon transitions for $\hbar\omega > D$ and hence show how the parameters $g_{\parallel}$, $g_{\perp}$ and $D$ could be obtained by measuring ESR for $\theta = 0$ and $\theta = \pi/2$.

(3.9) In the Mössbauer effect a 14.4 keV photon is emitted from an $^{57}$Fe atom. Calculate the velocity of recoil for a free $^{57}$Fe atom and a $^{57}$Fe atom that is rigidly held in the lattice of a 10 mg crystal. What is the Doppler shift of the 14.4 keV photon in each case? A sharp line in a Mössbauer experiment was obtained when the relative velocity between source and sample was 2 mm s$^{-1}$. Calculate the corresponding frequency shift in Hz and cm$^{-1}$ and also the energy shift in eV.

(3.10) The average muon lifetime is 2.2 μs; assuming that a measurable fraction of muons live for 20 μs, what lower limit does this put on the internal field that can be measured. In a typical experimental run, $10^7$ muons are measured. Of these, how many live for 20 μs or longer? Pulses of muons are produced by a synchrotron with a pulse-width of 50 ns. What upper limit does this put on the internal magnetic field in a specimen that can be measured? (Hint: consider the precession signal resulting from the muons at the back and the front of the pulse.) At a cyclotron, a continuous beam of muons can be produced which avoids this problem. An ordered magnetic oxide has an internal field of 0.4 T at the interstitial site which is occupied by an implanted muon. What precession frequency do you expect to be measured in this case?

(3.11) The saturation moment of an Fe$^{3+}$ ion ($^6$S$_{5/2}$) in a crystal is expected to be $5\mu_B$ (see Fig. 3.26). But the effective moment deduced from the susceptibility measurements is expected to be $5.92\mu_B$ (see Table 3.1). Why the difference?

# 4 Interactions

In this chapter we consider the different types of magnetic interaction which can be important in allowing the magnetic moments in a solid to communicate with each other and potentially to produce long range order.

## 4.1 Magnetic dipolar interaction

The first interaction which might be expected to play a rôle is the magnetic dipolar interaction. Two magnetic dipoles $\mu_1$ and $\mu_2$ separated by $\mathbf{r}$ have an energy equal to

$$E = \frac{\mu_0}{4\pi r^3}\left[\boldsymbol{\mu}_1 \cdot \boldsymbol{\mu}_2 - \frac{3}{r^2}(\boldsymbol{\mu}_1 \cdot \mathbf{r})(\boldsymbol{\mu}_2 \cdot \mathbf{r})\right] \qquad (4.1)$$

which therefore depends on their separation and their degree of mutual alignment. We can easily estimate the order of magnitude of this effect for two moments each of $\mu \approx 1\ \mu_{\mathrm{B}}$ separated by $r \approx 1$ Å to be approximately $\mu^2/4\pi r^3 \sim 10^{-23}$ J which is equivalent to about 1 K in temperature.[1] Since many materials order at much higher temperatures (some around 1000 K), the magnetic dipolar interaction must be too weak to account for the ordering of most magnetic materials. Nevertheless, it can be important in the properties of those materials which order at milliKelvin temperatures.

[1] See Fig. 3.8 for a handy conversion from energy to temperature.

## 4.2 Exchange interaction

Exchange interactions lie at the heart of the phenomenon of long range magnetic order. The exchange effect is subtle and not a little mysterious, since it seems surprising that one has to go to the bother of thinking about exchange operators and identical particles when all one is dealing with is a bar magnet and a pile of iron filings. But this, as so often with the subject of magnetism, is a demonstration of how quantum mechanics is at the root of many everyday phenomena. Exchange interactions are nothing more than electrostatic interactions, arising because charges of the same sign cost energy when they are close together and save energy when they are apart.

### 4.2.1 Origin of exchange

Consider a simple model with just two electrons which have spatial coordinates $\mathbf{r}_1$ and $\mathbf{r}_2$ respectively. The wave function for the joint state can be written as a product of single electron states, so that if the first electron is in state $\psi_a(\mathbf{r}_1)$

and the second electron is in state $\psi_b(\mathbf{r}_2)$, then the joint wave function is $\psi_a(\mathbf{r}_1)\psi_b(\mathbf{r}_2)$. However this product state does not obey exchange symmetry, since if we exchange the two electrons we get $\psi_a(\mathbf{r}_2)\psi_b(\mathbf{r}_1)$ which is not a multiple of what we started with. Therefore, the only states which we are allowed to make are symmetrized or antisymmetrized product states which behave properly under the operation of particle exchange. This was discussed in Section 1.3.4.

For electrons the overall wave function must be antisymmetric so the spin part of the wave function must either be an antisymmetric singlet state $\chi_S$ ($S = 0$) in the case of a symmetric spatial state or a symmetric triplet state $\chi_T$ ($S = 1$) in the case of an antisymmetric spatial state. Therefore we can write the wave function for the singlet case $\Psi_S$ and the triplet case $\Psi_T$ as

$$\Psi_S = \frac{1}{\sqrt{2}}[\psi_a(\mathbf{r}_1)\psi_b(\mathbf{r}_2) + \psi_a(\mathbf{r}_2)\psi_b(\mathbf{r}_1)]\,\chi_S$$

$$\Psi_T = \frac{1}{\sqrt{2}}[\psi_a(\mathbf{r}_1)\psi_b(\mathbf{r}_2) - \psi_a(\mathbf{r}_2)\psi_b(\mathbf{r}_1)]\,\chi_T, \qquad (4.2)$$

where both the spatial and spin parts of the wave function are included. The energies of the two possible states are

$$E_S = \int \Psi_S^* \hat{\mathcal{H}} \Psi_S \, d\mathbf{r}_1 \, d\mathbf{r}_2$$

$$E_T = \int \Psi_T^* \hat{\mathcal{H}} \Psi_T \, d\mathbf{r}_1 \, d\mathbf{r}_2,$$

with the assumption that the spin parts of the wave function $\chi_S$ and $\chi_T$ are normalized. The difference between the two energies is

$$E_S - E_T = 2 \int \psi_a^*(\mathbf{r}_1)\psi_b^*(\mathbf{r}_2)\hat{\mathcal{H}}\psi_a(\mathbf{r}_2)\psi_b(\mathbf{r}_1) \, d\mathbf{r}_1 \, d\mathbf{r}_2. \qquad (4.3)$$

Equation 1.70 shows how the difference between singlet and triplet states can be parametrized using $\mathbf{S}_1 \cdot \mathbf{S}_2$. For a singlet state $\mathbf{S}_1 \cdot \mathbf{S}_2 = -\frac{3}{4}$ while for a triplet state $\mathbf{S}_1 \cdot \mathbf{S}_2 = \frac{1}{4}$. Hence the Hamiltonian can be written in the form of an 'effective Hamiltonian'

$$\hat{\mathcal{H}} = \frac{1}{4}(E_S + 3E_T) - (E_S - E_T)\mathbf{S}_1 \cdot \mathbf{S}_2. \qquad (4.4)$$

This is the sum of a constant term and a term which depends on spin. The constant can be absorbed into other constant energy terms, but the second term is more interesting. The **exchange constant** (or **exchange integral**), J is defined by

$$J = \frac{E_S - E_T}{2} = \int \psi_a^*(\mathbf{r}_1)\psi_b^*(\mathbf{r}_2)\hat{\mathcal{H}}\psi_a(\mathbf{r}_2)\psi_b(\mathbf{r}_1) \, d\mathbf{r}_1 \, d\mathbf{r}_2. \qquad (4.5)$$

and hence the spin-dependent term in the effective Hamiltonian can be written

$$\hat{\mathcal{H}}^{\text{spin}} = -2J\mathbf{S}_1 \cdot \mathbf{S}_2. \qquad (4.6)$$

If $J > 0$, $E_S > E_T$ and the triplet state $S = 1$ is favoured. If $J < 0$, $E_S < E_T$ and the singlet state $S = 0$ is favoured. This equation is relatively simple

Werner Heisenberg (1901–1976)

to derive for two electrons, but generalizing to a many-body system is far from trivial. Nevertheless, it was recognized in the early days of quantum mechanics that interactions such as that in eqn 4.6 probably apply between all neighbouring atoms. This motivates the Hamiltonian of the **Heisenberg model**:

$$\hat{\mathcal{H}} = -\sum_{ij} J_{ij} \mathbf{S}_i \cdot \mathbf{S}_j, \tag{4.7}$$

where $J_{ij}$ is the exchange constant between the $i^{\text{th}}$ and $j^{\text{th}}$ spins. The factor of 2 is omitted because the summation includes each pair of spins twice. Another way of writing eqn 4.7 is

$$\hat{\mathcal{H}} = -2\sum_{i>j} J_{ij} \mathbf{S}_i \cdot \mathbf{S}_j, \tag{4.8}$$

where the $i > j$ avoids the 'double-counting' and hence the factor of two returns. Often it is possible to take $J_{ij}$ to be equal to a constant $J$ for nearest neighbour spins and to be 0 otherwise.

Note that in some books, J is replaced by twice the value used here so that eqns 4.6–4.8 would become

$$\hat{\mathcal{H}}^{\text{spin}} = -J\mathbf{S}_1 \cdot \mathbf{S}_2, \tag{4.9}$$

$$\hat{\mathcal{H}} = -\frac{1}{2}\sum_{ij} J_{ij}\mathbf{S}_i \cdot \mathbf{S}_j, \tag{4.10}$$

$$\hat{\mathcal{H}} = -\sum_{i>j} J_{ij}\mathbf{S}_i \cdot \mathbf{S}_j. \tag{4.11}$$

The calculation of the exchange integral can be complicated in general, but we here mention some general features. First, if the two electrons are on the same atom, the exchange integral is usually positive. This stabilizes the triplet state and ensures an antisymmetric spatial state which minimizes the Coulomb repulsion between the two electrons by keeping them apart. This is consistent with Hund's first rule.

When the two electrons are on neighbouring atoms, the situation is very different. Any joint state will be a combination of a state centred on one atom and a state centred on the other. It is worth remembering that the energy of a particle in a one-dimensional box of length $L$ is proportional to $L^{-2}$; this is a kinetic energy and hence demonstrates that there is a large kinetic energy associated with being squeezed into a small box. The electrons therefore can save kinetic energy by forming bonds because this allows them to wander around both atoms rather than just one (i.e. wander in a 'bigger box'). The correct states to consider are now not atomic orbitals but molecular orbitals (see Fig. 4.1). These can be **bonding** (spatially symmetric) or 'antibonding' (spatially antisymmetric), with the **antibonding** orbitals more energetically costly. This is because the antibonding orbital has a greater curvature and hence a larger kinetic energy. This favours singlet (antisymmetric) states and the exchange integral is therefore likely to be negative.

### 4.2.2  Direct exchange

If the electrons on neighbouring magnetic atoms interact via an exchange interaction, this is known as **direct exchange**. This is because the exchange interaction proceeds directly without the need for an intermediary. Though this seems the most obvious route for the exchange interaction to take, the reality in physical situations is rarely that simple.

Very often direct exchange cannot be an important mechanism in controlling the magnetic properties because there is insufficient direct overlap between neighbouring magnetic orbitals. For example, in rare earths the 4f electrons are strongly localized and lie very close to the nucleus, with little probability density extending significantly further than about a tenth of the interatomic

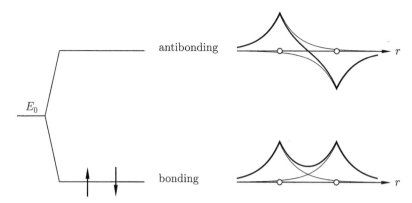

**Fig. 4.1** Molecular orbitals for a diatomic molecule. The bonding orbital, which corresponds to the sum of the two atomic orbitals (symmetric under exchange, as far as the spatial part of the wave function is concerned), is of lower energy than the antibonding orbital, which corresponds to the difference of the two atomic orbitals (antisymmetric under exchange). This therefore favours a singlet ground state in which two electrons fill the bonding state and the antibonding state is empty. This diagram is appropriate for the hydrogen molecule $H_2$ which has a lower energy than that of two isolated H atoms ($E_0$). Note that the diatomic form of helium, $He_2$, does not form because the four electrons from two He atoms would fill both the bonding and antibonding orbitals, corresponding to no net energy saving in comparison with two isolated He atoms.

spacing. This means that the direct exchange interaction is unlikely to be very effective in rare earths. Even in transition metals, such as Fe, Co and Ni, where the 3d orbitals extend further from the nucleus, it is extremely difficult to justify why direct exchange should lead to the observed magnetic properties. These materials are metals which means that the rôle of the conduction electrons should not be neglected, and a correct description needs to take account of both the localized and band character of the electrons.

Thus in many magnetic materials it is necessary to consider some kind of **indirect exchange interaction**.

### 4.2.3  Indirect exchange in ionic solids: superexchange

A number of ionic solids, including some oxides and fluorides, have magnetic ground states. For example, MnO (see Fig. 4.2) and $MnF_2$ are both antiferromagnets, though this observation appears at first sight rather surprising because there is no direct overlap between the electrons on $Mn^{2+}$ ions in each system. The exchange interaction is normally very short-ranged so that the longer-ranged interaction that is operating in this case must be in some sense 'super'.

The exchange mechanism which is operative here is in fact known as **superexchange**. It can be defined as an indirect exchange interaction between non-neighbouring magnetic ions which is mediated by a non-magnetic ion which is placed in between the magnetic ions. It arises because there is a kinetic energy advantage for antiferromagnetism, which can be understood by reference to Fig. 4.3 which shows two transition metal ions separated by an oxygen ion. For simplicity we will assume that the magnetic moment on the transition metal ion is due to a single unpaired electron (more complicated cases can be dealt with in analogous ways). Hence if this system were perfectly ionic, each metal ion would have a single unpaired electron in a d orbital

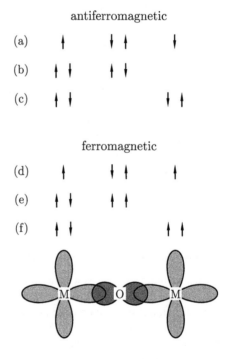

**Fig. 4.2** The crystal structure of MnO. Nearest neighbour pairs of $Mn^{2+}$ (manganese) ions are connected via $O^{2-}$ (oxygen) ions.

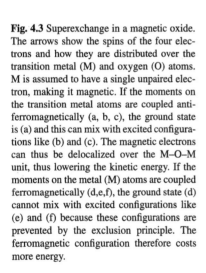

**Fig. 4.3** Superexchange in a magnetic oxide. The arrows show the spins of the four electrons and how they are distributed over the transition metal (M) and oxygen (O) atoms. M is assumed to have a single unpaired electron, making it magnetic. If the moments on the transition metal atoms are coupled antiferromagnetically (a, b, c), the ground state is (a) and this can mix with excited configurations like (b) and (c). The magnetic electrons can thus be delocalized over the M–O–M unit, thus lowering the kinetic energy. If the moments on the metal (M) atoms are coupled ferromagnetically (d,e,f), the ground state (d) cannot mix with excited configurations like (e) and (f) because these configurations are prevented by the exclusion principle. The ferromagnetic configuration therefore costs more energy.

and the oxygen would have two p electrons in its outermost occupied states. The figure demonstrates that antiferromagnetic coupling lowers the energy of the system by allowing these electrons to become delocalized over the whole structure, thus lowering the kinetic energy.

Because superexchange involves the oxygen orbitals as well as the metal atom, it is a second-order process and is derived from second-order perturbation theory. A general consequence of second-order perturbation theory is that the energy involved is approximately given by the square of the matrix element of the transition divided by the energy cost of making the excited state. Here the transition matrix element is controlled by a parameter called the hopping integral $t$, which is proportional to the energy width of the conduction band (i.e. the bandwidth) in a simple tight-binding approach. The energy cost of making an excited state is given by the Coulomb energy $U$. Thus we have that $J \sim -t^2/U$. (In fourth-order it is possible to have an interaction of the form $\Delta E \propto -t^4(\mathbf{S}_1 \cdot \mathbf{S}_2)^2/U^3$ which is known as **biquadratic exchange**.)

The exchange integral consists of two parts. The first is a potential exchange term which represents the electron repulsion and favours ferromagnetic ground states, but is small when the ions are well separated. The second is a kinetic exchange term which dominates here and is the effect discussed above. It depends on the degree of overlap of orbitals and thus superexchange is strongly dependent upon the angle of the M–O–M bond. The figure has been drawn for one type of d orbital only, but the effect of other d orbitals which can overlap with the oxygen orbitals may also need to be added.

In some circumstances, superexchange can actually be ferromagnetic. For example, imagine a situation in which there is a coupling, through an oxygen ion, between an occupied $e_g$ orbital on one magnetic ion and an unoccupied $e_g$ orbital on another magnetic ion. There is an energetic advantage to the $e_g$ electron hopping onto the unoccupied orbital, if when it arrives its spin is aligned with the spin of the $t_{2g}$ electrons because of the Hund's rule coupling. Thus the superexchange could be ferromagnetic in this case, but this is weaker interaction and less common than the usual antiferromagnetic superexchange.

### 4.2.4 Indirect exchange in metals

In metals the exchange interaction between magnetic ions can be mediated by the conduction electrons. A localized magnetic moment spin-polarizes the conduction electrons and this polarization in turn couples to a neighbouring localized magnetic moment a distance $r$ away. The exchange interaction is thus indirect because it does not involve direct coupling between magnetic moments. It is known as the **RKKY interaction** (or also as **itinerant exchange**). The name RKKY is used because of the initial letters of the surnames of the discoverers of the effect, Ruderman, Kittel, Kasuya and Yosida. The coupling takes the form of an $r$-dependent exchange interaction $J_{RKKY}(r)$ given by

$$J_{RKKY}(r) \propto \frac{\cos(2k_F r)}{r^3}. \tag{4.12}$$

at large $r$ (assuming a spherical Fermi surface of radius $k_F$). The interaction is long range and has an oscillatory dependence on the distance between the magnetic moments. Hence depending on the separation it may be either ferromagnetic or antiferromagnetic. The coupling is oscillatory with wavelength $\pi/k_F$ because of the sharpness of the Fermi surface. The RKKY interaction will be considered in more detail in chapter 7.

### 4.2.5 Double exchange

In some oxides, it is possible to have a ferromagnetic exchange interaction which occurs because the magnetic ion can show **mixed valency**, that is it can exist in more than one oxidation state. Examples of this include compounds containing the Mn ion which can exist in oxidation state 3 or 4, i.e. as $Mn^{3+}$

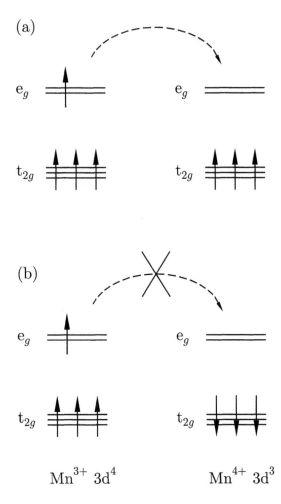

**Fig. 4.4** Double exchange mechanism gives ferromagnetic coupling between $Mn^{3+}$ and $Mn^{4+}$ ions participating in electron transfer. The single-centre exchange interaction favours hopping if (a) neighbouring ions are ferromagnetically aligned and not if (b) neighbouring ions are antiferromagnetically aligned.

or $Mn^{4+}$. One such is the material $La_{1-x}Sr_xMnO_3$ ($0 \leq x \leq 1$) which adopts a perovskite structure. Sr is divalent (it exists as $Sr^{2+}$) and La is trivalent (it exists as $La^{3+}$). This implies that a fraction $x$ of the Mn ions are $Mn^{4+}$ and $1 - x$ are $Mn^{3+}$. The end members of the series, with $x = 0$ and $x = 1$, are both antiferromagnetic insulators, as would be expected for an oxide material in which the magnetism is mediated by superexchange through the oxygen. $LaMnO_3$ contains only $Mn^{3+}$ ions and $Mn^{3+}$ is a Jahn–Teller ion. $LaMnO_3$ has A-type antiferromagnetic ordering (see Section 5.2). However when $LaMnO_3$ is doped with Sr up to a level of $x = 0.175$, the Jahn–Teller distortion vanishes and the system becomes ferromagnetic with a Curie temperature around room temperature, below which temperature the material becomes metallic.

The ferromagnetic alignment is due to the **double exchange** mechanism which can be understood with reference to Fig. 4.4. The $e_g$ electron on a $Mn^{3+}$ ion can hop to a neighbouring site only if there is a vacancy there of the same spin (since hopping proceeds without spin-flip of the hopping electron). If the neighbour is a $Mn^{4+}$ which has no electrons in its $e_g$ shell, this should present no problem. However, there is a strong single-centre (Hund's rule number 1) exchange interaction between the $e_g$ electron and the three electrons in the

$t_{2g}$ level which wants to keep them all aligned. Thus it is not energetically favourable for an $e_g$ electron to hop to a neighbouring ion in which the $t_{2g}$ spins will be antiparallel to the $e_g$ electron (Fig. 4.4(b)). Ferromagnetic alignment of neighbouring ions is therefore required to maintain the high-spin arrangement on both the donating and receiving ion. Because the ability to hop gives a kinetic energy saving, allowing the hopping process shown in Fig. 4.4(a) reduces the overall energy. Thus the system ferromagnetically aligns to save energy. Moreover, the ferromagnetic alignment then allows the $e_g$ electrons to hop through the crystal and the material becomes metallic. The issue of conductivity in double exchange ferromagnets will be further discussed in Section 8.9.5. Double exchange is, essentially, ferromagnetic superexchange in an extended system.[2]

Double exchange is also found in magnetite ($Fe_3O_4$) which contains an equal mixture of $Fe^{2+}$ ($3d^6$) and $Fe^{3+}$ ($3d^5$) ions on octahedral sites, together with the same number again of $Fe^{3+}$ ions on tetrahedral sites. A double exchange interaction ferromagnetically aligns the $Fe^{2+}$ and $Fe^{3+}$ ions on the octahedral sites. The $Fe^{3+}$ ions on the tetrahedral sites do not participate in this interaction and are coupled to the $Fe^{3+}$ ions on the octahedral sites by an antiferromagnetic superexchange interaction. Thus the two sets of $Fe^{3+}$ ions cancel out, leaving a net moment due to the $Fe^{2+}$ ions alone. The measured magnetic moment per formula unit is very close to the expected $4 \mu_B$ due to just the $Fe^{2+}$ ions.

[2] Ferromagnetic superexchange is normally applied to two isolated ions. The kinetic energy saved by ferromagnetic alignment corresponds to hopping into an excited state. Double exchange is applied to an extended system so the kinetic energy saved corresponds to a gain in electron bandwidth.

### 4.2.6   Anisotropic exchange interaction

It is also possible for the spin–orbit interaction to play a rôle in a similar manner to that of the oxygen atom in superexchange. Here the excited state is not connected with oxygen but is produced by the spin–orbit interaction in one of the magnetic ions. There is then an exchange interaction between the excited state of one ion and the ground state of the other ion. This is known as the **anisotropic exchange interaction**, or also as the **Dzyaloshinsky–Moriya interaction**. When acting between two spins $S_1$ and $S_2$ it leads to a term in the Hamiltonian, $\hat{\mathcal{H}}_{DM}$ equal to

$$\hat{\mathcal{H}}_{DM} = \mathbf{D} \cdot \mathbf{S}_1 \times \mathbf{S}_2. \tag{4.13}$$

The vector $\mathbf{D}$ vanishes when the crystal field has an inversion symmetry with respect to the centre between the two magnetic ions. However, in general $\mathbf{D}$ may not vanish and then will lie parallel or perpendicular to the line connecting the two spins, depending on the symmetry. The form of the interaction is such that it tries to force $\mathbf{S}_1$ and $\mathbf{S}_2$ to be at right angles in a plane perpendicular to the vector $\mathbf{D}$ in such an orientation as to ensure that the energy is negative. Its effect is therefore very often to cant (i.e. slightly rotate) the spins by a small angle. It commonly occurs in antiferromagnets and then results in a small ferromagnetic component of the moments which is produced perpendicular to the spin-axis of the antiferromagnet. The effect is known as **weak ferromagnetism**. It is found in, for example, $\alpha$-$Fe_2O_3$, $MnCO_3$ and $CoCO_3$.

### 4.2.7 Continuum approximation

In this section we return to the Heisenberg model given in eqn 4.7. For what follows later in the book it is useful to find an expression for this interaction in a continuum approximation in which the discrete nature of the lattice is ignored.

Let us first assume that $J_{ij}$ can be taken to be equal to a constant, $J$, if $i$ and $j$ are nearest neighbours, and to be zero otherwise. Hence we write

$$\hat{\mathcal{H}} = -\sum_{<ij>} J\mathbf{S}_i \cdot \mathbf{S}_j, \tag{4.14}$$

where the symbol $\langle ij \rangle$ below the $\sum$ denotes a sum over nearest neighbours only. Let us consider classical spins, and assume that the angle between nearest neighbour spins is $\phi_{ij}$ and that it is very small, i.e. $\phi_{ij} \ll 1$ for all $i$ and $j$. What we are essentially doing is assuming that the system shows ferromagnetism (see the next chapter) but the spins are not completely aligned.

With these assumptions, the energy of the system can be written as

$$E = -JS^2 \sum_{<ij>} \cos\phi_{ij} = \text{constant} + \frac{JS^2}{2} \sum_{<ij>} \phi_{ij}^2, \tag{4.15}$$

where the last equality is obtained using $\cos\phi_{ij} \approx 1 - \phi_{ij}^2/2$ for $\phi_{ij} \ll 1$. We will now ignore the constant term which just refers to the energy of the fully aligned state. We now define the reduced moment by $\mathbf{m} = \mathbf{M}/M_s$, where $\mathbf{M}$ is the magnetization and $M_s$ is the saturation magnetization. The unit vector $\mathbf{m}$ therefore follows the direction of the spins and $m_x$, $m_y$ and $m_z$ can be thought of as the direction cosines of the spin at lattice point $r_{ij}$. Using the notation of Fig. 4.5, we can write

$$|\phi_{ij}| \approx |\mathbf{m}_i - \mathbf{m}_j| \approx |(\mathbf{r}_{ij} \cdot \nabla)\mathbf{m}|, \tag{4.16}$$

and so the energy can be written

$$E = JS^2 \sum_{<ij>} [(\mathbf{r}_{ij} \cdot \nabla)\mathbf{m}]^2. \tag{4.17}$$

In the continuum limit, we ignore the discrete nature of the lattice and therefore write

$$E = A \int_V [(\nabla m_x)^2 + (\nabla m_y)^2 + (\nabla m_z)^2]\mathrm{d}^3r, \tag{4.18}$$

where $A$ is given by

$$A = 2JS^2 z/a, \tag{4.19}$$

$a$ is the nearest neighbour distance and $z$ is the number of sites in the unit cell ($z = 1$ for simple cubic, $z = 2$ for body-centred cubic (bcc) and $z = 4$ for face-centred cubic (fcc)).

This result can be seen from another perspective. If we assert that exchange arises from a non-uniform magnetization distribution, then if the non-uniformities are relatively smooth we can derive a result based on symmetry. The expression must be invariant with respect to spin rotations and also under change of sign of the magnetization components. Therefore we look for an expression of the lowest even orders of the derivatives of $\mathbf{M}$, consistent with the

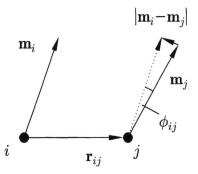

**Fig. 4.5** The magnetic moments are represented by the reduced moments $\mathbf{m}_i$ and $\mathbf{m}_j$ at neighbouring sites $i$ and $j$ separated by a vector $\mathbf{r}_{ij}$. The angle between the moments is $\phi_{ij}$. The reduced moments $\mathbf{m}_i$ and $\mathbf{m}_j$ are unit vectors by definition.

symmetry of the crystal. Since terms proportional to $\partial M_\alpha/\partial x_\beta$ would change sign if **M** were reversed, we are left with terms quadratic in the gradient of the magnetization. The most general expression is

$$E = \sum_{\alpha\beta\gamma} C_{\alpha\beta} \frac{\partial M_\gamma}{\partial x_\alpha} \frac{\partial M_\gamma}{\partial x_\beta}, \qquad (4.20)$$

where $C_{\alpha\beta}$ is a tensor with the symmetry of the crystal. In a cubic crystal, this reduces to

$$E = C \sum_{\alpha\gamma} \frac{\partial M_\gamma}{\partial x_\alpha} \frac{\partial M_\gamma}{\partial x_\alpha} = C\left[(\nabla M_x)^2 + (\nabla M_y)^2 + (\nabla M_z)^2\right], \qquad (4.21)$$

The expression in eqn 4.21 is sometimes written $C|\nabla \mathbf{M}|^2$ but note that you have to be a bit careful when taking the gradient of a vector.

which is equivalent to the result we have in eqn 4.18 above.

# Further reading

- Further information on exchange interactions may be found in C. Herring in *Magnetism*, ed. G. Rado and H. Suhl, vol. **2B**, p.1, Academic Press, New York 1966.
- A theoretical account of interactions in magnetic systems may be found in K. Yosida, *Theory of magnetism*, Springer 1996.

- Various exchange interactions in real systems are reviewed in P. A. Cox, *Transition metal oxides*, OUP 1995.
- A very thorough and helpful reference on exchange and exchange interactions is D. C. Mattis, *The theory of magnetism I*, Springer 1981.

# Exercises

(4.1) Show that two magnetic dipoles $\mu_1$ and $\mu_2$ separated by **r** have a dipolar energy equal to

$$E = \frac{\mu_0}{4\pi r^3}\left[\mu_1 \cdot \mu_2 - \frac{3}{r^2}(\mu_1 \cdot \mathbf{r})(\mu_2 \cdot \mathbf{r})\right]. \qquad (4.22)$$

(4.2) Calculate the magnitude of the magnetic field 1 Å and 10 Å from a proton in a direction (a) parallel and (b) perpendicular to the proton spin direction.

(4.3) Estimate the ratio of the exchange and dipolar coupling of two adjacent Fe atoms in metallic Fe. (The exchange constant in Fe can be crudely estimated by setting it equal to $k_B T_C$ where $T_C$ is the Curie temperature. For Fe, $T_C = 1043$ K.)

(4.4) Provide a rough estimate of the size of the exchange constant in a magnetic oxide which is coupled by superexchange using the measured value of the electronic bandwidth (determined by inelastic neutron scattering) of

0.05 eV. Take the Coulomb energy to be $\sim 1$ eV. Hence estimate the antiferromagnetic ordering temperature.

(4.5) Consider the case of two interacting spin-$\frac{1}{2}$ electrons. The good quantum numbers are $S = 0$ and 1 so that there is a triplet state and a singlet state which will be separated by an energy gap $\Delta$. We define the sign of $\Delta$ so that when $\Delta > 0$ the singlet state ($S = 0$) is the lower state and when $\Delta < 0$ the triplet state is the lower state. These situations are shown in Fig. 4.6(a) and (b). Show that the susceptibility in this model is given by

$$\chi = \frac{2ng\mu_B^2}{k_B T(3 + e^{\Delta/k_B T})}, \qquad (4.23)$$

which is known as the Bleaney–Bowers equation. It is plotted in Fig. 4.6(c).

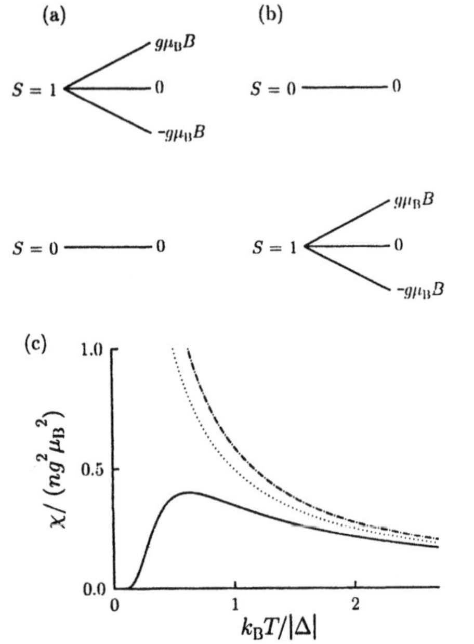
(a)
(b)
S = 1
$g\mu_B B$
0
$-g\mu_B B$
S = 0
(c)
1.0
0.5
0.0
$\chi/(ng^2\mu_B^2)$
0
1
2
$k_BT/|\Delta|$

**Fig. 4.6** Two spins coupled by $\Delta$ give rise to a singlet and triplet state. The energy levels for the case when (a) $\Delta > 0$ and (b) $\Delta < 0$. (c) The susceptibility which is given by the Bleaney–Bowers equation. It is plotted for $\Delta > 0$ (lower curve), $\Delta = 0$ (middle curve), and $\Delta < 0$ (upper curve).

(4.6) Two nuclei with spins $\mathbf{I}_1$ and $\mathbf{I}_2$, separated by a vector $\mathbf{r}$, are coupled by a dipole–dipole interaction and the Hamiltonian is

$$\hat{\mathcal{H}} = \frac{\mu_0}{4\pi r^3}\left[\hat{\mu}_1 \cdot \hat{\mu}_2 - \frac{3}{r^2}(\hat{\mu}_1 \cdot \mathbf{r})(\hat{\mu}_2 \cdot \mathbf{r})\right], \quad (4.24)$$

where $\mu_i = g_I \mu_N \mathbf{I}_i$, $i = 1, 2$. Show that the Hamiltonian can be expressed as

$$\hat{\mathcal{H}} = \frac{\mu_0 g_I^2 \mu_N^2}{4\pi r^3}\left[A + B + C + D + E + F\right]. \quad (4.25)$$

where

$$A = (1 - 3\cos^2\theta) I_{1z} I_{2z}$$
$$B = -\frac{1}{4}(1 - 3\cos^2\theta)(I_1^+ I_2^- + I_1^- I_2^+)$$
$$C = -\frac{3}{2}\sin\theta\cos\theta e^{-i\phi}(I_{1z} I_2^+ + I_1^+ I_{2z}) \quad (4.26)$$
$$D = -\frac{3}{2}\sin\theta\cos\theta e^{i\phi}(I_{1z} I_2^- + I_1^- I_{2z})$$
$$E = -\frac{3}{4}\sin\theta e^{-2i\phi} I_1^+ I_2^+$$
$$F = -\frac{3}{4}\sin\theta e^{2i\phi} I_1^- I_2^-,$$

the angles $\theta$ and $\phi$ relate to the vector $\mathbf{r}$ in spherical polars, and the raising and lowering operators are $I_j^{\pm} = I_{jx} \pm i I_{jy}$. For two protons ($I_1 = I_2 = \frac{1}{2}$) in the triplet state (i.e. for $\mathbf{I} = \mathbf{I}_1 + \mathbf{I}_2$ with magnitude $I = 1$) show that the terms in the Hamiltonian above induce transitions as follows: $\Delta M = 0$ for $A$ and $B$, $\Delta M = +1$ for $C$, $\Delta M = -1$ for $D$, $\Delta M = +2$ for $E$ and $\Delta M = -2$ for $F$.

(4.7) A diatomic molecule is formed when two atoms bond together. Two electrons are in a molecule and have coordinates $\mathbf{r}_1$ and $\mathbf{r}_2$. There are two nuclei, one fixed at $\mathbf{R}_1$ and the other fixed at $\mathbf{R}_2$, each with charge $Ze$. Ignoring the repulsion between the nuclei, the Hamiltonian can be written as $\hat{\mathcal{H}}_0 + \hat{\mathcal{H}}'$ where

$$\hat{\mathcal{H}}_0 = -\frac{\hbar^2}{2m}(\nabla_1^2 + \nabla_2^2) - \frac{Ze^2}{4\pi\epsilon_0|\mathbf{r}_1 - \mathbf{R}_1|} - \frac{Ze^2}{4\pi\epsilon_0|\mathbf{r}_2 - \mathbf{R}_2|}$$
$$= \hat{\mathcal{H}}_1 + \hat{\mathcal{H}}_2, \quad (4.27)$$

which is a sum of single particle Hamiltonians $\hat{\mathcal{H}}_1$ and $\hat{\mathcal{H}}_2$ (each electron orbiting its own nucleus) and

$$\hat{\mathcal{H}}' = \frac{e^2}{4\pi\epsilon_0|\mathbf{r}_1 - \mathbf{r}_2|} - \frac{Ze^2}{4\pi\epsilon_0|\mathbf{r}_1 - \mathbf{R}_2|} - \frac{Ze^2}{4\pi\epsilon_0|\mathbf{r}_2 - \mathbf{R}_1|}, \quad (4.28)$$

which contains terms for the electron repulsion energy and the attraction energy for each electron being attracted by its partner's nucleus. Show that the energy of the molecule is given by

$$E = E_1 + E_2 + \frac{K \pm J}{1 \pm S} \quad (4.29)$$

where $E_1$ and $E_2$ are the eigenvalues of $\hat{\mathcal{H}}_1$ and $\hat{\mathcal{H}}_2$ respectively,

$$K = \int \psi_a^*(\mathbf{r}_1)\psi_b^*(\mathbf{r}_2)\hat{\mathcal{H}}'\psi_a(\mathbf{r}_1)\psi_b(\mathbf{r}_2)\,d\mathbf{r}_1\,d\mathbf{r}_2.$$
$$= \int |\psi_a(\mathbf{r}_1)|^2 \hat{\mathcal{H}}' |\psi_b(\mathbf{r}_2)|^2\,d\mathbf{r}_1\,d\mathbf{r}_2 \quad (4.30)$$

which is known as the Coulomb integral,

$$J = \int \psi_a^*(\mathbf{r}_1)\psi_b^*(\mathbf{r}_2)\hat{\mathcal{H}}'\psi_a(\mathbf{r}_2)\psi_b(\mathbf{r}_1)\,d\mathbf{r}_1\,d\mathbf{r}_2 \quad (4.31)$$

which is known as the exchange integral,

$$S = \int \psi_a^*(\mathbf{r}_1)\psi_b^*(\mathbf{r}_2)\psi_a(\mathbf{r}_2)\psi_b(\mathbf{r}_1)\,d\mathbf{r}_1\,d\mathbf{r}_2 \quad (4.32)$$

which is known as the overlap integral, and where the eigenfunctions are assumed to be given by

$$\Psi_{\pm} = \frac{1}{\sqrt{2}}[\psi_a(\mathbf{r}_1)\psi_b(\mathbf{r}_2) \pm \psi_a(\mathbf{r}_2)\psi_b(\mathbf{r}_1)]. \quad (4.33)$$

(4.8) Three $S = 1$ atoms are placed on the corners of an equilateral triangle and can be described by a Hamiltonian

$$\hat{\mathcal{H}} = -2J(\hat{\mathbf{S}}_1 \cdot \hat{\mathbf{S}}_2 + \hat{\mathbf{S}}_2 \cdot \hat{\mathbf{S}}_3 + \hat{\mathbf{S}}_3 \cdot \hat{\mathbf{S}}_1). \quad (4.34)$$

Show that the energy eigenvalues of the system are $-6J, 0, 4J$ and $6J$.

# Order and magnetic structures

5

In the previous chapter the different types of magnetic interaction which operate between magnetic moments in a solid have been presented. In this chapter we will consider the different types of magnetic ground state which can be produced by these interactions. Some of these ground states are illustrated in Fig. 5.1. The different ground states include ferromagnets in which all the magnetic moments are in parallel alignment, antiferromagnets in which adjacent magnetic moments lie in antiparallel alignment, spiral and helical structures in which the direction of the magnetic moment precesses around a cone or a circle as one moves from one site to the next, and spin glasses in which the magnetic moments lie in frozen random arrangements. This chapter will be concerned with showing how, in broad terms, the interactions discussed in the previous chapter lead to these differing ground states. In the following chapter the phenomenon of order will be examined in a more general context and it will be seen that order is a consequence of broken symmetry.

Notation reminder: In this book J refers to the exchange constant, $J$ is total angular momentum.

## 5.1 Ferromagnetism

A ferromagnet has a spontaneous magnetization even in the absence of an applied field. All the magnetic moments lie along a single unique direction.[1] This effect is generally due to exchange interactions which were described in the previous chapter. For a ferromagnet in an applied magnetic field **B**, the appropriate Hamiltonian to solve is

[1] In fact in many ferromagnetic samples this is not true throughout the sample because of domains. In each domain there is a uniform magnetization, but the magnetization of each domain points in a different direction from its neighbours. See Section 6.7 for more on magnetic domains.

$$\hat{\mathcal{H}} = -\sum_{ij} J_{ij}\mathbf{S}_i \cdot \mathbf{S}_j + g\mu_B \sum_j \mathbf{S}_j \cdot \mathbf{B}, \qquad (5.1)$$

and the exchange constants for nearest neighbours will be positive in this case, to ensure ferromagnetic alignment. The first term on the right is the Heisenberg exchange energy (see eqn 4.7). The second term on the right is the Zeeman energy (see eqn 1.35). To keep things simple to begin with, let us assume[2] that we are dealing with a system in which there is no orbital angular momentum, so that $L = 0$ and $J = S$.

[2] We will relax this assumption later in Section 5.1.4.

### 5.1.1 The Weiss model of a ferromagnet

To make progress with solving eqn 5.1 it is necessary to make an approximation. We define an effective molecular field at the $i^{\text{th}}$ site by

$$\mathbf{B}_{\text{mf}} = -\frac{2}{g\mu_B}\sum_j J_{ij}\mathbf{S}_j. \qquad (5.2)$$

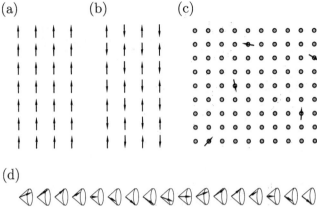

**Fig. 5.1** Various spin arrangements in ordered systems: (a) ferromagnets, (b) antiferromagnets, (c) spin glasses and (d) spiral and (e) helical structures.

Now focus on the $i^{\text{th}}$ spin. Its energy is due to a Zeeman part $g\mu_B\mathbf{S}_i \cdot \mathbf{B}$ and an exchange part. The total exchange interaction between the $i^{\text{th}}$ spin and its neighbours is $-2\sum_j J_{ij}\mathbf{S}_i \cdot \mathbf{S}_j$, where the factor of 2 is because of the double counting.[3] This term can be written as

$$-2\mathbf{S}_i \cdot \sum_j J_{ij}\mathbf{S}_j = -g\mu_B\mathbf{S}_i \cdot \mathbf{B}_{\text{mf}}. \tag{5.3}$$

Hence the exchange interaction is replaced by an effective molecular field $\mathbf{B}_{\text{mf}}$ produced by the neighbouring spins. The effective Hamiltonian can now be written as

$$\hat{\mathcal{H}} = g\mu_B \sum_i \mathbf{S}_i \cdot (\mathbf{B} + \mathbf{B}_{\text{mf}}) \tag{5.4}$$

which now looks like the Hamiltonian for a paramagnet in a magnetic field $\mathbf{B} + \mathbf{B}_{\text{mf}}$. The assumption underpinning this approach is that all magnetic ions experience the same molecular field. This may be rather questionable, particularly at temperatures close to a magnetic phase transition, as will be discussed in the following chapter. For a ferromagnet the molecular field will act so as to align neighbouring magnetic moments. This is because the dominant exchange interactions are positive. (For an antiferromagnet, they will be negative.)

Since the molecular field measures the effect of the ordering of the system, one can assume that

$$\mathbf{B}_{\text{mf}} = \lambda\mathbf{M} \tag{5.5}$$

where $\lambda$ is a constant which parametrizes the strength of the molecular field as a function of the magnetization. For a ferromagnet, $\lambda > 0$. Because of the large Coulomb energy involved in the exchange interaction, the molecular field is often found to be extremely large in ferromagnets.

We are now able to treat this problem as if the system were a simple paramagnet placed in a magnetic field $\mathbf{B} + \mathbf{B}_{\text{mf}}$. At low temperature, the moments can be aligned by the internal molecular field, even without any applied field being present. Notice that the alignment of these magnetic moments gives rise to the internal molecular field that causes the alignment in the first place, so that this is something of a 'chicken-and-egg' scenario. At low temperature the magnetic order is self-sustaining. As the temperature is raised, thermal fluctuations begin to progressively destroy the magnetization and at a critical temperature the order will be destroyed. This model is known as the **Weiss model of ferromagnetism.**

Pierre Weiss (1865–1946)

To find solutions to this model, it is necessary to solve simultaneously the equations

$$\frac{M}{M_s} = B_J(y) \tag{5.6}$$

(see eqn 2.38) and

$$y = \frac{g_J \mu_B J (B + \lambda M)}{k_B T} \tag{5.7}$$

Reminder: we are assuming $J = S$ and $L = 0$ at this stage.

(see eqn 2.37). Without the $\lambda M$ term due to the molecular field, this would be identical to our treatment of a paramagnet in Section 2.4.3.

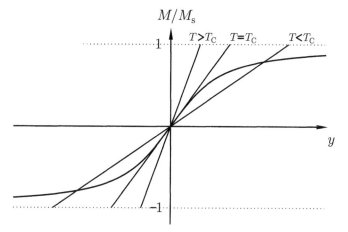

**Fig. 5.2** The graphical solution of eqns 5.6 and 5.7 for $B = 0$.

These equations can be solved graphically. First, we restrict our attention to the case of $B = 0$, so that $M = k_B T y / g_J \mu_B J \lambda$. Hence the straight line produced by plotting $M$ against $y$ has a gradient which is proportional to temperature $T$ as illustrated in Fig. 5.2. For high temperature, there is no simultaneous solution of eqns 5.6 and 5.7 except at the origin where $y = 0$ and $M_s = 0$. This situation changes when the gradient of the line is less than that of the Brillouin function at the origin. At low temperatures there are then three solutions, one at $M_s = 0$ and another two for $M_s$ at $\pm$ some non-zero value. It turns out that when the curve is less steep than the Brillouin function at the origin, the non-zero solutions are stable and the zero-solution is unstable. (If the system has $M_s = 0$ for $T < T_C$, any fluctuation, no matter how small, will cause the system to turn into either one of the two stable states.) Thus below a certain temperature, non-zero magnetization occurs and this grows as the material is cooled. The substance thus becomes magnetized, even in the absence of an external field. This **spontaneous magnetization** is the characteristic of ferromagnetism.

The temperature at which the transition occurs can be obtained by finding when the gradients of the line $M = k_B T y / g_J \mu_B J \lambda M_s$ and the curve $M = M_s B_J(y)$ are equal at the origin. For small $y$, $B_J(y) = (J+1)y/3J + O(y^3)$.

The transition temperature, known as the **Curie temperature** $T_C$, is then defined by

$$T_C = \frac{g_J \mu_B (J+1) \lambda M_s}{3k_B} = \frac{n\lambda \mu_{\text{eff}}^2}{3k_B}. \tag{5.8}$$

The molecular field $B_{\text{mf}} = \lambda M_s$ is thus $3k_B T_C / g_J \mu_B (J+1)$ and so for a ferromagnet with $J = \frac{1}{2}$ and $T_C \sim 10^3$ K, $B_{\text{mf}} = k_B T_C / \mu_B \sim 1500$ T. This is an enormous effective magnetic field and reflects the strength of the exchange interaction.

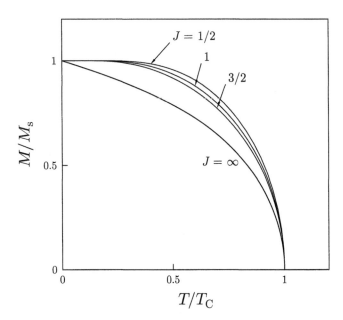

**Fig. 5.3** The mean-field magnetization as a function of temperature, deduced for different values of $J$.

The solutions of these equations as a function of temperature are shown in Fig. 5.3 for a range of values of $J$. Although the form of the curves is slightly different in each case, some general features persist. The magnetization is zero for temperatures $T \geq T_C$ and is non-zero for $T < T_C$. The magnetization is continuous at $T = T_C$, but its gradient is not. This classifies the phase transition between the non-magnetic and ferromagnetic phases in this molecular field model as a **second-order phase transition**. The order of a phase transition is the order of the lowest differential of the free energy which shows a discontinuity at the transition. A first-order phase transition would have a discontinuous jump in the first derivative of the free energy, i.e. in quantities like the volume, entropy or the magnetization. The jump in the entropy gives a latent heat. A second-order phase transition has a discontinuity in the second derivative of the free energy, i.e. in quantities like the compressibility or the heat capacity. In the present case the discontinuity is in the gradient of the magnetization, i.e. in the second derivative of the free energy, so the transition is second order. Phase transitions and critical exponents will be considered in more detail in Section 6.4. The properties of some common ferromagnets are listed in Table 5.1.

**Table 5.1** Properties of some common ferromagnets.

| Material | $T_C$ (K) | magnetic moment ($\mu_B$ /formula unit) |
|---|---|---|
| Fe | 1043 | 2.22 |
| Co | 1394 | 1.715 |
| Ni | 631 | 0.605 |
| Gd | 289 | 7.5 |
| MnSb | 587 | 3.5 |
| EuO | 70 | 6.9 |
| EuS | 16.5 | 6.9 |

## 5.1.2    Magnetic susceptibility

Applying a small $B$ field at $T \geq T_C$ will lead to a small magnetization, so that the $y \ll 1$ approximation for the Brillouin function can be used. Thus

$$\frac{M}{M_s} \approx \frac{g_J \mu_B (J+1)}{3k_B} \left( \frac{B + \lambda M}{T} \right) \tag{5.9}$$

so that

$$\frac{M}{M_s} \approx \frac{T_C}{\lambda M_s} \left( \frac{B + \lambda M}{T} \right). \tag{5.10}$$

This can be rearranged to give

$$\frac{M}{M_s} \left( 1 - \frac{T_C}{T} \right) \approx \frac{T_C B}{\lambda M_s} \tag{5.11}$$

so that

$$\chi = \lim_{B \to 0} \frac{\mu_0 M}{B} \propto \frac{1}{T - T_C} \tag{5.12}$$

which is known as the **Curie Weiss law**.

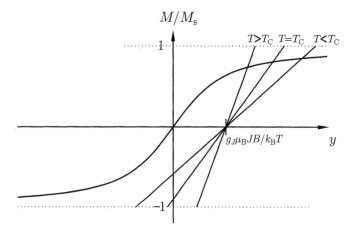

**Fig. 5.4** The graphical solution of eqns 5.6 and 5.7 for $B \neq 0$.

## 5.1.3    The effect of a magnetic field

The effect of adding a magnetic field is to shift to the right the straight line in the graphical solution of the equations (see Fig. 5.4). This results in a solution with $M \neq 0$ for all temperatures and so the phase transition is removed. For ferromagnets in a non-zero magnetic field there is always an energetic advantage to have a non-zero magnetization with the moments lining up along the magnetic field. This removal of the phase transition can be seen in Fig. 5.5 which shows graphical solutions to eqns 5.6 and 5.7 for a range of magnetic fields. In this model it is not necessary to consider the effect of applying a magnetic field in different directions. Whichever direction the magnetic field is applied, the magnetization will rotate round to follow it. The model contains no special direction associated with the ferromagnet itself. In a real ferromagnet this is not the case, and the effect of magnetic anisotropy associated with the material will need to be considered (see the following chapter).

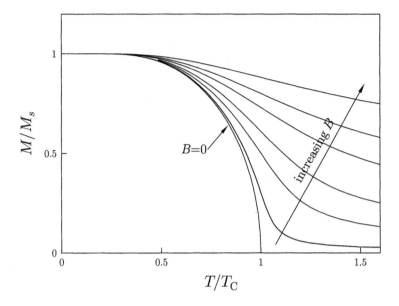

**Fig. 5.5** The mean-field magnetization as a function of temperature for $J = \frac{1}{2}$, calculated for different values of the applied field $B$. The phase transition is only present when $B = 0$.

At $T = T_{\mathrm{C}}$, the effect of the magnetic field is simple to work out analytically. At this temperature, the magnetization is given by $M \propto B^{1/3}$ for small magnetic fields. To prove this, it is necessary to take the next term in the Taylor expansion of $\mathrm{B}_J(y)$. Writing $\mathrm{B}_J(y) = (J+1)y/3J - \zeta y^3 + O(y^5)$, where $\zeta$ is a constant, we must solve simultaneously $M = M_{\mathrm{s}}\mathrm{B}_J(y)$ and

$$y = \frac{g_J \mu_{\mathrm{B}} J (B + \lambda M)}{k_{\mathrm{B}} T_{\mathrm{C}}} = \frac{(B + \lambda M)}{\zeta M_{\mathrm{s}}} \tag{5.13}$$

which yields

$$M = M_{\mathrm{s}} \zeta \frac{(B + \lambda M)}{\lambda} - \zeta M_{\mathrm{s}} \left( \frac{3J(B + \lambda M)}{\lambda(J+1)M_{\mathrm{s}}} \right)^3 \tag{5.14}$$

and hence

$$B \propto (B + \lambda M)^3 \tag{5.15}$$

and given that $\lambda M \gg B$, the right-hand side is dominated by the $M^3$ term so that $M \propto B^{1/3}$.

### 5.1.4   Origin of the molecular field

When Weiss proposed his molecular field model in 1907 he was disappointed that the constant $\lambda$ needed to be very large to agree with the large values of $T_{\mathrm{C}}$ found in nature. Considering only dipole fields, it was not possible to account for an internal field which, as discussed above, needs to be $\sim 10^3$ T to account for the Curie temperature of Fe. Thirty years later, Heisenberg showed that it was the exchange interaction, which involves large Coulomb energies, which is responsible for the large molecular field.[4]

The molecular field, parametrized by $\lambda$, can be related to the size of the exchange interaction, parametrized by $J_{ij}$. Assuming that the exchange interaction is effective only over the $z$ nearest neighbours of an ion, where it takes

[4]The molecular field is a convenient fiction and one shouldn't think that magnetic fields $\sim 10^3$ T are experienced by electrons in ferromagnets. The exchange interaction is purely an electrostatic effect. When we talk about the molecular field we are pretending that the exchange interaction is actually internal magnetic field. The point is that, if it really existed and Heisenberg exchange didn't, the molecular field would have to be $\sim 10^3$ T to explain the effects we see.

a value $J$, then using eqns 5.1, 5.4 and 5.5, one can easily find that

$$\lambda = \frac{2zJ}{ng^2\mu_{\rm B}^2}.$$

(5.16)

Using eqn 5.8, the Curie temperature can then be written

$$T_{\rm C} = \frac{2zJ J(J+1)}{3k_{\rm B}}.$$

(5.17)

In our discussion so far we have assumed that $L = 0$ and $J = S$. This works for most 3d ions. Exchange is between spin degrees of freedom and therefore depends on $S$. The magnetic moment of an ion depends on $J$, the total (spin + orbital) angular momentum. For 3d ions they are the same thing, because $L$ is quenched.

For 4f ions however, $\mathbf{S}$ is not a good quantum number, but $\mathbf{J}$ is. It follows that the component of $\mathbf{S}$ which is perpendicular to $\mathbf{J}$ must average to zero. The component of $\mathbf{S}$ which is parallel to $\mathbf{J}$ is conserved. Thus one must project $\mathbf{S}$ onto $\mathbf{J}$. Now $\mathbf{J} = \mathbf{L} + \mathbf{S}$ and $\mathbf{L} + 2\mathbf{S}$ is equal to $g_J\mathbf{J}$ plus a component perpendicular to $\mathbf{J}$. Hence the component of $\mathbf{S}$ that is a good quantum number is $(g_J - 1)\mathbf{J}$. Values of $(g_J - 1)$ for various 4f ions are listed in Table 5.2. From this, it is clear that $\mathbf{S}$ and $\mathbf{J}$ are parallel for the so-called 'heavy rare earths' (Gd to Yb), but antiparallel for the so-called 'light rare earths' (Ce to Sm).

**Table 5.2** The g-factors for 4f ions using Hund's rules.

| ion | shell | $S$ | $L$ | $J$ | $g_J$ | $g_J - 1$ | $(g_J - 1)^2 J(J+1)$ |
|-----|-------|-----|-----|-----|-------|-----------|----------------------|
| $Ce^{3+}$ | $4f^1$ | $\frac{1}{2}$ | 3 | $\frac{5}{2}$ | $\frac{6}{7}$ | $-\frac{1}{7}$ | 0.18 |
| $Pr^{3+}$ | $4f^2$ | 1 | 5 | 4 | $\frac{4}{5}$ | $-\frac{1}{5}$ | 0.80 |
| $Nd^{3+}$ | $4f^3$ | $\frac{3}{2}$ | 6 | $\frac{9}{2}$ | $\frac{72}{99}$ | $-\frac{27}{99}$ | 1.84 |
| $Pm^{3+}$ | $4f^4$ | 2 | 6 | 4 | $\frac{3}{5}$ | $-\frac{2}{5}$ | 3.20 |
| $Sm^{3+}$ | $4f^5$ | $\frac{5}{2}$ | 5 | $\frac{5}{2}$ | $\frac{2}{7}$ | $-\frac{5}{7}$ | 4.46 |
| $Eu^{3+}$ | $4f^6$ | 3 | 3 | 0 | – | – | – |
| $Gd^{3+}$ | $4f^7$ | $\frac{7}{2}$ | 0 | $\frac{7}{2}$ | 2 | 1 | 15.75 |
| $Tb^{3+}$ | $4f^8$ | 3 | 3 | 6 | $\frac{3}{2}$ | $\frac{1}{2}$ | 10.50 |
| $Dy^{3+}$ | $4f^9$ | $\frac{5}{2}$ | 5 | $\frac{15}{2}$ | $\frac{4}{3}$ | $\frac{1}{3}$ | 7.08 |
| $Ho^{3+}$ | $4f^{10}$ | 2 | 6 | 8 | $\frac{5}{4}$ | $\frac{1}{4}$ | 4.50 |
| $Er^{3+}$ | $4f^{11}$ | $\frac{3}{2}$ | 6 | $\frac{15}{2}$ | $\frac{6}{5}$ | $\frac{1}{5}$ | 2.55 |
| $Tm^{3+}$ | $4f^{12}$ | 1 | 5 | 6 | $\frac{7}{6}$ | $\frac{1}{6}$ | 1.17 |
| $Yb^{3+}$ | $4f^{13}$ | $\frac{1}{2}$ | 3 | $\frac{7}{2}$ | $\frac{8}{7}$ | $\frac{1}{7}$ | 0.32 |
| $Lu^{3+}$ | $4f^{14}$ | 0 | 0 | 0 | – | – | – |

Using $(g_J - 1)\mathbf{J}$ for the conserved part of $\mathbf{S}$, the expression $-\sum_{ij} J_{ij}\mathbf{S}_i \cdot \mathbf{S}_j$ can be replaced with $-\sum_{ij}(g_J - 1)^2 J_{ij}\mathbf{J}_i \cdot \mathbf{J}_j$. The magnetic moment $\mu = -g_J\mu_{\rm B}\mathbf{J}$ can then be used to repeat the calculation leading to eqn 5.16, resulting in

$$\lambda = \frac{2zJ(g_J - 1)^2}{ng_J^2\mu_{\rm B}^2}.$$

(5.18)

(For the special case of transition metal ions with orbital quenching, $g_J = 2$ and eqn 5.18 reduces to eqn 5.16.) The Curie temperature can then be written

$$T_C = \frac{2z(g_J - 1)^2 J}{3k_B} J(J+1). \tag{5.19}$$

The critical temperature is therefore expected to be proportional to the de Gennes factor $(g_J - 1)^2 J(J+1)$. Values of the de Gennes factor are also listed in Table 5.2. Gd, with the largest de Gennes factor, is a ferromagnet but the rare earth metals show a variety of different ground states, including antiferromagnetism and helimagnetism which will be considered in the next sections.

Pierre-Gilles de Gennes (1932–)

## 5.2   Antiferromagnetism

If the exchange interaction is negative, $J < 0$, the molecular field is oriented such that it is favourable for nearest neighbour magnetic moments to lie antiparallel to one another. This is antiferromagnetism. Very often this occurs in systems which can be considered as two interpenetrating sublattices (see Fig. 5.6), on one of which the magnetic moments point up and on the other of which they point down. The nearest neighbours of each magnetic moment in Fig. 5.6 will then be entirely on the other sublattice. Initially we will therefore assume that the molecular field on one sublattice is proportional to the magnetization of the other sublattice. We will also assume that there is no applied magnetic field.

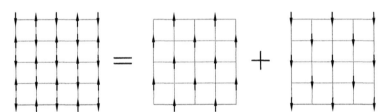

**Fig. 5.6** An antiferromagnet can be decomposed into two interpenetrating sublattices.

### 5.2.1   Weiss model of an antiferromagnet

If we label the 'up' sublattice $+$ and the 'down' sublattice $-$ then the molecular field on each sublattice is

$$B_+ = -|\lambda|M_-$$
$$B_- = -|\lambda|M_+, \tag{5.20}$$

where $\lambda$ is the molecular field constant which is now negative. On each sublattice, the molecular field is therefore given by

$$M_\pm = M_s B_J \left( -\frac{g_J \mu_B J |\lambda| M_\mp}{k_B T} \right). \tag{5.21}$$

The two sublattices are equivalent in everything except the direction of the moments so that

$$|M_+| = |M_-| \equiv M, \tag{5.22}$$

and hence

$$M = M_s B_J \left( \frac{g_J \mu_B J |\lambda| M}{k_B T} \right). \tag{5.23}$$

This is almost identical to the corresponding equation for ferromagnetism (eqns 5.6 and 5.7), and so the molecular field on each sublattice will follow exactly the form shown in Fig. 5.3 and will disappear for temperatures above a transition temperature, known as the **Néel temperature** $T_N$, which is then defined by

$$T_N = \frac{g_J \mu_B (J + 1) |\lambda| M_s}{3k_B} = \frac{n |\lambda| \mu_{eff}^2}{3k_B}. \tag{5.24}$$

Although the magnetization on each sublattice will follow the form shown in Fig. 5.3, the two magnetizations will be in oppositedirections so that the net magnetization $M_+ + M_-$ of the antiferromagnet will be zero. One can define a quantity known as the **staggered magnetization** as the *difference* of the magnetization on each sublattice, $M_+ - M_-$; this is then non-zero for temperatures below $T_N$ and hence can be used as an order parameter[5] for antiferromagnets.

## 5.2.2 Magnetic susceptibility

For temperatures above $T_N$ the effect of a small applied magnetic field can be calculated in the same way as for the ferromagnet, by expanding the Brillouin function $B_J(y) = (J + 1)y/3J + O(y^3)$, and results in the magnetic susceptibility $\chi$ being given by

$$\chi = \lim_{B \to 0} \frac{\mu_0 M}{B} \propto \frac{1}{T + T_N}, \tag{5.25}$$

which is the Curie Weiss law again but with the term $-T_C$ replaced by $+T_N$.

This result gives a ready means of interpreting susceptibility data in the paramagnetic state (i.e. for temperatures above the transition to magnetic order). The magnetic susceptibility can be fitted to a Curie Weiss dependence

$$\chi \propto \frac{1}{T - \theta}, \tag{5.26}$$

where $\theta$ is the **Weiss temperature**. If $\theta = 0$, the material is a paramagnet (see eqn 2.44). If $\theta > 0$ the material is a ferromagnet and we expect $\theta = T_C$ (see eqn. 5.12). If $\theta < 0$ the material is a antiferromagnet and we expect $\theta = -T_N$ (see eqn. 5.25). These possibilities are shown in Fig. 5.7.

Experimentally determined Weiss temperatures in antiferromagnets are often a long way from $-T_N$ (see Table 5.3 which contains data for some common antiferromagnets). This discrepancy is largely due to the assumption we have made that the molecular field on one sublattice depends only on the magnetization of the other sublattice. A more realistic calculation is considered in Exercise 5.3.

Applying a magnetic field to an antiferromagnet at temperatures below $T_N$ is more complicated than the case of a ferromagnet below $T_C$ because the *direction* in which the magnetic field is applied is crucial. There is no longer an energetic advantage for the moments to line up along the field because any energy saving on one sublattice will be cancelled by the energy cost for

Louis E. F. Néel (1904–2000)

[5]An order parameter will be defined in Section 6.1.

(a)

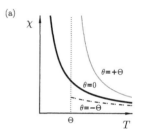

(b)

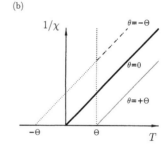

(c)

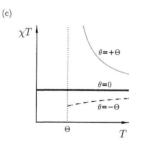

**Fig. 5.7** The Curie Weiss law states that $\chi \propto 1/(T - \theta)$ for $T > \theta$. This is shown in (a) for three cases: $\theta = 0$ (paramagnet), $\theta = \Theta > 0$ (ferromagnet) and $\theta = -\Theta < 0$ (antiferromagnet). Straight-line graphs are obtained by plotting $1/\chi$ against $T$ as shown in (b) with the intercept on the temperature axis yielding $\theta$. A graph of $\chi T$ against $T$ can be constant ($\theta = 0$), increasing for decreasing $T$ ($\theta > 0$) or decreasing for decreasing $T$ ($\theta < 0$), as shown in (c).

**Table 5.3** Properties of some common antiferromagnets.

| Material | $T_N$ (K) | $\theta$ (K) | $J$ |
|---|---|---|---|
| $MnF_2$ | 67 | −80 | $\frac{5}{2}$ |
| $MnO$ | 122 | −610 | $\frac{5}{2}$ |
| $CoO$ | 292 | −330 | $\frac{3}{2}$ |
| $FeO$ | 198 | −507 | 2 |
| $Cr_2O_3$ | 307 | −485 | $\frac{3}{2}$ |
| $\alpha\text{-}Fe_2O_3$ | 950 | −2000 | $\frac{5}{2}$ |

the other sublattice, if the magnetization on the two sublattices is equal and opposite.

Consider first the case of absolute zero ($T = 0$), so that thermal agitation effects can be ignored. $|M_+|$ and $|M_-|$ are both equal to $M_s$. If a small magnetic field is applied *parallel* to the magnetization direction of one of the sublattices (and hence antiparallel to the magnetization direction of the other sublattice), a small term is added or subtracted to the local field of each sublattice. Since both sublattices are already saturated, this has no effect and the net magnetization induced in the material is zero so that $\chi_\parallel = 0$. If instead the small magnetic field is applied *perpendicular* to the magnetization direction of one of the sublattices, this causes the magnetization of both sublattices to tilt slightly so that a component of magnetization is produced along the applied magnetic field (see Fig. 5.8). Thus $\chi_\perp \neq 0$.

If the temperature is now increased, but still kept below $T_N$, the thermal fluctuations decrease the molecular field at each sublattice. This greatly affects the case of applying a small magnetic field *parallel* to the magnetization direction of one of the sublattices, since the field enhances the magnetization of one sublattice and reduces it on the other. In the perpendicular case, raising the temperature has little effect since the $M_+$ and $M_-$ are reduced equally and are also symmetrically affected by the small magnetic field. $\chi_\perp$ is independent of temperature, whereas $\chi_\parallel$ rises from 0 up to $\chi_\perp$ as $T \rightarrow T_N$. These characteristics are shown in Fig. 5.9.

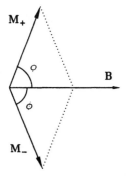

**Fig. 5.8** The origin of $\chi_\perp$. A small magnetic field $B$ is applied perpendicular to the magnetization direction of the sublattices, causing the magnetization of both sublattices to tilt slightly so that a component of magnetization is produced along the applied magnetic field.

### 5.2.3 The effect of a strong magnetic field

Let us first consider the effect of a strong magnetic field on an antiferromagnet with $T = 0$ to avoid any complications from thermal fluctuations. If the magnetic field is large enough, it must eventually dominate over any internal molecular field and force all the magnetic moments to lie parallel to each other. But as the field is increased, although the final end result is clear, the route to that destination depends strongly on the direction of the applied field with respect to the initial direction of sublattice magnetization.

If the applied magnetic field is perpendicular to the sublattice magnetizations, all that happens is that as the field increases the magnetic moments bend round more and more ($\phi$ gets progressively smaller, see Fig. 5.8) until the moments line up with the applied magnetic field.

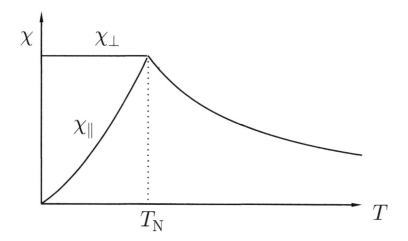

**Fig. 5.9** The effect of temperature on $\chi_\parallel$ and $\chi_\perp$.

If the applied magnetic field is parallel to the sublattice magnetizations, the case is more interesting. At small magnetic fields the moments don't rotate round but stay in line (Fig. 5.10(a)). However, at a critical field the system suddenly snaps into a different configuration (Fig. 5.10(b)); this is called a **spin-flop transition**. For further increases of magnetic field the angle $\theta$ gets progressively smaller until eventually the magnetic moments line up with the applied magnetic field.

These effects can be calculated quantitatively. Let $\mathbf{M}_+$ lie at an angle of $\theta$ to the magnetic field (measured counterclockwise) and let $\mathbf{M}_-$ lie at an angle of $\phi$ to the magnetic field (measured clockwise). We will apply the magnetic field along the crystallographic $z$ axis. The antiferromagnetic phase corresponds to $\theta = 0$ and $\phi = \pi$ (see Fig. 5.10(a)) and the spin-flop phase corresponds to $\theta = \phi$. It is necessary to determine which phase has lower energy.

We assume that the total energy $E$ is due to the sum of the Zeeman energies of the individual sublattices and a term representing the exchange coupling which will depend on the relative orientation between the two sublattice moments. This leads to

$$E = -MB\cos\theta - MB\cos\phi + AM^2\cos(\theta + \phi), \qquad (5.27)$$

where $A$ is a constant connected with the exchange coupling. To model the magnetic anisotropy, it is necessary to add on a term of the form

$$-\frac{1}{2}\Delta(\cos^2\theta + \cos^2\phi) \qquad (5.28)$$

where $\Delta$ is a small constant. This accounts for the fact that the magnetizations actually do prefer to lie along a certain crystallographic axis (in this case the $z$ axis) so that that $\theta$ and $\phi$ prefer to be 0 or $\pi$ but not somewhere in between. In the antiferromagnetic case ($\theta = 0, \phi = \pi$) we have $E = -AM^2 - \Delta$ which is independent of field. In the spin-flop case ($\phi = \theta$) we have

$$E = -2MB\cos\theta + AM^2\cos 2\theta - \Delta\cos^2\theta. \qquad (5.29)$$

The condition $\partial E/\partial\theta = 0$ shows that there is a minimum energy when $\theta = \cos^{-1}[B/2AM]$, ignoring the anisotropy term. Substituting this back into $E$

(a)

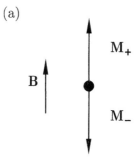

(b)

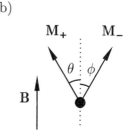

**Fig. 5.10** A magnetic field is applied parallel to the sublattice magnetizations. (a) For small fields nothing happens and the system remains in the antiferromagnetic phase. (b) Above a critical field the system undergoes a spin-flop transition into a spin-flop phase.

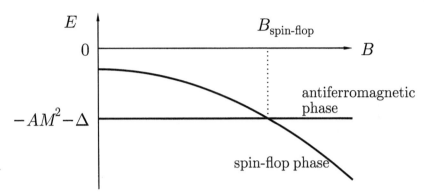

**Fig. 5.11** The energy of the antiferromagnetic phase and the spin-flop phase as a function of $B$.

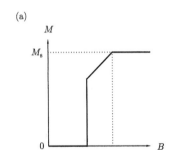

(a)

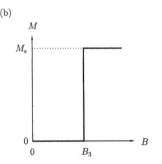

(b)

**Fig. 5.12** (a) Magnetization for applying a parallel magnetic field to an antiferromagnet. Initially nothing happens but then there is a spin-flop transition to a spin-flop phase at $B_1$. The magnetic field then rotates the moments until saturation is achieved at the field $B_2$. (b) If there is a strong preference for the spins to lie along the parallel direction, no spin-flop occurs. Instead there is a spin-*flip* transition at $B_3$. Both figures show the expected curves for absolute zero. Finite temperature will round off the sharp corners. This is also known as a metamagnetic transition.

and plotting the result leads to the graph shown in Fig. 5.11. Below the critical field $B_{\text{spin-flop}}$ the antiferromagnetic case has the lowest energy. At the critical field $B_{\text{spin-flop}}$ the system switches from one state to the other and we have a spin-flop transition. Above this field the spin-flop phase has the lowest energy.

The magnetization for the antiferromagnet in a large parallel magnetic field is shown in Fig. 5.12(a). There is no effect until the spin-flop transition, above which the magnetization increases steadily until saturation is reached. If the anisotropy effect is very strong ($\Delta$ is large), another effect can occur. In this case, if the external field is along $z$, no spin-flop occurs. Instead we get a **spin-flip transition**, i.e. the magnetization of one sublattice suddenly reverses when $B$ reaches a critical value, and the system moves in a single step to the ferromagnetic state. This is illustrated in Fig. 5.12(b).

### 5.2.4   Types of antiferromagnetic order

Another complication with antiferromagnetism is that there is a large number of ways of arranging an equal number of up and down spins on a lattice. The different possible arrangements also depend on the kind of crystal lattice on which the spins are to be arranged. A selection of possible arrangements is shown in Figures 5.13 and 5.14.

In cubic perovskites, which have the magnetic atoms arranged on a simple cubic lattice, G-type ordering (see Fig. 5.13(d)) is very common because superexchange interactions through oxygen atoms force all nearest-neighbour magnetic atoms to be antiferromagnetically aligned. This is the case for G-type ordering only and is, for example, found in $LaFeO_3$ and $LaCrO_3$. $LaMnO_3$ is also a cubic perovskite but shows A-type ordering (see Fig. 5.13(a)), with alternately aligned ferromagnetic (100) planes. This occurs because of the Jahn–Teller distortion of the $Mn^{3+}$ ions which gives alternate long and short Mn–O bonds within the (100) planes. The orbitals on adjacent $Mn^{3+}$ ions are differently oriented, and the superexchange leads to an interaction between an occupied orbital on one atom with an unoccupied orbital on its neighbour. The in-plane interaction is thus ferromagnetic while the out-of-plane interaction is antiferromagnetic because of the conventional operation of the superexchange interaction.

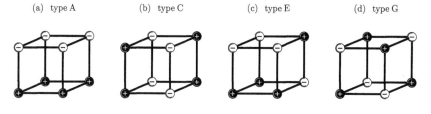

(a) type A    (b) type C    (c) type E    (d) type G

**Fig. 5.13** Four types of antiferromagnetic order which can occur on simple cubic lattices. The two possible spin states are marked + and −.

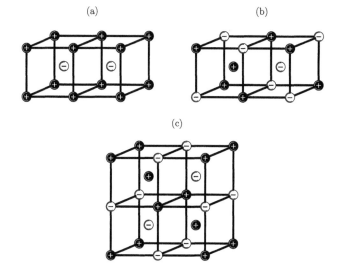

(a)    (b)

(c)

**Fig. 5.14** Three types of antiferromagnetic order which can occur on body-centred cubic lattices.

## 5.3 Ferrimagnetism

The above treatment of antiferromagnetism assumed that the two sublattices were equivalent. But what if there is some crystallographic reason for them not to be equivalent? In this case the magnetization of the two sublattices may not be equal and opposite and therefore will not cancel out. The material will then have a net magnetization. This phenomenon is known as **ferrimagnetism**. Because the molecular field on each sublattice is different, the spontaneous magnetizations of the sublattices will in general have quite different temperature dependences. The net magnetization itself can therefore have a complicated temperature dependence. Sometimes one sublattice can dominate the magnetization at low temperature but another dominates at higher temperature; in this case the net magnetization can be reduced to zero and change sign at a temperature known as the **compensation temperature**. The magnetic susceptibilities of ferrimagnets therefore do not follow the Curie Weiss law.

**Ferrites** are a family of ferrimagnets. They are a group of compounds with the chemical formula $MO \cdot Fe_2O_3$ where M is a divalent cation such as $Zn^{2+}$, $Co^{2+}$, $Fe^{2+}$, $Ni^{2+}$, $Cu^{2+}$ or $Mn^{2+}$. The crystal structure is the **spinel** structure which contains two types of lattice sites, tetrahedral sites (with four oxygen

**Table 5.4** Properties of some common ferrimagnets.

| Material | $T_C$ (K) | magnetic moment ($\mu_B$/formula unit) | Compensation temperature (K) |
|---|---|---|---|
| $Fe_3O_4$ | 858 | 4.1 | – |
| $CoFe_2O_4$ | 793 | 3.7 | – |
| $NiFe_2O_4$ | 858 | 2.3 | – |
| $CuFe_2O_4$ | 728 | 1.3 | – |
| $Y_3Fe_5O_{12}$ | 560 | 5.0 | – |
| $Gd_3Fe_5O_{12}$ | 564 | 16.0 | 290 |
| $Dy_3Fe_5O_{12}$ | 563 | 18.2 | 220 |
| $Ho_3Fe_5O_{12}$ | 567 | 15.2 | 137 |

neighbours, these are known as A sites) and octahedral sites (with six oxygen neighbours, these are known as B sites). There are twice as many B sites as A sites. The two sublattices are non-equivalent because there are two types of crystallographic site and they contain two types of different ion. In **normal spinels**, the $M^{2+}$ cations sit at the A sites and the $Fe^{3+}$ ($^6S_{5/2}$ and therefore a moment of 5 $\mu_B$) cations sit at the B sites. In **inverse spinels** the $M^{2+}$ cations sit at half of the B sites, while the $Fe^{3+}$ cations occupy the other half of the B sites and all the A sites. In inverse spinels, the moments of the $Fe^{3+}$ cations on the A and the B sites are antiparallel, so that the total moment of the sample is due to the $M^{2+}$ ions only.

The case of M=Fe, i.e. $Fe_3O_4$ (which is a semiconductor, in contrast to the other ferrites which are insulators), has already been discussed in Section 4.2.5.

Another family of ferrimagnets is the **garnets** which have the chemical formula $R_3Fe_5O_{12}$ where $R$ is a trivalent rare earth atom. The crystal structure is cubic, but the unit cell is quite complex. Three of the $Fe^{3+}$ ions are on tetrahedral sites, two are on octahedral sites and the $R^{3+}$ ions are on sites of dodecahedral symmetry. In yttrium iron garnet (YIG), $Y_3Fe_5O_{12}$, the $Y^{3+}$ has no magnetic moment (it is $4d^0$) and the moments of the $Fe^{3+}$ ions on the tetrahedral sites are antiparallel to those on the octahedral sites, so that the net moment is $5\mu_B$.

Barium ferrite ($BaFe_{12}O_{19} = BaO \cdot 6Fe_2O_3$) has a hexagonal structure. Eight of the $Fe^{3+}$ ions are antiparallel to the other four, so that the net moment is equivalent to four $Fe^{3+}$ ions, i.e. $20\mu_B$. In powder form, it is used in magnetic recording since it has a high coercivity (see Section 6.7.9). The properties of some common ferrimagnets are listed in Table 5.4.

Most ferrimagnets are electrical insulators and this fact is responsible for many of their practical applications. Ferromagnets are often metallic and thus are unsuitable in applications in which an oscillating magnetic field is involved; a rapidly changing magnetic field induces a voltage and causes currents (known as eddy currents) to flow in conductors. These currents cause resistive heating in a metal (eddy current losses). Many ferrimagnets therefore can be used when a material with a spontaneous magnetization is required to operate at high frequencies, since the induced voltage will not be able to cause any significant eddy currents to flow in an insulator. Solid ferrite cores are used in many high frequency applications including aerials and transformers requiring high permeability and low energy loss, as well as applications in microwave components. Also many ferrimagnets are more corrosion resistant than metallic ferromagnets since they are already oxides.

## 5.4   Helical order

In many rare earth metals, the crystal structure is such that the atoms lie in layers. Consider first the case (relevant for dysprosium) in which there is ferromagnetic alignment of atomic moments within the layers and that the interaction between the layers can be described by a nearest-neighbour exchange constant $J_1$ and a next-nearest-neighbour exchange constant $J_2$. If the angle between the magnetic moments in successive basal planes (i.e. the planes corresponding to the layers) is $\theta$ (see Fig. 5.15(a)), then the energy of the system can be written

$$E = -2NS^2(J_1 \cos\theta + J_2 \cos 2\theta), \qquad (5.30)$$

where $N$ is the number of atoms in each plane. The energy is minimized when $\partial E/\partial\theta = 0$ which yields

$$(J_1 + 4J_2 \cos\theta) \sin\theta = 0. \qquad (5.31)$$

Solutions to this are either $\sin\theta = 0$, which implies $\theta = 0$ or $\theta = \pi$ (ferromagnetism or antiferromagnetism), or

$$\cos\theta = -\frac{J_1}{4J_2}. \qquad (5.32)$$

This last solution corresponds to helical order (also known as helimagnetism) and is favoured over either ferromagnetism or antiferromagnetism when $J_2 < 0$ and $|J_1| < 4|J_2|$ (see Fig. 5.15(b)). The pitch of the spiral will not in general be commensurate with the lattice parameter and so no two layers in a crystal will have exactly the same spin directions.

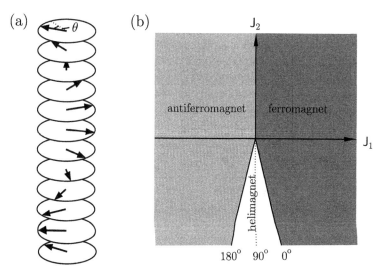

**Fig. 5.15** (a) Helimagnetic ordering. (b) The phase diagram for the model of planes coupled by a nearest-neighbour exchange constant $J_1$ and a next-nearest-neighbour exchange constant $J_2$.

Helical structures are found in many magnetic systems, most famously in rare earth metals. Many of these rare earth metals have hexagonally close packed crystal structures. The axis of the helix is perpendicular to the hexagonally close packed planes, along what is usually defined as the $c$ axis. The plane in which the spins rotate in Tb, Dy and Ho is the hexagonally close packed plane, but in Er and Tm the easy axis for spins is the $c$ axis so that the $c$ component of the spins is modulated sinusoidally over a certain temperature range. Spiral structures (Fig. 5.1(d)) are also possible in which the spins have a component along the easy axis, perpendicular to the plane, but also a component in the plane which precesses around a helix.

The exchange interaction in rare earth metals is an indirect RKKY interaction mediated by the conduction electrons. It has much longer range than superexchange interactions and changes its sign as a function of distance. The model presented above, which contains just nearest and next-nearest neighbour interactions, is therefore an over-simplification. Details of the Fermi surface of each rare earth metal are needed to compute the wave vector-dependent exchange interaction $J(\mathbf{q})$. If this takes a maximum at a certain wave vector $\mathbf{q}$, then helimagnetism can be induced with wave vector $\mathbf{q}$. A further feature is the large effect of the crystal field in rare earth metals. The crystal field splits the electronic states and usually one finds that not only the ground state but also some excited states of the system are thermally populated, further complicating the analysis that needs to be performed to understand these systems.

## 5.5 Spin glasses

So far we have considered materials in which there is one or more magnetic moments in each unit cell. What happens if we start with a non-magnetic lattice and sparsely populate it with a dilute, random distribution of magnetic atoms? One's intuition might suggest that the end result would be something which is entirely random which would not be likely to exhibit a phase transition from a high temperature disordered state to a low temperature ordered state. This is only partly right because such systems, although inherently random, do show something approximating to a phase transition at a particular temperature to a state which, while not ordered, is distinctly different from the high temperature disordered state. We can define a spin glass as a random, magnetic system with mixed interactions characterized by a random, yet cooperative, freezing of spins at a well defined temperature $T_f$ (the freezing temperature) below which a metastable frozen state appears without the usual magnetic long-range ordering.

### Example 5.1

A well studied example of a spin glass is the alloy CuMn in which the concentration of Mn is a few atomic percent. The Mn ions are therefore present only in dilute quantities and their magnetic moments interact with each other via a RKKY (see sections 4.2.4 and 7.7.3) interaction mediated by the conduction electrons in the Cu. The RKKY interaction oscillates in sign so that the interactions can be either antiferromagnetic or ferromagnetic. The net result is a system with a great deal of in built frustration with no well defined ground state, but a large number of alternative possible ground states. At high temperature the Mn moments thermally fluctuate, but as the temperature is reduced, locally correlated clusters build up as the moments slow down. At $T_f$ the moments get stuck in one of these degenerate ground states and become frozen. In CuMn site-randomness implies a distribution of distances between spins and thus provides the frustration; spin glasses can also be found in systems which have bond-randomness, for which the nearest–neighbour interactions vary between $+J$ and $-J$. Spin glasses and frustration will be discussed in more detail in Section 8.2.

## 5.6   Nuclear ordering

So far only electronic moments have been considered. But some nuclei possess magnetic moments; can they order? The answer is yes, but the effect is very small and only important at low temperature in materials with no electronic moments. Nuclear spins are well localized inside the atoms and do not couple strongly with each other; there can be no direct exchange interaction between them. Nuclear moments are also about three orders of magnitude smaller than electronic moments and since the dipolar interaction is proportional to the square of the magnetic moment, any dipolar interaction will be six orders of magnitude lower than in the electronic case. If electronic moments are present, they will completely dwarf any nuclear effect. The small size of the nuclear magnetic moments means that nuclear spin systems can only show magnetic order at temperatures below $\sim 1$ $\mu$K. There is however also the possibility of RKKY coupling between nuclear moments in metallic samples in which exchange can be mediated by the conduction electrons. However this effect is about the same order of size as the nuclear dipolar interaction.

Some remarkable experiments have nevertheless been performed on a number of materials which have no electronic moments and nuclear spin ordering has actually been observed. In copper there is a first order phase transition to an antiferromagnetically ordered nuclear ground state at 58 nK. In silver, the Néel temperature for the nuclear spins is found to be 560 pK. How may these extremely low temperatures be obtained? The experiments rely on the fact that the lattice and the nuclei can have different temperatures, even though they are in the same sample. Using adiabatic demagnetization (see Section 2.6) the nuclei can be cooled and reach thermal equilibrium among themselves in a time characterized by the spin–spin relaxation time $T_2$ which is typically a few milliseconds. The nuclear spins only relax back up to the lattice temperature in a time characterized by $T_1$ which can be as long as a few hours.

There is a further cunning trick which may be used: if the nuclei are placed in a magnetic field, the nuclear levels are split and populated according to their Boltzmann probability. If the magnetic field is then suddenly reversed (in a time which is short in comparison to $T_2$) the nuclear spins do not immediately move relative to their levels and one obtains a population inversion. Moreover, one has the correct Boltzmann distribution between the levels appropriate for a *negative* temperature. The negative temperature state is a distinct thermodynamic state with a distinct magnetic phase diagram. Experiments on silver show that at low negative temperatures the silver nuclei order ferromagnetically at $-1.9$ nK.

## 5.7   Measurement of magnetic order

In the preceding sections, the various types of magnetic order have been described. But how do you measure them? In this section, various techniques for measuring magnetic order are discussed.

### 5.7.1   Magnetization and magnetic susceptibility

The most obvious experimental technique to try is a conventional measurement of magnetization. This is essentially a measurement of the sample's net magnetic moment $\mu$ (dividing this by the sample volume yields the magnetization).

In an **extraction magnetometer**, the sample is placed at the centre of a coil and then removed to a large distance, inducing a voltage $V$ in the coil. The magnetic flux $\Phi$ produced by the sample is equal to $\int V \, dt$. Usually the magnetometer is constructed so that there are two counterwound coils, arranged so that voltages due to changes in the applied field are cancelled out and only the signal from the sample remains. A **vibrating sample magnetometer** (commonly known as a VSM) vibrates a sample sinusoidally up and down, and then an electrical signal can be induced in a stationary pick-up coil by the movement of the magnetic moment of the sample. The signal is proportional to the magnetic moment, as well as to the amplitude and frequency of the vibration. In the **alternating gradient magnetometer**, the sample is mounted on a piezoelectric strip and an alternating field is provided by counterwound coils, so that at the sample there is an alternating field gradient. This results in a force on the sample equal to $(\mu \cdot \nabla)\mathbf{B}$. The piezoelectric strip therefore deflects in time with the driving frequency and by an amount that can be measured by the piezoelectric signal, allowing $\mu$ to be deduced. In a **torque magnetometer**, the sample is suspended on a torsion fibre. The application of a magnetic field $\mathbf{B}$ on the sample produces a torque equal to $\mu \times \mathbf{B}$. One of the most sensitive techniques uses a **SQUID** (superconducting quantum interference device). This is a superconducting ring which contains a 'weak-link', so that there is a Josephson junction in it, and the ring is therefore able to act like a very sensitive quantum interferometer. If a sample is passed through the ring, the persistent current induced is proportional to the magnetization of the sample. These techniques are of course useful only if the sample has a non-zero magnetization. For an antiferromagnet the magnetization is zero unless the applied magnetic field is large.

An alternative strategy is to measure the magnetic susceptibility which gives a good indication of the type of magnetic order present. This can be measured and data fitted to the Curie Weiss law. The magnetic susceptibility can be obtained by a number of methods. It is equal to the magnetization induced in a sample for a small applied magnetic field, and so the techniques described

above which measure magnetization can be used to measure magnetic susceptibility. In addition, a method which is mainly used for extracting small magnetic
susceptibilities is the Gouy method. In this method a long cylindrical sample
is suspended between the pole pieces of a magnet so that one end of the sample
is between the pole pieces and the other end is well out of the magnet in a
region of zero field. When the magnet is turned on, the apparent weight of the
sample changes. In the related Faraday method, a smaller sample can be used
but this time the force is produced using a magnetic field gradient obtained
with appropriately shaped pole pieces.

Louis George Gouy (1854–1926)

Michael Faraday (1791–1867)

### 5.7.2   Neutron scattering

Neutrons have been found to be extremely useful in studying magnetism in
condensed matter, and neutron scattering has significant advantages over other
experimental techniques in the study of magnetic structure and dynamics. They
give the most direct information on the arrangement of magnetic moments in
a specimen.

   The neutron is a spin-$\frac{1}{2}$ particle (see Table 2.3) and has a non-zero magnetic
moment. Neutrons are produced in great quantities by fission reactions inside
the fuel elements of nuclear reactors. The beam of neutrons which emerges
from a reactor has a spectrum of energies determined by the temperature $T$ of
the moderator, usually room temperature but $T$ can be higher or lower if the
moderator is heated or cooled. Neutrons emerging from the reactor therefore
have a distribution of velocities given by

$$n(v) \propto v^3 \exp\left(-\frac{\frac{1}{2}m_n v^2}{k_B T}\right), \tag{5.33}$$

where $n(v)\,dv$ is the number of neutrons through unit area per second and $m_n$
is the mass of the neutron. The distribution contains a Boltzmann factor and a
$v^3$ term which is the product of a phase space factor $v^2$ (as for the Maxwellian
distribution of velocities in kinetic theory) and a factor of $v$ due to effusion
through the hole. This function is plotted in Fig. 5.16(a). The maximum of this
distribution is at

$$v = \sqrt{\frac{3k_B T}{m_n}} \tag{5.34}$$

which corresponds to the condition $\frac{1}{2}m_n v^2 = \frac{3}{2}k_B T$. The de Broglie wavelength $\lambda$ of a neutron with velocity $v$ is

$$\lambda = \frac{h}{m_n v}. \tag{5.35}$$

The distribution function is plotted as a function of wavelength in Fig. 5.16(b).
This demonstrates that thermal neutrons have wavelengths similar to atomic
spacings, thus permitting diffraction measurements to be performed. The
ability to tune the wavelength means that diffraction experiments range in
lengthscale from directly probing the wave function of the small atoms to the
low-resolution study of macromolecules.

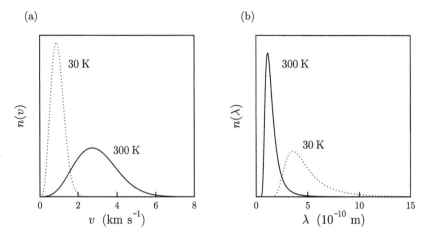

**Fig. 5.16** (a) The distribution function of neutrons as a function of velocity $v$ for temperatures 30 K and 300 K. (b) The same distribution functions plotted as a function of de Broglie wavelength $\lambda = h/m_\mathrm{n}v$. The curves are normalized to have the same area.

Neutrons scatter from the nucleus via the strong nuclear force and also from variations in magnetic field within a crystal via the electromagnetic interaction. The latter interaction enables the neutron to probe magnetic properties of materials since the neutron couples to the magnetic moment of the electrons.

We first consider the nuclear scattering. The nuclear forces which cause scattering of neutrons are very short range ($10^{-15}$–$10^{-14}$ m), much smaller than the wavelength of typical thermal neutrons. Hence an incident neutron wave $\psi_i = \mathrm{e}^{\mathbf{ik \cdot r}}$ produces a spherically symmetric elastically scattered wave $\psi_f = -(b/r)\mathrm{e}^{\mathrm{i}kr}$. The quantity $b$ is known as the **scattering length** and the minus sign in our expression for $\psi_f$ ensures that $b > 0$ for a repulsive nuclear potential. The scattering length depends on the particular nucleus and the spin state of the nucleus–neutron system (which can be $I + \frac{1}{2}$ or $I - \frac{1}{2}$ for a nucleus of spin $I$).

In an **elastic neutron scattering** experiment, elastically scattered neutrons produce strong Bragg reflections when the scattering vector is equal to a reciprocal lattice vector (Fig. 5.17(a, b)). These reflections can be studied in samples by a number of experimental methods, including (i) rotating a crystal and measuring the scattering from a monochromatic beam, looking for strong reflections, (ii) using a range of incident wavelengths (the Laue method, see Fig. 5.17(c)) and (iii) measuring the diffraction of monochromatic neutrons from a powder sample (see Fig. 5.17(d)). A possible experimental arrangement at a reactor source is shown schematically in Fig. 5.17(e). Neutrons can also be produced at a **spallation source**. High energy protons from a synchrotron source strike a heavy metal target (e.g. $^{238}$U) resulting in the production of neutrons. The proton beam from the synchrotron is pulsed, so that the neutron beam is also pulsed. The spread of energies in the pulse of neutrons can be exploited by building spectrometers that use the time-of-flight technique. By measuring the time dependence of the scattered neutrons after the start of the pulse, the scattering of a range of incident neutron wave vectors can be extracted from a single pulse (see Fig. 5.17(f)).

The nuclear scattering of neutrons results from interaction with the nucleus, rather than with the electron cloud, and so the scattering power of an atom is not strongly related to its atomic number. This is in sharp contrast with the cases of both X-ray and electron scattering. This has some advantages:

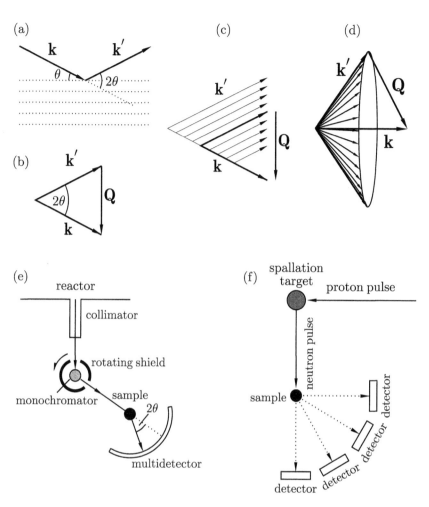

**Fig. 5.17** (a) Incident neutrons with wave vector **k** are scattered to **k′** through a scattering angle of $2\theta$ by the crystal planes. Elastic scattering implies that $|\mathbf{k}| = |\mathbf{k}'|$. The incident neutron wavelength is $\lambda = 2\pi/|\mathbf{k}|$. (b) Constructive interference of the scattered neutrons from crystal planes occurs when the scattering vector $\mathbf{Q} = \mathbf{k} - \mathbf{k}'$ equals a reciprocal lattice vector. (c) The Laue method in which a beam of neutrons with a fixed incident direction but with a range of wavelengths gives rise to a Bragg reflection when **Q** is a reciprocal lattice vector. (d) In a powder diffraction method, a monochromatic beam of incident neutrons is incident upon a powder sample. The scattered neutrons lie on a cone (the Debye–Scherrer cone) of semiangle $2\theta$. (e) Neutron diffraction at a reactor source. The monochromator and rotating shield are used to select the wavelength of the incident neutrons and the scattered neutrons from the sample are measured by a multidetector. (f) At a spallation source, a pulse of neutrons is produced by spallation from a target. The detector signals are recorded as a function of time, data from the faster (shorter wavelength) neutrons arriving earlier.

it is easier to sense light atoms, such as hydrogen, in the presence of heavier ones, and neighbouring elements in the periodic table generally have substantially different scattering cross sections and can be distinguished. The nuclear dependence of scattering allows isotopes of the same element to have substantially different scattering lengths for neutrons, so that isotopic substitution can thus be employed. Neutrons are a highly penetrating probe, allowing the investigation, again non-destructively, of the interior of materials, rather than just the surface layers probed by techniques such as X-ray scattering, electron microscopy or optical methods.

Neutrons have no charge, which is why the electron clouds are invisible to them. However, if an atom has a net magnetic moment, the neutron's magnetic moment can directly couple to it. This is the second type of scattering mechanism which is important for neutrons: magnetic scattering. The result is that spin-up and spin-down electronic moments 'look' different to neutrons. Hence, when a sample becomes magnetic, new peaks can appear in the neutron diffraction pattern. An example of this is shown in Figures 5.18 and 5.19. which show the magnetic structure of the antiferromagnet MnO. Above $T_N = 116$ K there are neutron diffraction peaks resulting from the face–centred cubic

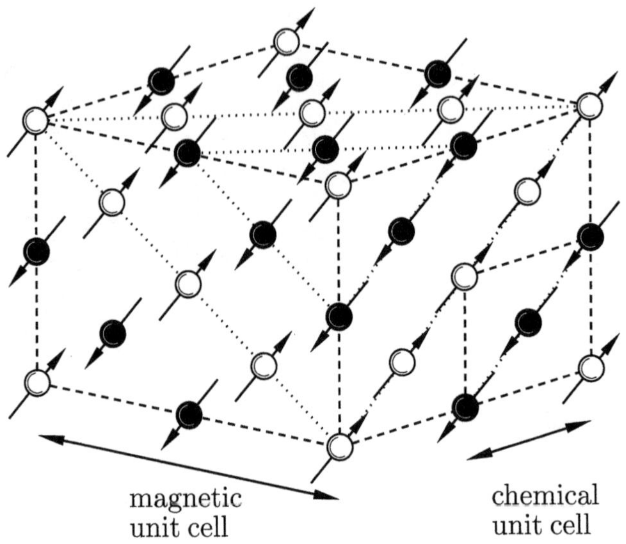

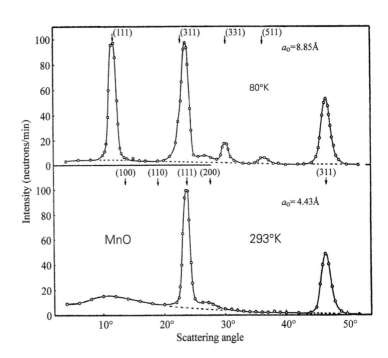

**Fig. 5.18** The magnetic structure of MnO. The Mn ions are on a face–centred cubic lattice. The $O^{2-}$ ions are not shown (see Fig. 4.2). The Mn ions are shaded black or white depending on their spin state.

magnetic unit cell

chemical unit cell

**Fig. 5.19** Neutron diffraction patterns for MnO below and above $T_N$. After C. G. Shull, W. A. Strauser and E. O. Wollan, *Phys. Rev.*, **83**, 333 (1951).

(fcc) lattice of Mn ions, all of which are equivalent (see Fig. 5.19). The fcc lattice gives rise to systematic absences in the diffraction pattern. Below $T_N$, the magnetic structure is as shown in Fig. 5.18 with spins all being parallel within each (111) plane, but the moments in any two adjacent (111) planes are antiparallel. The magnetic unit cell is thus double the size of the chemical unit cell and so a few peaks suddenly appear in the neutron diffraction pattern below

$T_N$. The amplitude of the magnetic Bragg peaks can be used as a measure of the strength of the magnetic order, and hence the magnetic order can be followed as a function of temperature.

The amplitude of the scattering from a magnetic moment depends on the direction of alignment of the moment. Hence neutron diffraction can be used to determine the arrangement of the atomic magnetic moments in a magnetically ordered crystal.

In a periodic system, sharp Bragg reflections occur when the scattering vector $\mathbf{Q}$ equals a reciprocal lattice vector $\mathbf{G}$. In a helimagnet, there is an additional periodicity in the system because of the helix which can be described by the wave vector of the helix $\mathbf{q}$. The helix hence has its axis along the vector $\mathbf{q}$ and the pitch of the helix is equal to $2\pi/|\mathbf{q}|$. Bragg reflections are then observed also at $\mathbf{G} \pm \mathbf{q}$. Since the helix pitch is larger than the lattice spacing, $|\mathbf{q}| < |\mathbf{G}|$ so that these Bragg peaks observed above the magnetic transition temperature are still observed but together with smaller satellite peaks on either side of each Bragg peak below the transition temperature.

Neutron scattering is the most direct method available for determining the details of the magnetic state of a sample. Neutrons are however expensive and because neutrons interact relatively weakly with matter, large sample sizes are needed (although this drawback is most serious for inelastic experiments, as described in Section 6.6.4).

### 5.7.3  Other techniques

The internal field can be measured by the techniques described in chapter 3. NMR, Mössbauer and $\mu$SR experiments can all measure the temperature dependences of the magnetization of ferromagnets. Because they are local probes, i.e. they do not measure an average magnetization but rather the magnetization at a particular crystallographic site, they can be used to measure the magnetization of individual sublattices in antiferromagnets or in ferrimagnets. They can also give information about the spontaneous magnetization from within a magnetic domain, even if the sample as a whole is unmagnetized.

A further technique to be considered is X-ray scattering. Magnetic X-ray scattering is typically a rather weak effect (in diffraction experiments the magnetic scattering is nominally five orders of magnitude smaller than the non-magnetic scattering) so X-rays were, until recently, mainly used for structural rather than magnetic investigations. This has changed because of the development of **X-ray resonant exchange scattering** (XRES) in which the magnetic scattering is enhanced as the energy of the X-rays is tuned through an absorption edge. Another major factor has been the development of high brilliance X-ray synchrotron sources and the provision of good beams of polarized photons. These factors counter the intrinsic weakness of the effect and have allowed X-rays to be used in the study of magnetism. X-rays have a number of advantages. They can be used to study small samples (because the diameter of the beam can be very small), they are element specific (because you tune the energy to the absorption edge of a particular atom in your sample and therefore measure a signal which is a signature of a particular element in the sample) and can give better resolution than neutrons. They can also be used to separately measure the spin and orbital contributions to the magnetization density.

Dichroism is the polarization dependence of the absorption of light (a Polaroid sheet which consists of oriented long-chain molecules is a well known example of linear dichroism at optical frequencies). Magnetic dichroism works in a similar way, and the effect is driven by asymmetries in the magnetization distribution around an atom caused by unpaired electrons. This is essentially an orbital phenomenon since a photon transfers helicity ±1 to an absorber, but the spin–orbit interaction can lead to a dependence of the dichroism on the spin angular momentum. In the technique of **magnetic X-ray circular dichroism** (MXCD for short) the difference in attenuation between right-handed and left-handed circularly polarized X-rays is measured and the results used to separately infer the orbital and spin magnetic moments.

# Further reading

- A review of experiments on nuclear magnetic ordering may be found in A. S. Oja and O. V. Lounsmaa, *Reviews of Modern Physics* **67**, 1 (1997).

- Neutron scattering is described in M. Dove, *Structure and dynamics*, OUP, forthcoming, G. L. Squires, *Introduction to the theory of thermal neutron scattering*, CUP 1978, and W. Marshall and S. W. Lovesey, *Theory of*

*thermal neutron scattering*, OUP 1971.

- A more advanced treatment may be found in *Theory of neutron scattering from condensed matter* by S. W. Lovesey, OUP 1984.

- *X-ray scattering and absorption by magnetic materials* by S. W. Lovesey and S. P. Collins, OUP 1996, provides useful information on magnetic X-ray scattering.

# Exercises

(5.1) Estimate the size of the molecular field, $B_{mf}$ in iron in units of Tesla. Compare this with $\mu_0 M$, the contribution to the $B$-field due to the magnetization. Hence explain why the concept of exchange is necessary to explain the ferromagnetism of iron. The density of Fe is 7873 kg m$^{-3}$, the relative atomic mass is 55.847, the Curie temperature $T_C$ is 1043 K and each atom carries a moment of approximately $2.2\mu_B$.

(5.2) Generalize the Weiss model for spins $S > \frac{1}{2}$. Show that the magnetization $M$ just below $T_C$ is given by

$$\frac{M}{M_s} = \left(\frac{10(S+1)^2(T_C - T)}{3[(S+1)^2 + S^2]T_C}\right)^{1/2} \quad (5.36)$$

and that there is a discontinuity in the heat capacity at $T_C$ equal to

$$\Delta C = \frac{5}{2}nk_B\frac{(2S+1)^2 - 1}{(2S+1)^2 + 1}. \quad (5.37)$$

You will need the relation

$$B_S(y) = \left[\frac{(2S+1)^2 - 1}{(2S)^2}\right]\frac{y}{3}$$

$$- \left[\frac{(2S+1)^4 - 1}{(2S)^4}\right]\frac{y^3}{45} + O(y^5) \ y \ll 1.$$
$$(5.38)$$

(5.3) The molecular field on each sublattice of an antiferromagnet is given by

$$B_+ = -\Gamma M_+ - |\lambda| M_-$$
$$B_- = -\Gamma M_- - |\lambda| M_+; \quad (5.39)$$

where $\Gamma$ is a constant which expresses the contribution to the molecular field from the same sublattice. Generalize the treatment given in Section 5.2 to show that on each sublattice the magnetization is

$$M_\pm = M_s B_J \left(-\frac{g_J\mu_B J(\Gamma M_\pm + |\lambda| M_\mp)}{k_B T}\right). \quad (5.40)$$

Hence show that the Néel temperature, the temperature at which the spontaneous magnetization on each sublattice disappears, is

$$T_N = \frac{n(|\lambda| - \Gamma)\mu_{eff}^2}{3k_B}. \tag{5.41}$$

Also show that the susceptibility is $\chi \propto (T-\theta)^{-1}$ where $\theta$ is given by

$$\theta = -\frac{n(|\lambda| + \Gamma)\mu_{eff}^2}{3k_B}, \tag{5.42}$$

so that $|\theta| = T_N$ only if $\Gamma = 0$.

(5.4) For the helimagnet described in Section 5.4, the energy is given by eqn 5.30. Show that the energies for ferromagnetism, antiferromagnetism and helimagnetism are given by $E_{FM}$, $E_{AFM}$ and $E_{HM}$ respectively where

$$E_{FM} = -2NS^2(J_1 + J_2) \tag{5.43}$$

$$E_{AFM} = -2NS^2(-J_1 + J_2) \tag{5.44}$$

$$E_{HM} = -2NS^2\left(\frac{-J_1^2}{8J_2} - J_2\right) \tag{5.45}$$

so that helimagnetism is favoured over either ferromagnetism or antiferromagnetism when $J_2 < 0$ and $|J_1| < 4|J_2|$.

(5.5) Show that for $T \ll T_C$ the Weiss model of a ferromagnet leads to a magnetization $M$ given by

$$M = M_s(1 - 2e^{-2T_C/T}). \tag{5.46}$$

Note that this is in disagreement with the experimentally determined power law behaviour for $T \ll T_C$ which follows

$$M = M_s(1 - AT^{3/2}) \tag{5.47}$$

where $A$ is a constant. This low temperature behaviour can be explained by the effect of spin waves (see Section 6.6).

(5.6) The susceptibility $\chi$ of a ferrimagnet can be treated using mean-field theory. Consider a ferrimagnet with two inequivalent sublattices such that the molecular fields $B_1$ and $B_2$ on sublattice 1 and 2 are given by

$$B_1 = \mu_0 H - \lambda M_1 \tag{5.48}$$

$$B_2 = \mu_0 H - \lambda M_2, \tag{5.49}$$

where $M_1$ and $M_2$ are the magnetization on each sublattice and $H$ is the applied field. However, the two sublattices have different Curie constants $C_1$ and $C_2$ (i.e. different values of $g_J J$). Show that the magnetic susceptibility is given by

$$\chi = \frac{1}{T^2 - \theta^2}\left[(C_1 + C_2)T - \frac{2\lambda C_1 C_2}{\mu_0}\right], \tag{5.50}$$

and find a value for the transition temperature $\theta$. Show that if $C_1 = C_2 = C$ the susceptibility reduces to that for an antiferromagnet with $\theta = \lambda C/\mu_0$.

(5.7) Show that the Hamiltonian for the Heisenberg model

$$\hat{\mathcal{H}} = -\sum_{<ij>} J\mathbf{S}_i \cdot \mathbf{S}_j, \tag{5.51}$$

can be rewritten as

$$\hat{\mathcal{H}} = -\sum_{<ij>} J\left[S_i^z S_j^z + \frac{1}{2}(S_i^+ S_j^- + S_i^- S_j^+)\right]. \tag{5.52}$$

Hence show that for a Heisenberg ferromagnet ($J > 0$) the state $|\Phi\rangle$, which consists of spins on every site pointing up (say), is an eigenstate of the Hamiltonian and has energy $E_0$ given by

$$E_0 = -NS^2 J. \tag{5.53}$$

Consider the Heisenberg antiferromagnet ($J < 0$) with the spins residing on two sublattices, each spin interacting only with those on the other sublattice. Show that the 'obvious' ground state, namely one with each sublattice ferromagnetically aligned but with oppositely directed sublattice magnetizations, is not an eigenstate of the Hamiltonian. This emphasizes that the Heisenberg antiferromagnet is a complex and difficult problem.

(5.8) $MnF_2$ has a tetragonal crystal structure in which the Mn ions are situated at the corners of the tetragonal unit cell $a = b = 0.5$ nm and $c = 0.3$ nm, and at the body–centred position in the unit cell. Below 70 K the spins of the Mn ions become antiferromagnetically aligned along the $c$ axis with the spins of an ion at the centre of the unit cell aligned opposite to those at the corners. The neutron scattering from a powdered sample of $MnF_2$ is measured using an incident neutron wavelength of 0.3 nm and an angle of scattering between $0°$ and $90°$. Sketch the results you would expect to observe at (a) 100 K and (b) 10 K. You can neglect the scattering from the F ions.

(5.9) It is possible to find exact solutions to the Weiss model. With no applied field the problem reduces to solving simultaneously the equations

$$\tilde{m} = \frac{M}{M_s} = B_J(y) \tag{5.54}$$

and

$$y = \frac{g_J \mu_B J\lambda M}{k_B T} \tag{5.55}$$

where $\tilde{m}$ is the reduced magnetization and $M_s = ng_J\mu_B J$. It is helpful to define $x = e^y$.

(a) For $J = \frac{1}{2}$, $g_J = 2$ and $B_J(y) = \tanh y$. Show that this implies that

$$\tilde{m} = \frac{x - \frac{1}{x}}{x + \frac{1}{x}} \tag{5.56}$$

and hence that

$$\log x = \frac{1}{2}[\log(1 + \tilde{m}) - \log(1 - \tilde{m})]. \qquad (5.57)$$

Writing $T_C = n\mu_B^2 \lambda / k_B$, show that

$$\frac{T}{T_C} = \frac{2\tilde{m}}{\log(1 + \tilde{m}) - \log(1 - \tilde{m})}. \qquad (5.58)$$

(b) For $J = 1$, show that

$$B_1(y) = \frac{2 \sinh y}{2 \coth y + 1} \qquad (5.59)$$

and hence

$$\tilde{m} = \frac{x - \frac{1}{x}}{x + 1 + \frac{1}{x}} \qquad (5.60)$$

with

$$x = \exp(2\lambda M \mu_B / k_B T). \qquad (5.61)$$

Show that this implies that

$$x = \frac{\tilde{m} \pm \sqrt{4 - 3\tilde{m}^2}}{2(1 - \tilde{m})}, \qquad (5.62)$$

and hence (think carefully about the $\pm$ sign) that this implies that

$$\frac{T}{T_C} = \frac{\frac{3}{2}\tilde{m}}{\log(\tilde{m} + \sqrt{4 - 3\tilde{m}^2}) - \log(2(1 - \tilde{m}))}, \qquad (5.63)$$

with $T_C = 8n\mu_B^2 \lambda / 3 k_B$. The approach given in this question is detailed in M.A.B. Whittaker, *American Journal of Physics* **57**, 45 (1989).

# Order and broken symmetry

**6**

The appearance of spontaneous order at low temperature is a fundamental phenomenon of condensed matter physics. Ferromagnets, antiferromagnets, liquid crystals and superconductors are all ordered phases, as is the solid state itself. All these phenomena share some fundamental properties and characteristics. For example, they are all characterized by a temperature dependence in which some relevant physical property shows a marked difference above and below a critical temperature $T_c$. For each phase one can define an order parameter which is zero for $T > T_c$ and non-zero for $T < T_c$. This quantity therefore acts as an indicator of whether or not the system is ordered. In the case of ferromagnetism, the order parameter is simply the magnetization. In this chapter we will show that each type of ordered phase is associated with a broken symmetry, a concept which is explained in the following section.

## 6.1 Broken symmetry

In ferromagnets a single unique direction has been chosen along which all the atomic magnetic moments have lined up. They have all chosen 'up' rather than 'down'. This is actually a rather surprising effect since the underlying physical equations do not distinguish between 'up' and 'down'. Thus the microscopic physics has a symmetry not possessed by the experimentally observed ground state. To delve further into this mystery, we must consider symmetry in greater detail.

A useful example to consider is the Euler strut, shown in Fig. 6.1. This consists of a vertical rod which is clamped into the ground at the bottom and uniformly loaded from the top. As the weight of the load increases, the rod compresses, and above a critical weight it buckles. Whether it buckles to the left or the right depends on the precise details of exactly how symmetrically one places the load on the top. But even in the idealization that it is placed perfectly centrally on the top, the rod still has to choose which way to buckle. Balanced precariously, as if on a knife-edge, a random thermal fluctuation might be enough to send it buckling one way or the other. The left–right symmetry, inherent in the lightly loaded rod, is said to be broken by the buckling.

Many similar features are observed in condensed matter systems. The parameter which drives the symmetry-breaking transition can be a force, such as an applied pressure, but it is very often temperature. Figure 6.2 shows atoms in a liquid and in a solid. As a liquid cools there is a very slight contraction of the system but it retains a very high degree of symmetry. However, below a critical temperature, the melting temperature, the liquid becomes a solid and that symmetry is broken. This may at first sight seem surprising because the picture

Terminology concerning symmetry

A system possesses a particular symmetry if the Hamiltonian describing it is invariant with respect to the transformations associated with elements of a symmetry group. A group is a collection of elements and an operation which combines them, that follow a particular set of rules: the group is 'closed', so that combinations of elements are also members of the group; the group combination is associative; an inverse and identity are well defined.

A discrete symmetry refers to a symmetry group with countable elements, such as the rotational symmetry group of a cube. A continuous symmetry has an uncountable continuum of elements, such as the rotational symmetry group of a sphere. A system possesses a global symmetry if it is invariant under the symmetry elements of the group being applied globally to the entire system. A local symmetry applies to the Hamiltonian if it is unchanged after the symmetry operations are applied differently to different points in space. The gauge symmetry of superconductivity is such a local symmetry.

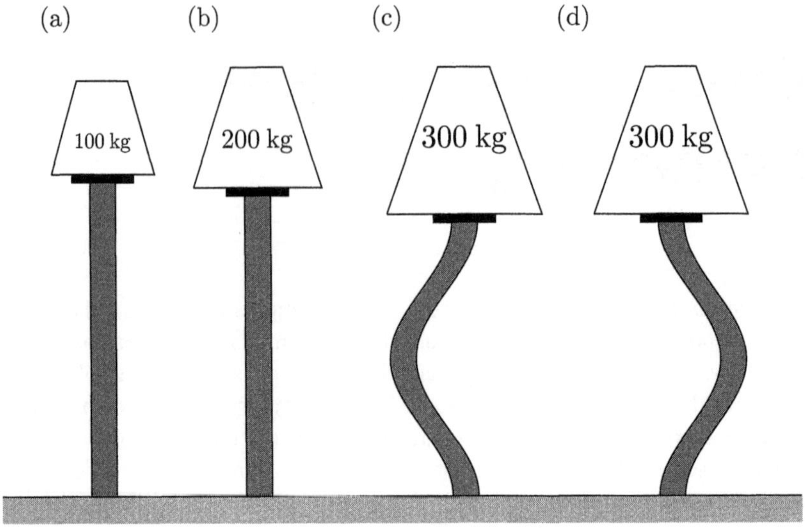

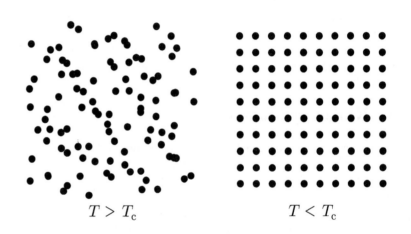

**Fig. 6.1** (a) The Euler strut. (b) As an increasing vertical force is applied to the strut, the strut is compressed but retains left–right symmetry. (c) Once the force exceeds a critical value, the strut buckles, breaking the left–right symmetry. (d) The strut could also break the symmetry in the opposite sense.

**Fig. 6.2** Liquid–solid phase transition. Left: The high temperature state (statistically averaged) has complete translational and rotational symmetry. Right: These symmetries are broken as the system becomes a solid below the critical temperature $T_c$.

of the solid 'looks' more symmetrical than that of the liquid. The atoms in the solid are all symmetrically lined up while in the liquid they are all over the place. The crucial observation is that any point in a liquid is exactly the same as any other. If you average the system over time, each position is visited by atoms as often as any other. There are no unique directions or axes along which atoms line up. In short, the system possesses complete translational and rotational symmetry. In the solid however this high degree of symmetry is nearly all lost. The solid drawn in Fig. 6.2 still possesses some residual symmetry: rather than being invariant under arbitrary rotations, it is invariant under four-fold rotations ($\pi/2$, $\pi$, $3\pi/2$, $2\pi$); rather than being invariant under arbitrary translations, it is now invariant under a translation of an integer combination of lattice basis vectors. Therefore not all symmetry has been lost but the high symmetry of the liquid state has been broken.

The situation is similar for a ferromagnet (Fig. 6.3). A ferromagnet above the Curie temperature $T_C$ possesses complete rotational symmetry. All directions are equivalent and each magnetic moment can point in any direction. Below $T_C$ the system 'chooses' a unique direction for all the spins to point. The

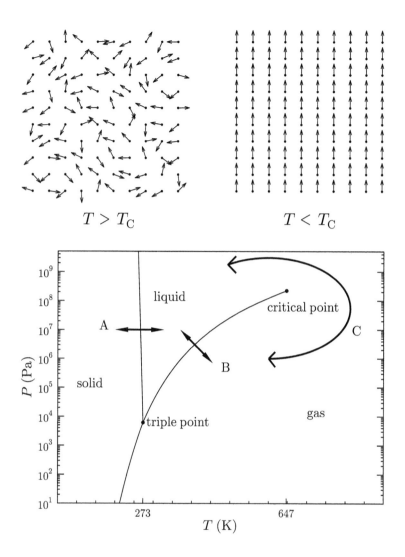

**Fig. 6.3** Paramagnet–ferromagnet phase transition. Left: The high temperature state (statistically averaged) has rotational symmetry. Right: This symmetry is broken as the system becomes a ferromagnet below the Curie temperature $T_C$.

**Fig. 6.4** The phase diagram of water. Paths A and B in the phase diagram cross a phase boundary, while C does not. Path A involves a change of symmetry, while B and C do not.

higher rotational symmetry of the high temperature state is broken. The lower temperature state has a reduced rotational symmetry (only rotations about the magnetization axis, the up-direction in Fig. 6.3, are allowed).

An important point to notice is that it is impossible to change symmetry *gradually*. Either a particular symmetry is present or it is not. Hence, phase transitions are sharp and there is a clear delineation between the ordered and disordered states. The appearance of order at low temperatures can then be understood from quite general thermodynamic considerations. The free energy $F$ is related to the energy $E$ and the entropy $S$ by $F = E - TS$. In order to minimize the free energy, at low temperature a system will choose its lowest energy ground state which is usually ordered. Thus it will minimize $E$. As the temperature increases, it becomes more important in minimizing $F$ to find a state which maximizes $S$ and so a disordered state is favoured.

Not all phase transitions involve a change of symmetry. Figure 6.4 shows the phase diagram of water. The boundary line between the liquid and gas regions is terminated by a **critical point**. Hence it is possible to 'cheat' the

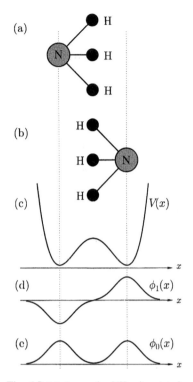

**Fig. 6.5** (a) Ammonia, $NH_3$, is a tetrahedral molecule which can exist in one of two forms. The other form is shown in (b). There are two stable positions of the N atom, and we can thus define a potential energy $V(x)$ associated with this system. (d) The first excited state wave function $\phi_1(x)$. (e) The ground state wave function $\phi_0(x)$.

*To preserve the analogy with the ammonia molecule, we here consider a uniaxial ferromagnet (see Section 6.7.2).

sharp phase transition by taking a path through the phase diagram which avoids a discontinuous change (path C in Fig. 6.4). For temperatures above the critical temperature (647 K) a gas and a liquid are distinguished only by their density. The transition between a gas and a liquid involves no change of symmetry. In contrast, the solid–liquid transition involves a change of symmetry and consequently there is no critical point for the melting curve.

A puzzle about the nature of symmetry-breaking transitions is that the broken-symmetry ground state does not possess the symmetry of the Hamiltonian. On cooling a ferromagnet through the transition temperature, the system has to choose a particular direction along which all the spins will point, even though no particular direction is singled out in the underlying physics. How can this come about?

To understand this, it is useful to consider a non-magnetic example, that of the molecule ammonia ($NH_3$) shown in Fig. 6.5. The three hydrogen atoms form three corners of an equilateral triangle (viewed edge-on in Figures 6.5(a) and 6.5(b)) and the nitrogen atom can be just above the plane of the triangle or just below it. This pyramidal shape is due to the lone pair of electrons on the nitrogen which jockeys for space with the three nitrogen-hydrogen bonds. This gives the molecule a dipole moment, but also defines a particular direction in space and breaks the symmetry of the underlying Hamiltonian. Therefore this pyramid cannot be a stable state of the system.

In fact we can imagine transforming between two pyramidal states shown in Figures 6.5(a) and (b) by moving the nitrogen through the middle of the plane, thus inverting the tetrahedron much in the same way as an umbrella is inverted when it is blown out by a strong wind. If we denote the position of the nitrogen atom by $x$, the distance from the plane, we could imagine a potential energy curve $V(x)$ which has two minima associated with the two stable configurations. This is shown in Fig. 6.5(c). The two lowest energy eigenstates of this potential are shown in Figures 6.5(d) and (e) for the first excited state $\phi_1(x)$ and ground state $\phi_0(x)$ respectively.

The configurations represented in Fig. 6.5(a) and (b) do not therefore represent stable configurations of the system, but are superpositions of energy eigenstates. They therefore represent metastable states. Thus the state in Fig. 6.5(a) is represented by $(\phi_0(x) - \phi_1(x))/\sqrt{2}$ and the state in Fig. 6.5(b) is represented by $(\phi_0(x) + \phi_1(x))/\sqrt{2}$. If you prepared the system in the metastable state in Fig. 6.5(a), you would find that it would oscillate between that and the metastable state in Fig. 6.5(b) at a frequency given by the difference between the energy of $\phi_0(x)$ and $\phi_1(x)$. This frequency is 24 GHz in ammonia and this transition is the one used in the ammonia maser. If you replace the nitrogen for the more massive phosphorous to make $PH_3$, the frequency goes down by about a factor of ten. Thus as we make the system heavier we *increase* the time constant of the metastable state in Fig. 6.5(a).

In a ferromagnet* we have a more massive state yet, involving not just heavier atoms, but more atoms, maybe $10^{23}$ of them. The ground state of the system with all spins pointing in a single direction is actually not a true stationary state of the system. However it is a metastable state but with a stability time which is greater than the age of the universe. To access the other state, the one with all spins reversed, one would have to flip every spin in the system simultaneously, which is extremely unlikely.

## 6.2  Models

In order to discuss some of the consequences of symmetry breaking it is convenient to think about some simple *models* of magnetism and try to solve them.

### 6.2.1  Landau theory of ferromagnetism

A convenient model which simply produces a phase transition was provided by
the Russian physicist Lev Landau  and arises from some very general considerations. We write down the free energy for a ferromagnet with magnetization $M$
as a power series in $M$. Because there is no energetic difference between 'up'
or 'down', this power series cannot contain any odd power of $M$. Therefore we
can write for the free energy $F(M)$ the expression

Lev D. Landau (1908–1968)

$$F(M) = F_0 + a(T)M^2 + bM^4 \qquad (6.1)$$

where $F_0$ and $b$ are constants (we assume $b > 0$) and $a(T)$ is temperature
dependent. We can show that this system yields an appropriate phase transition
if we allow $a(T)$ to change sign at the transition temperature $T_C$. Thus in the
region of interest, near the transition, we write $a(T) = a_0(T - T_C)$ where $a_0$
is a positive constant. To find the ground state of the system, it is necessary
to minimize the free energy so we look for solutions of $\partial F/\partial M = 0$. This
condition implies

$$2M[a_0(T - T_C) + 2bM^2] = 0. \qquad (6.2)$$

The left-hand side of this equation is a product of two terms, so either of them
could be zero. This means

$$M = 0 \ \text{ or } \ M = \pm \left[ \frac{a_0(T_C - T)}{2b} \right]^{1/2}. \qquad (6.3)$$

The second condition is only valid when $T < T_C$, otherwise one is trying to
take the square root of a negative number. The first condition applies above or
below $T_C$ but below $T_C$ it only produces a position of unstable equilibrium
(which can be deduced by evaluating $\partial^2 F/\partial M^2$). Thus the magnetization
follows the curve shown in Fig. 6.6(b); it is zero for temperatures $T \geq T_C$
and is non-zero and proportional to $(T_C - T)^{1/2}$ for $T < T_C$.

Landau's approach to studying phase transitions is called a **mean-field
theory** which means that it assumes that all spins 'feel' an identical average
exchange field produced by all their neighbours. This field is proportional
to the magnetization. This approach is identical to the Weiss model outlined
in Section 5.1.1. Mean–field theories are the simplest type of theory that
can be constructed to describe many different types of phase transition and
give similar results in each case. They go under different names in different
cases: e.g. Bragg–Williams theory for order–disorder transitions in alloys.
Mean-field theories fail to explain the critical region accurately because
the assumption that all regions of the sample are the same then becomes
particularly misplaced.

Mean-field theories ignore **correlations** and **fluctuations** which become very
important near $T_C$. Very near to the critical temperature, large fluctuations are
seen in the order parameter. The critical region is characterized by fluctuations,

(a)

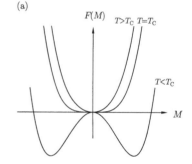

(b)

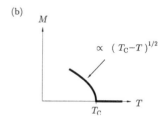

**Fig. 6.6** (a) The free energy $F(M)$ of a
ferromagnet. (b) The corresponding magnetization as a function of temperature.

at all lengthscales. When a saucepan of water is heated, the water warms quite quietly and unobtrusively until near the boiling point when it makes a great deal of noise and it bubbles and foams.[1] The same type of effect occurs near $T_C$ in a ferromagnet. The dominant length scale which characterizes the fluctuations is the correlation length $\xi$. As $T \to T_C$, $\xi \to \infty$ i.e., the correlation length becomes infinite at the critical point. The importance of fluctuations is notable also from the form of $F(M)$. The curvature at $M = 0$, $(\partial^2 F / \partial M^2)_{M=0}$, tends to zero at $T = T_C$ so that the minimum at $M = 0$ becomes broad and flat-bottomed (see Fig. 6.6). Hence predictions about the critical region obtained from mean-field theories, though the theories are straightforward to solve, must be regarded with some caution.

### 6.2.2   Heisenberg and Ising models

An alternative approach to understanding the magnetic behaviour of solids is to consider particular microscopic models of the magnetic interaction. A commonly studied model is the **nearest neighbour Heisenberg model** which has a Hamiltonian

$$\hat{\mathcal{H}} = -\sum_{\langle ij \rangle} \mathsf{J} \mathbf{S}_i \cdot \mathbf{S}_j, \qquad (6.4)$$

where the constant $\mathsf{J}$ is the exchange integral and the symbol $\langle ij \rangle$ below the $\sum$ denotes a sum over nearest neighbours. The spins $\mathbf{S}_i$ are treated as three-dimensional vectors because we allow them to point in any direction in three-dimensional space. However the sum can be taken over a lattice of 1, 2 or 3 dimensions. Here it is important to distinguish between the dimensionality $d$ of the lattice on which the spins sit and the dimensionality $D$ of the spins themselves (in general $D$ is known as the dimensionality of the *order parameter*). For the Heisenberg model $D = 3$ (because the spins are three-dimensional vectors). However we could be considering a lattice of these spins in 1, 2 or 3 dimensions (or 4 dimensions if we wanted to!) so $d$ can be 1, 2, 3, . . .

A related model is the **Ising model** in which the spins are only allowed to point up or down, i.e. we only consider the $z$ component of the spin. The Hamiltonian of this model is

$$\hat{\mathcal{H}} = -\sum_{\langle ij \rangle} \mathsf{J} S_i^z S_j^z. \qquad (6.5)$$

Here the dimensionality of the order parameter $D$ is equal to 1 (the spins are only allowed to point along $\pm\mathbf{z}$). Nevertheless we could arrange these one-dimensional spins on a lattice with $d = 1, 2, \ldots$

### 6.2.3   The one-dimensional Ising model ($D = 1, d = 1$)

If the Ising spins are placed on a one-dimensional lattice, we will show that there is no phase transition. First, consider a chain with $N + 1$ spins (and hence we need to consider $N$ 'bonds' between each neighbour). The Hamiltonian is

$$\hat{\mathcal{H}} = -2\mathsf{J} \sum_{i=1}^{N} S_i^z S_{i+1}^z \qquad (6.6)$$

Ernst Ising (1900–1998)

In Chapter 1, the operator for the $z$ component of the spin was written as $\hat{S}_z$. Because we now also need to label the site of the spin, we will write the $z$ as a superscript and put the site label as a subscript. Hence $\hat{S}_i^z$ is the operator for the $z$ component of the spin at site $i$.

**Fig. 6.7** The one-dimensional Ising model. (a) The ground state contains all $N + 1$ spins aligned ferromagnetically. (b) A single defect is added.

and assume $J > 0$ so that the ground state is obtained by having all adjacent spins lined up ferromagnetically (this is illustrated in Fig. 6.7(a)). The ground state thus has energy $-NJ/2$ because $S_i^z = +\frac{1}{2}$ for all values of $z$. Now consider adding one 'mistake', a single defect (see Fig. 6.7(b)). This costs an extra energy $E = J$ because we have to turn one favourable interaction (energy saving $J/2$) into an unfavourable one (energy cost $J/2$, so the change in energy is $J$). However, there is an entropy gain equal to $S = k_B \ln N$ because we can put the defect in any one of $N$ places. As we let the chain get very large ($N \to \infty$) the energy cost of a defect remains the same ($J$) but the entropy gain becomes infinite. The properties are determined by the free energy $F = E - TS$ so that as long as the temperature is not zero, the entropy consideration means that the presence of the defect causes the free energy to plummet to $-\infty$. This means that defects can spontaneously form and in fact no long range order occurs for $T > 0$. Another way of saying this is that the critical temperature is zero. This consideration is valid for most models on one-dimensional lattices (because entropy usually wins in one dimension) and we conclude that long range order is not usually possible in one-dimension.

### 6.2.4 The two-dimensional Ising model ($D = 1, d = 2$)

If the Ising spins are placed on a two-dimensional lattice, a phase transition to a magnetically ordered state below a non-zero critical temperature will result. This is because the energy cost and the entropy gain of making a defect both scale with the perimeter size of the defect (see Fig. 6.8). Energy and entropy therefore can have a fair fight with neither having an overwhelming advantage. The detailed solution of this model was one of the outstanding achievements of twentieth century statistical physics (Lars Onsager solved it in 1944) and his solution is beyond the scope of this book. This illustrates that even problems which are simple to state are by no means easy to solve.

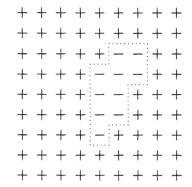

**Fig. 6.8** The two-dimensional Ising model. Up spins are labelled $+$ and down spins are labelled $-$. The energy and entropy of the down spin defect scale with the perimeter of the defect boundary.

Lars Onsager (1903–1976)

## 6.3 Consequences of broken symmetry

When you break symmetry there are various consequences:

- **Phase transitions**: The system will have a sharp change of behaviour at a temperature $T_c$. We describe this by saying the system has changed *phase* (e.g. liquid $\to$ solid, paramagnet $\to$ ferromagnet, etc). The region near the phase transition is called the **critical region** (see Section 6.4).
- **Rigidity**: Having broken the symmetry, the system will have a strong energetic preference for staying in that broken-symmetry state, and attempts by us to change the way the system has broken the symmetry meet with resistance. Thus crystals don't bend easily and ferromagnets show permanent magnetism (see Section 6.5).
- **Excitations**: At $T = 0$ the system is perfectly ordered apart from quantum fluctuations. At finite temperature this order is weakened by excitations in the order parameter. In crystals these excitations are called lattice waves, quantized into phonons. In ferromagnets the analogous modes are called spin waves, quantized into magnons (see Section 6.6).

• **Defects**: If symmetry is broken differently in two adjacent parts of a macroscopic sample, the boundary will conain a defect: e.g. a grain boundary in a crystal or a domain wall in a ferromagnet (see Section 6.7).

A summary of the properties of a number of different broken symmetry phases is contained in Table 6.1. Each is characterized by a high temperature disordered state and a low temperature ordered state. Each has an order parameter, a rigidity phenomenon, a set of excitations representing wave-like departures from the ordered state, and defects associated with breaking the symmetry differently in different spatial regions.

A crystalline solid gives rise to Bragg peaks in a diffraction experiment. Thus we can use as an order parameter $\rho_G$, the Fourier component of the charge density corresponding to a spatial frequency equal to a reciprocal lattice vector $\mathbf{G}$. The case of an antiferromagnet is similar to that of a ferromagnet, except that the order parameter is the magnetization on one sublattice of the antiferromagnet. A superconductor has a complex order parameter given by $\psi = |\psi| e^{i\phi}$. Here $|\psi|^2$ represents the density of superconducting electrons (for a superfluid or Bose condensate, it represents the condensate fraction). The phase $\phi$ of the wave function is the broken symmetry; it is free to take any value at any point in the normal state, but in the superconducting state it becomes uniform over the whole sample. For a nematic liquid crystal (see Fig. 6.9), the order parameter is given by $S = \langle \frac{1}{2}(3\cos^2\theta - 1) \rangle$, which is the average value of the function $\frac{1}{2}(3\cos^2\theta - 1)$ where $\theta$ is the angle between the long axis of the molecule and the **director**, $\hat{\mathbf{n}}$, a unit vector which points along the mean orientation of the axis of the liquid crystal molecule. An isotropic (liquid) state corresponds to $S = 0$. Full alignment corresponds to $S = 1$.

**Table 6.1** The properties of broken symmetry phases. Here $\rho_G$ is the Fourier component of the charge density corresponding to a spatial frequency equal to a reciprocal lattice vector $\mathbf{G}$. The complex wave function in a superconductor is $\psi = |\psi| e^{i\phi}$. The electric polarization $\mathbf{P}$ is the electric dipole moment per unit volume.

| Phenomenon | High $T$ Phase | Low $T$ Phase | Order parameter | Excitations | Rigidity phenomenon | Defects |
|---|---|---|---|---|---|---|
| crystal | liquid | solid | $\rho_G$ | phonons | rigidity | dislocations, grain boundaries |
| ferromagnet | paramagnet | ferromagnet | $\mathbf{M}$ | magnons | permanent magnetism | domain walls |
| antiferromagnet | paramagnet | antiferromagnet | $\mathbf{M}$ (on sublattice) | magnons | (rather subtle) | domain walls |
| nematic (liquid crystal) | liquid | oriented liquid | $S = \langle \frac{1}{2}(3\cos^2\theta - 1) \rangle$ | director fluctuations | various | disclinations, point defects |
| ferroelectric | non-polar crystal | polar crystal | $\mathbf{P}$ | soft modes | ferroelectric hysteresis | domain walls |
| superconductor | normal metal | superconductor | $|\psi| e^{i\phi}$ | – | superconductivity | flux lines |

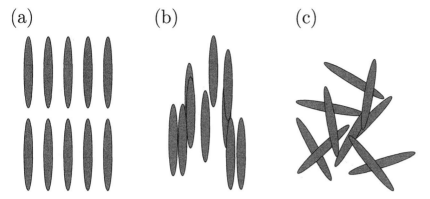

**Fig. 6.9** A nematic liquid crystal contains rod-shaped molecules. (a) At low temperatures these order as a solid. (c) At high temperatures they are in a liquid state. (b) There is an intermediate temperature region where they are in the nematic state, in which the long-axis of the molecules become aligned, even though the positions of the centre of mass of the molecules are not ordered.

**Table 6.2** Critical exponents for various models.

| Model | Mean-field | Ising | Ising | Heisenberg |
|-------|-----------|-------|-------|------------|
| $D$ | any | 1 | 1 | 3 |
| $d$ | any | 2 | 3 | 3 |
| $\beta$ | $\frac{1}{2}$ | $\frac{1}{8}$ | 0.326 | 0.367 |
| $\gamma$ | 1 | $\frac{7}{4}$ | 1.2378(6) | 1.388(3) |
| $\delta$ | 3 | 15 | 4.78 | 4.78 |

## 6.4 Phase transitions

In Section 6.2 the Landau theory of ferromagnetism was presented. This is a mean-field theory in which an identical exchange field is felt by all spins. This leads to the magnetization behaving as $(T_C - T)^{1/2}$ below the transition. In real systems it is found that the magnetization does behave as $(T_C - T)^\beta$ close to the transition, but the exponent $\beta$ is not necessarily equal to $\frac{1}{2}$. The exponent therefore gives important information about the nature of the phase transition. A number of other similar exponents, known as **critical exponents**, can be defined. Thus near the phase transition temperature $T_C$ it is found experimentally that

$$
\begin{aligned}
\chi &\propto (T - T_C)^{-\gamma} & T > T_C \\
M &\propto (T_C - T)^\beta & T < T_C \\
M &\propto H^{1/\delta} & T = T_C
\end{aligned}
\qquad (6.7)
$$

where $\beta$, $\gamma$, and $\delta$ are the critical exponents. The values taken by these critical exponents for various models are shown in Table 6.2.

As stated earlier, correlations and fluctuations are ignored in mean-field theories. This means that they cannot hope to give a correct description close to $T_C$, precisely where one might wish to calculate critical exponents. It turns out however that mean-field theory *is* a correct description of systems with four or more dimensions (i.e. $d \geq 4$) and predicts the correct critical exponents.

Despite the failure of mean-field theory to account successfully for critical behaviour in systems with dimensionality below four, it is only necessary to consider a small representative set of ideal statistical models to calculate critical exponents of any physical system if the hypothesis of universality is accepted. This hypothesis is based on the observation that critical exponents do seem to be surprisingly independent of the type of phase transition, whether liquid–gas, ferromagnetic–paramagnetic, superconducting–non-superconducting, or any other. It is supposed that for a continuous phase transition, the critical exponents depend only on

(1) The dimensionality of the *system*, $d$.
(2) The dimensionality of the *order parameter*, $D$. (Actually the symmetry of the order parameter.)
(3) Whether the forces are *short* or *long* range.

Hence, it is only important to look at particular universality classes, i.e. particular values of $D$ and $d$ for both short and long range forces, and pick the simplest models in each class and calculate the exponents for those. (Actually the assumptions stated above apply to static critical exponents, rather than the dynamic critical exponents which characterize time-dependent properties.)

We now need to list those models for which solutions are known. There are a number of exactly solved models which include the following:

(1) Most cases for $d = 1$. As we have seen, such systems do not exhibit continuous phase transitions.
(2) All cases for $d \geq 4$, which give mean-field solutions.
(3) Many cases for long range interactions, which give mean-field solutions.
(4) The case $d = 2$, $D = 1$. This is the 2D Ising model (see Table 6.2).
(5) The case $D = \infty$ for any $d$. This is known as the spherical model.

Unfortunately, most real situations correspond to $d = 3$ and short-range interactions, which have not been solved exactly.

A method of calculating critical behaviour, even in models which cannot be exactly solved, was discovered in the early 1970s, in which a block of spins with volume $L^d$ is considered, where $L$ is much less than the correlation length $\xi$. If we increase (scale) the linear dimensions of this block by a factor $n$, the Hamiltonian will be suitably scaled, and it is possible to examine how the order parameter is thus changed (renormalized) by this transformation. If such a transformation is denoted by $\tau$, then the Hamiltonian $\mathcal{H}$ transforms as $\mathcal{H}' = \tau(\mathcal{H})$. At the critical point, the correlation length $\xi \to \infty$, so that there must exist a limiting function $\mathcal{H}^*$ such that $\mathcal{H}^*$ is a fixed point of $\tau$, i.e. $\tau(\mathcal{H}^*) = \mathcal{H}^*$. This approach, namely that of looking for fixed points of scaling transformations, is known as the renormalization group method, since the set of scaling transformations $\tau$ form a group. It has proved to be particularly successful in condensed matter physics, and has deep connections with renormalization in quantum field theory. A more detailed treatment of the renormalization group is however beyond the scope of this book.

Cluster models can be useful in describing the paramagnetic regime of ferromagnets, i.e. the region $T > T_C$. They retain some of the simplicity (and hence predictive power) of mean-field approaches but make some attempt to model the fluctuations. The idea is to evaluate the configuration of a single finite-size cluster of spins exactly, and then to couple that to the mean field from the rest of the sample. Of course the correlation length diverges at $T_C$ and so the size of the cluster needs to increase as you get closer to the transition. Very good agreement with data can be obtained if a cluster model is chosen in which the size of the clusters is unrestricted, and is allowed to vary.[2]

[2] See R. V. Chamberlin, *Nature*, **408**, 337 (2000).

## 6.5 Rigidity

Breaking symmetry involves a choice of ground state: should the spins all point up, or down, or left or right? The energy of a macroscopic sample is minimized when the symmetry is broken the same way throughout its volume. If you try to make different parts of a macroscopic sample break symmetry differently, forces will appear reflecting an additional energy cost. This gives rise to a generalized rigidity or stiffness. This rigidity is intimately connected to the order, as can be deduced from the following argument for a crystal.

Consider a cubic crystal with lattice parameter $a$ and imagine stretching one plane of bonds by increasing $a$ to $a + u$ inducing a strain $\epsilon = u/a$ (see Fig. 6.10). This produces a stress equal to $G\epsilon$ (where $G$ is an elastic modulus) so that the force on an individual bond is equal to $G\epsilon a^2 = Gau$. The energy stored in this bond is then $\frac{1}{2}Gau^2$. At high temperatures (large compared with the Debye temperature[3]) the mean stored energy in each bond is equal to $\frac{1}{2}k_B T$ (by the equipartition theorem, see Appendix E) so that

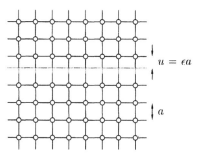

**Fig. 6.10** A cubic crystal of lattice parameter $a$ is stretched producing a strain $\epsilon = u/a$.

$$\langle u^2 \rangle = \frac{k_B T}{Ga} \qquad (6.8)$$

and hence $\langle u^2 \rangle \propto 1/G$. Thus if $G$ vanishes, the fluctuations diverge and *vice versa*. Hence the existence of a non-zero elastic modulus (i.e. a rigidity) is connected with the finiteness of the fluctuations (i.e. the stability of the crystal).

This means that because we have an ordered crystal, we must have a rigid crystal. Crystals do not deform easily (in contrast to liquids) because there is an elastic energy proportional to the elastic modulus and to $(\nabla u)^2$ where $u$ is the lattice displacement, and the lattice transmits this force to the other end of a macroscopic sample. Solids are therefore rigid.

These ideas are also applicable to isotropic ferromagnets. The spins are aligned in a ferromagnet and it costs energy to turn them with respect to each other. Hence we have the phenomenon of permanent magnetism. If the magnetization is non-uniform there is an exchange cost proportional to $(\nabla \mathbf{M})^2$ (see Section 4.2.7, and eqn 4.18 in particular).

As a final example, consider a superconductor in which the phase of the wave function, $\phi$, is the order parameter. In a superconductor, $\phi$ is uniform across a sample and there is a 'phase rigidity' energy proportional to $(\nabla \phi)^2$. (In fact twisting this phase across a sample produces a supercurrent $\propto \nabla \phi$.)

[3] The Debye temperature $\Theta_D$ gives the energy of the highest occupied phonon mode, divided by $k_B$, under the assumption that the phonons are non-dispersive. Of course there is always some dispersion, but $\Theta_D$ nevertheless characterizes a typical phonon energy of the system. Most Debye temperatures are in the range 100–1000 K. When $T \gg \Theta_D$, the phonon modes can be treated classically.

## 6.6 Excitations

A solid is ordered at $T = 0$, although zero-point fluctuations mean that, even then, atoms are not purely static. At non-zero temperature, the order is disrupted by thermally excited lattice vibrations, which are quantized as phonons. The behaviour of the phonons is characterized by a dispersion relation, i.e.

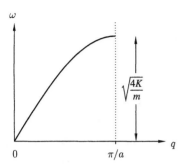

**Fig. 6.11** The phonon dispersion for a one-dimensional monatomic chain of atoms of mass $m$ connected by bonds with spring constant $K$. These phonons are acoustic. There is no gap ($\omega = 0$ at $q = 0$) and at small $q$, $\omega \approx v_s q$ where $v_s$ is the speed of sound. The dispersion is massless because $\omega = 0$ at $q = 0$ (i.e. energy $E = h\omega = 0$ at momentum $p = hq = 0$). A relativistic particle has energy $E^2 = p^2 c^2 + m^2 c^4$, so that the energy at zero momentum is equal to $mc^2$, i.e. proportional to the mass.

a relationship between angular frequency $\omega$ and wave vector $q$ (equivalently between energy $\hbar\omega$ and momentum $\hbar q$). An example for a one-dimensional monatomic chain is shown in Fig. 6.11. The crucial feature of this is that $\omega = 0$ at $q = 0$ for acoustic phonons, so that it costs a vanishingly small energy to produce a phonon of wave vector $q$, provided its wavelength $\lambda = 2\pi/q$ is long enough. The fact that an acoustic phonon can be thermally generated, as long as the temperature is non-zero, is because there is no **energy gap** to leap across from the ground state of the system ($q = 0$, $\omega = 0$) to the lowest acoustic phonon level. There is no energy gap in the phonon dispersion relation at $q = 0$ (in contrast to the case for optic phonons). Acoustic phonons give rise to a $T^3$ heat capacity at low temperature in a three-dimensional crystal.

Whenever you have broken a continuous global symmetry (as you do when you make a solid from a liquid or a ferromagnet from a paramagnet) it is possible to produce long-wavelength excitations in the order parameter for vanishingly small energy cost. Such excitations are called **Goldstone modes** (or sometimes Goldstone bosons). Because they cost no energy they are 'massless'. In particle physics an example of this is the photon which is a Goldstone boson. For a superconductor, the situation is rather different because it turns out that you are breaking a continuous *local* symmetry and you don't get Goldstone modes. The reason for this is rather subtle and is connected with the Higgs mechanism.

A ferromagnet is perfectly ordered at $T = 0$ but at non-zero temperature the order is disrupted by **spin waves**, quantized as **magnons**. The crucial feature of these is that it costs a vanishingly small energy to produce a magnon, provided its wavelength is long enough. Thus the magnons play the same rôle in ferromagnets as phonons do in solids and are the Goldstone modes of the system. We will show below that, for an isotropic ferromagnet, there is no energy gap in the magnon dispersion relation at $q = 0$.

### 6.6.1 Magnons

In this section we will derive the magnon dispersion relation for an isotropic ferromagnet. This is an important problem so two alternative derivations will be presented, the first using a semiclassical approach and the second using a quantum mechanical approach.

(1) We begin with a semiclassical derivation of the spin wave dispersion. First, recall the Hamiltonian for the Heisenberg model,

$$\hat{\mathcal{H}} = -\sum_{\langle ij \rangle} J\hat{\mathbf{S}}_i \cdot \hat{\mathbf{S}}_j \qquad (6.9)$$

(which is eqn 6.4). In a one-dimensional chain each spin has two neighbours, so the Hamiltonian reduces to

$$\hat{\mathcal{H}} = -2J\sum_i \hat{\mathbf{S}}_i \cdot \hat{\mathbf{S}}_{i+1}. \qquad (6.10)$$

We can calculate the time dependence of $\langle \hat{\mathbf{S}}_j \rangle$ using eqn C.7, with the

result that

$$\frac{d\langle\hat{\mathbf{S}}_j\rangle}{dt} = \frac{1}{i\hbar}\langle[\hat{\mathbf{S}}_j, \hat{\mathcal{H}}]\rangle \tag{6.11}$$

$$= -\frac{2J}{i\hbar}\langle[\hat{\mathbf{S}}_j, \cdots + \hat{\mathbf{S}}_{j-1}\cdot\hat{\mathbf{S}}_j + \hat{\mathbf{S}}_j\cdot\hat{\mathbf{S}}_{j+1} + \cdots]\rangle \tag{6.12}$$

$$= -\frac{2J}{i\hbar}\langle[\hat{\mathbf{S}}_j, \hat{\mathbf{S}}_{j-1}\cdot\hat{\mathbf{S}}_j] + [\hat{\mathbf{S}}_j, \hat{\mathbf{S}}_j\cdot\hat{\mathbf{S}}_{j+1}]\rangle \tag{6.13}$$

$$= \frac{2J}{\hbar}\langle\hat{\mathbf{S}}_j \times (\hat{\mathbf{S}}_{j-1} + \hat{\mathbf{S}}_{j+1})\rangle. \tag{6.14}$$

We now treat the spins at each site as classical vectors. The ground state of the system has all the spins aligned, say along the $z$ axis, so that $S_j^z = S$, $S_j^x = S_j^y = 0$. Consider a state which is a small departure from this state with $S_j^z \approx S$, $S_j^x$, $S_j^y \ll S$, so that

$$\frac{dS_j^x}{dt} \approx \frac{2JS}{\hbar}(2S_j^y - S_{j-1}^y - S_{j+1}^y) \tag{6.15}$$

$$\frac{dS_j^y}{dt} \approx -\frac{2JS}{\hbar}(2S_j^x - S_{j-1}^x - S_{j+1}^x) \tag{6.16}$$

$$\frac{dS_j^z}{dt} \approx 0. \tag{6.17}$$

We now look for normal mode solutions, so put

$$S_j^x = Ae^{i(qja-\omega t)} \tag{6.18}$$

$$S_j^y = Be^{i(qja-\omega t)} \tag{6.19}$$

where $q$ is a wave vector. A little algebra leads to $A = iB$ (showing that the $x$ and $y$ motion are $\pi/2$ out of phase) and hence that

$$\hbar\omega = 4JS(1 - \cos qa), \tag{6.20}$$

which is the dispersion relation for the spin waves and which is plotted in Fig. 6.12. Thus the spin waves are as depicted in Fig. 6.13.

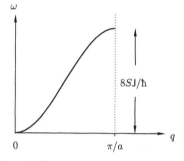
Fig. 6.12 The spin wave dispersion relation for a one-dimensional chain of spins.

(a)

(b)

Fig. 6.13 A spin wave on a line of spins. (a) perspective view. (b) view from above.

(2) It is perhaps more satisfying to derive eqn 6.20 using a quantum mechanical method, and so we now repeat the derivation.[4] We consider the ground state of the system, $|\Phi\rangle$, which consists of all the spins lying along the $+z$ direction. In one dimension the Hamiltonian for the Heisenberg model can be written

$$\hat{\mathcal{H}} = -2J\sum_i\left[\hat{S}_i^z\hat{S}_{i+1}^z + \frac{1}{2}(\hat{S}_i^+\hat{S}_{i+1}^- + \hat{S}_i^-\hat{S}_{i+1}^+)\right], \tag{6.21}$$

[4]We take $S = \frac{1}{2}$ for now, but it can be easily generalized.

$$|\Phi\rangle = \cdots \uparrow\uparrow\uparrow\uparrow\uparrow\uparrow\uparrow\uparrow\uparrow\uparrow \cdots$$

$$j$$

$$|j\rangle = \cdots \uparrow\uparrow\downarrow\uparrow\uparrow\uparrow\uparrow\uparrow\uparrow\uparrow \cdots$$

**Fig. 6.14** The state $|\Phi\rangle$ consists of all the spins lying along the $+z$ direction. The state $|j\rangle$ is the ground state with the spin at site $j$ flipped.

so that $\hat{\mathcal{H}}|\Phi\rangle = -NS^2 J|\Phi\rangle$. Now to create an excitation, flip a spin at site $j$, so let us consider a state $|j\rangle = \hat{S}_j^-|\Phi\rangle$ which is the ground state with the spin at site $j$ flipped (see Fig. 6.14). By flipping a spin, we have changed the total spin of the system by $\frac{1}{2} - (-\frac{1}{2}) = 1$. This excitation therefore has integer spin and is a boson. If we apply the Hamiltonian to this new state, we get

$$\hat{\mathcal{H}}|j\rangle = 2\left[(-NS^2 J + 2SJ)|j\rangle - SJ|j+1\rangle - SJ|j-1\rangle\right], \quad (6.22)$$

which is not a constant multiplied by $|j\rangle$, so this state is not an eigenstate of the Hamiltonian. Nevertheless, we can diagonalize the Hamiltonian by looking for plane wave solutions of the form

$$|q\rangle = \frac{1}{\sqrt{N}}\sum_j e^{iqR_j}|j\rangle. \quad (6.23)$$

The state $|q\rangle$ is essentially a flipped spin delocalized (smeared out) across all the sites. Since it is composed of states representing a single flipped spin, the total spin of $|q\rangle$ itself has the value $NS - 1$. It is then straightforward to show that

$$\hat{\mathcal{H}}|q\rangle = E(q)|q\rangle, \quad (6.24)$$

where

$$E(q) = -2NS^2 J + 4JS(1 - \cos qa). \quad (6.25)$$

The energy of the excitation is then $\hbar\omega = 4JS(1 - \cos qa)$ which is the same as the result we had above in eqn 6.20.

### 6.6.2   The Bloch $T^{3/2}$ law

At small $q$, eqn 6.20 yields

$$\hbar\omega \approx 2JSq^2 a^2, \quad (6.26)$$

so that $\omega \propto q^2$. In three dimensions, the density of states is given by

$$g(q)\,dq \propto q^2\,dq \quad (6.27)$$

which leads to

$$g(\omega)d\omega \propto \omega^{1/2}\,d\omega \quad (6.28)$$

at low temperature where only small $q$ and small $\omega$ are important. The spin waves are quantized in the same way as lattice waves. The latter are termed phonons, and so in the same way the former are termed **magnons**. As shown in Section 6.6.1, magnons have a spin of one and are bosons.

The number of magnon modes excited at temperature $T$, $n_{\text{magnon}}$, is calculated by integrating the magnon density of states over all frequencies after multiplying by the Bose factor, $(\exp(\hbar\omega/k_B T) - 1)^{-1}$, which must be included because magnons are bosons. Thus the result is given by

$$n_{\text{magnon}} = \int_0^\infty \frac{g(\omega)\,d\omega}{\exp(\hbar\omega/k_B T) - 1}, \quad (6.29)$$

which can be evaluated using the substitution $x = \hbar\omega/k_{\rm B}T$. At low temperature, where $g(\omega) \propto \omega^{1/2}$ in three dimensions, this yields the result

$$n_{\rm magnon} = \left(\frac{k_{\rm B}T}{\hbar}\right)^{3/2} \int_0^\infty \frac{x^{1/2}\,{\rm d}x}{e^x - 1} \propto T^{3/2}. \qquad (6.30)$$

Since each magnon mode which is thermally excited reduces the total magnetization[5] by $S = 1$, then at low temperature the reduction in the spontaneous magnetization from the $T = 0$ value is given by

$$\frac{M(0) - M(T)}{M(0)} \propto T^{3/2}. \qquad (6.31)$$

This result is known as the **Bloch $T^{3/2}$ law** and it fits experimental data in the low temperature regime (see Fig. 6.15).[6] The energy of the magnon modes is given by

$$E_{\rm magnon} = \int_0^\infty \frac{\hbar\omega g(\omega)\,{\rm d}\omega}{\exp(\hbar\omega/k_{\rm B}T) - 1} \propto T^{5/2}, \qquad (6.32)$$

so that the heat capacity $C = \partial E_{\rm magnon}/\partial T$ is also proportional to $T^{3/2}$.

[5]Because each magnon mode is a delocalized single reversed spin, see the previous section.

[6]The Bloch $T^{3/2}$ law is only really correct for the spontaneous magnetization within a domain. The data shown in Fig. 6.15 actually measure this because they were obtained using $\mu$SR which measures the local internal field with zero applied field.

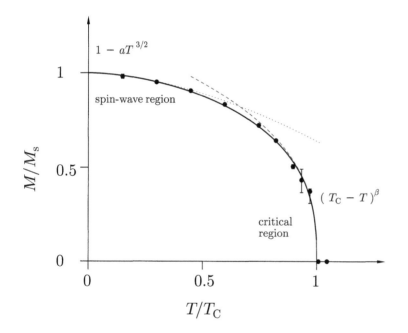

**Fig. 6.15** The spontaneous magnetization in a ferromagnet. At low temperatures this can be fitted using the spin-wave model and follows the Bloch $T^{3/2}$ law. Near the critical temperature, the magnetization is proportional to $(T - T_{\rm C})^\beta$ where $\beta$ is a critical exponent. Neither behaviour fits the real data across the whole temperature range. The data are for an organic ferromagnet which has $T_{\rm C} \approx 0.67$ K for which $\beta \approx 0.36$, appropriate for the three-dimensional Heisenberg model.

### 6.6.3 The Mermin–Wagner–Berezinskii theorem

In one and two dimensions, however, the integral in eqn 6.29 diverges, so that $M \to 0$ for all $T > 0$ for the isotropic 1-D and 2-D Heisenberg models.[7] The number of spin waves which are generated at finite temperature diverges, and hence spontaneous ferromagnetism is not possible. This result was first proved by Mermin and Wagner in 1966 and independently by Berezinskii. The absence of long range order in two-dimensional systems (with a continuous symmetry) is often referred to as the Mermin–Wagner–Berezinskii theorem.

[7]For different dimensions, the function $g(\omega)$ in eqn 6.29 has a different dependence on $\omega$.

[8]If the wavelength of the spin wave is sufficiently long, the exchange energy cost $\propto \int (|\nabla M_x|^2 + |\nabla M_y|^2)\, dx\, dy$ (in two dimensions) is minimized. See Exercise 6.7.

[9]This idea is explored in Exercise 6.7

This result only applies to an isotropic Heisenberg ferromagnet. This possesses rotational symmetry so that all of the spin directions can be globally rotated without any additional energy cost. This means that long wavelength excitations, in which the spin state may deviate from its ground state value over a considerable distance, cost very little energy.[8] Thus a fluctuation of the spins can be excited with very little energy cost; in one and two dimensions they destroy the long range order. If, however, there is significant anisotropy, there will be an energy cost associated with rotating the spins from their ground state value.

It turns out that the anisotropy energy penalty incurred by allowing these fluctuations increases with the square of $R$, the radius of the excitation, and hence the anisotropy energy will suppress all but the smallest of these non-linear fluctuations.[9] It is the presence of such symmetry-breaking fields which can stabilize long range order in two-dimensional systems. There is also a dipolar interaction between spins in real systems which, although much weaker than the exchange interaction, is anisotropic and can act in a similar way to suppress the growth of fluctuations.

It is observed in experiment that some ultra-thin magnetic films can show spontaneous ferromagnetism. Sometimes the anisotropy is such that it is energetically favourable for the spins to be perpendicular to the plane of the film. (The system is said to have a perpendicular easy axis for magnetization.) In this case the anisotropy leads to a gap in the spin wave excitation spectrum (between the ground state and the excited states) and long range magnetic order can exist at finite temperature. Dipolar interactions, or anisotropy of the energy of the spins in the plane of the film, can lead to a similar effect and stabilize the magnetization.

### 6.6.4 Measurement of spin waves

Elastic neutron scattering, which is used for magnetic structure determination, was discussed in Section 5.7.2. Spin wave dispersions can be measured using a technique known as inelastic neutron scattering. The magnitude of the incident neutron wave vector $\mathbf{k}$ is now no longer equal to the magnitude of the scattered neutron wave vector $\mathbf{k}'$. The energy of the neutron also changes from

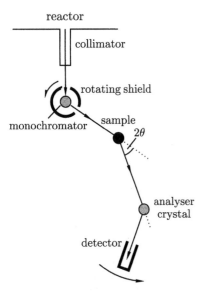

**Fig. 6.16** Schematic diagram of a neutron triple-axis spectrometer. By setting the angle of the rotating shield and the monochromator crystal, the incident neutron energy can be selected. The sample can also be rotated. The analysing crystal can also be rotated, allowing the scattered neutron energy to be measured.

$$E = \frac{\hbar^2 k^2}{2m_n} \tag{6.33}$$

to

$$E' = \frac{\hbar^2 k'^2}{2m_n} \tag{6.34}$$

This is because the neutron has produced in the sample an excitation of energy $\hbar\omega$ and wave vector $\mathbf{q}$. Conservation of energy and momentum implies that

$$E = E' + \hbar\omega \tag{6.35}$$

$$\mathbf{k} = \mathbf{k}' + \mathbf{q} + \mathbf{G}, \tag{6.36}$$

[10]The addition of $\mathbf{G}$ is necessary because the dispersion relation of magnons, like that of phonons, is periodic in the reciprocal lattice.

where $\mathbf{G}$ is a reciprocal lattice wave vector,[10] so that a measurement of $\mathbf{k}$, $\mathbf{k}'$, $E$ and $E'$ allows a determination of $\omega$ and $\mathbf{q}$. Neutrons have energies similar to the energies of atomic and electronic processes, i.e. in the meV to eV range.

Thus we can probe energy scales from the $\mu eV$ of quantum tunnelling, through molecular translations, rotations, vibrations and lattice modes, to eV transitions within the electronic structure of materials; magnon energies are typically in the range $10^{-3}$–$10^{-2}$ eV and therefore can be effectively measured using inelastic neutron scattering.

Experiments are typically performed using a **neutron triple-axis spectrometer** (see Fig. 6.16), so called because the angles of the monochromator, sample and analyser crystals can all be separately varied. This allows scattered neutrons corresponding to a large range of possible values of $\omega$ and $\mathbf{q}$ to be measured. In particular, it allows scans to be performed in which $q$ varies with fixed $\omega$ (or vice versa). Example data obtained by this technique are shown in Figures 6.17 for $Co_{0.92}Fe_{0.08}$ and 6.18 for the ferromagnetic oxide $La_{0.7}Pb_{0.3}MnO_3$ (which exhibits colossal magnetoresistance, see Section 8.9.5).

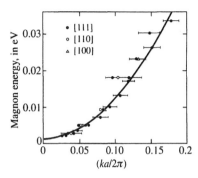

Fig. **6.17** The spectrum of magnon energy for spin waves in an alloy of $Co_{0.92}Fe_{0.08}$ obtained at room temperature (Sinclair and Brockhouse 1960). In this figure, $k$ is the magnon wave vector corresponding to $q$ in the text.

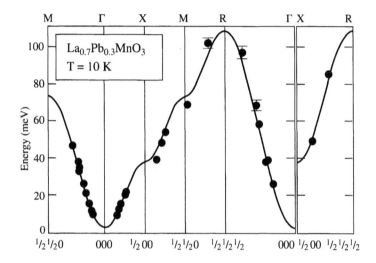

Fig. **6.18** Spin wave dispersion in the ferromagnetic oxide $La_{0.7}Pb_{0.3}MnO_3$ measured at 10 K by inelastic neutron scattering (though with a time-of-flight instrument, not a triple-axis spectrometer). The horizontal axis corresponds to the magnon wave vector. After Perring *et al.* 1996.

## 6.7 Domains

If different regions of a macroscopic system break symmetry in different ways, then in the interface between these regions the rigidity can break down. In general we expect domain walls, defects, vortices, dislocations and other singularities. In ferromagnets the most important singularity is the domain wall.

Weiss first proposed that a ferromagnet contains a number of small regions called **domains**, within each of which the local magnetization reaches the saturation value. The direction of the magnetization of different domains need not be parallel. Domains are separated by **domain walls**. The existence of domains explains the surprising observation that in some ferromagnetic specimens[11] it can be possible, at room temperature, to attain saturation magnetization of the whole sample (corresponding to $\mu_0 M \sim 1$ T) by the application of a very weak magnetic field (as low as $10^{-6}$ T). Such low applied fields would

[11]Which are magnetically 'soft', see Section 6.7.9.

[12]If $\chi \sim 10^{-3}$ for a particular paramagnet, see Table 2.1, then an applied field of $10^{-6}$ T would produce a magnetization of $\mu_0 M \sim 10^{-9}$ T.

[13]As we shall see, in some ferromagnetic materials the movement of domain walls can in fact be energetically costly. Such substances would require large applied magnetic fields to switch them from one magnetic state to the other.

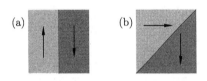

**Fig. 6.19** (a) 180° domain wall. (b) 90° domain wall.

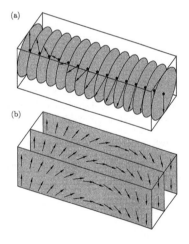

**Fig. 6.20** (a) A Bloch wall. (b) A Néel wall.

**Fig. 6.21** Two spins at an angle $\theta$ with respect to each other, producing an energy cost $\frac{1}{2}JS^2\theta^2$.

[14]Using $\cos\theta \approx 1 - \theta^2/2$ for $\theta \ll 1$.

have negligible effect on a paramagnet.[12] The large effect in the ferromagnetic specimen is because the applied field does not have to order the magnetic moments macroscopically (they are all already ordered!) but merely to cause the domains to align. This can be achieved by the energetically painless process of domain wall motion.[13]

In the same ferromagnetic specimen, it is also possible for the magnetization to be zero in zero applied field. This is also a manifestation of domains; the magnetization is saturated within each domain, but the directions of the magnetization of each domain are such that the net magnetization of the specimen is zero.

### 6.7.1   Domain walls

Between adjacent domins there is a boundary called a domain wall. The domain walls can be classified according to the angle between the magnetization in the two domains (see Fig. 6.19). A 180° domain wall separates domains of opposite magnetization. A 90° domain wall separates domains of perpendicular magnetization.

The most common type of 180° wall is a **Bloch wall** (see Fig. 6.20(a)) in which the magnetization rotates in a plane parallel to the plane of the wall. Another possible configuration is the **Néel wall** (see Fig. 6.20(b)) in which the magnetization rotates in a plane perpendicular to the plane of the wall. Let us attempt to calculate the domain wall width in a Bloch wall.

In a ferromagnet it costs energy to rotate neighbouring spins. Two spins, $\mathbf{S}_1$ and $\mathbf{S}_2$, which are at angle of $\theta$ with respect to each other (Fig. 6.21) have an energy $-2J\mathbf{S}_1 \cdot \mathbf{S}_2 = -2JS^2\cos\theta$. If $\theta = 0$, their energy is $-2JS^2$. Hence the energy cost of having $\theta \neq 0$ is approximately[14] $JS^2\theta^2$ if $\theta \ll 1$. In a Bloch wall, spins rotate over $N$ sites by an angle $\pi$ (see Fig. 6.20(a)). Hence the energy cost of a line of spins is equal to $N$ contributions of $JS^2\theta^2$ where $\theta = \pi/N$, i.e. to $JS^2\pi^2/N$. In a Bloch wall we have planes of spins (see Fig. 6.20) and so we are interested in $\sigma_{BW}$, the energy per unit area of the Bloch wall. In a square metre of wall, there are $1/a^2$ lines of spins like the one we have calculated. Hence

$$\sigma_{BW} = JS^2 \frac{\pi^2}{Na^2} \tag{6.37}$$

which tends to zero as $N \to \infty$. This result would seem to indicate that if a domain wall formed it would just unwind itself, growing in size throughout the entire system. This is because it costs energy to have spins twisted with respect to each other, and therefore they will all untwist unless some other interaction stops them. This other interaction is the **magnetocrystalline anisotropy**.

### 6.7.2   Magnetocrystalline anisotropy

Crystals possess a magnetic **easy axis** and a **hard axis**. Along certain crystallographic directions it is easy to magnetize the crystal, along others it is harder (this is shown for single crystals of Fe, Co and Ni in Fig. 6.22). In Co, for example, this anisotropy leads to an additional energy of the form

$$E = K_1 \sin^2\theta + K_2 \sin^4\theta \tag{6.38}$$

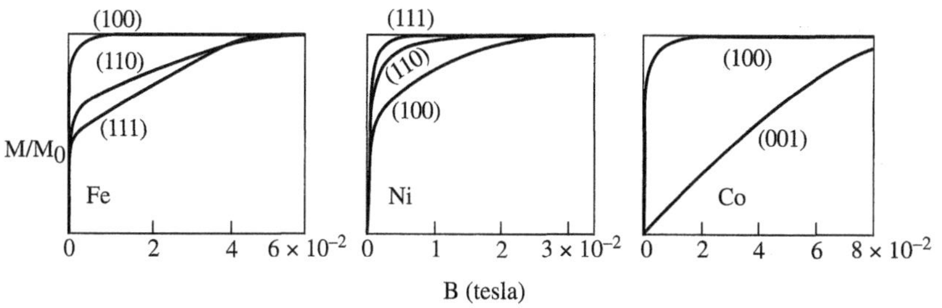

**Fig. 6.22** Magnetization in Fe, Co and Ni for applied fields in different directions showing anisotropy. After Honda and Kaya 1926, Kaya 1928.

where $K_1$ and $K_2$ are anisotropy constants and $\theta$ is the angle between the magnetization and the stacking direction of the hexagonally close packed planes. Because these constants are positive the energy is minimized when the magnetization lies along the stacking direction. In eqn 6.38, $E$, $K_1$ and $K_2$ are energy densities (i.e. they are measured in $J\,m^{-3}$). The anisotropy constants are found to be strongly temperature dependent.

Equation 6.38 is appropriate for **uniaxial anisotropy**, in which the energy depends on the angle to a single axis (in Co this is the stacking axis of the hexagonally close packed planes). In a cubic system, the appropriate expression is

$$E = K_1(m_x^2 m_y^2 + m_y^2 m_z^2 + m_z^2 m_x^2) + K_2 m_x^2 m_y^2 m_z^2 + \cdots, \qquad (6.39)$$

where $\mathbf{m} = (m_x, m_y, m_z) = \mathbf{M}/|\mathbf{M}|$. In spherical coordinates this is

$$E = K_1 \left( \frac{1}{4} \sin^2 \theta \sin^2 2\phi + \cos^2 \theta \right) \sin^2 \theta + \frac{K_2}{16} \sin^2 2\phi \sin^2 2\theta \sin^2 \theta + \cdots.$$
$$(6.40)$$

The anisotropy energy arises from the spin–orbit interaction and the partial quenching of the angular momentum. Anisotropy energies are usually in the range $10^2$–$10^7$ $Jm^{-3}$. This corresponds to an energy per atom in the range $10^{-8}$–$10^{-3}$ eV. The anisotropy energy is larger in lattices (of magnetic ions) of low symmetry and smaller in lattices of high symmetry. For example, cubic Fe and Ni have $K_1$ equal to $4.8 \times 10^4$ $Jm^{-3}$ and $-5.7 \times 10^3$ $Jm^{-3}$ respectively, but hexagonal Co has $K_1 = 5 \times 10^5$ $Jm^{-3}$. Low symmetry permanent magnet materials $Nd_2Fe_{14}B$ and $SmCo_5$ have $K_1$ equal to $5 \times 10^6$ $Jm^{-3}$ and $1.7 \times 10^7$ $Jm^{-3}$ respectively.

An additional energy term is due to the demagnetizing energy associated with the sample shape and is referred to as **shape anisotropy**. In thin films, a shape anisotropy $\frac{1}{2}\mu_0 M^2 \cos^2 \theta$ (where $\theta$ is the angle between the film normal and $\mathbf{M}$) leads to an energetic saving for keeping the magnetization in the plane of the film.

### 6.7.3 Domain wall width

In the magnetic domains of a ferromagnet the magnetization will prefer to lie along the easy direction but between domains, in the domain wall, it will have to rotate and a component will lie along the hard axis which will cost energy. If we assume a simple form for the anisotropy energy density, namely $E = K \sin^2 \theta$ where $K$ is an anisotropy constant, then we can easily find an expression for the anisotropy energy contribution to a Bloch wall. We take $K > 0$ so that spins prefer to line up along $\theta = 0$ or $\theta = \pi$. We add up the contribution from each of the $N$ spins and replace the sum by an integral for simplicity (the continuum limit). Thus the energy density contribution is

$$\sum_{i=1}^{N} K \sin^2 \theta_i \approx \frac{N}{\pi} \int_0^{\pi} K \sin^2 \theta \, d\theta = \frac{NK}{2}. \qquad (6.41)$$

The energy contribution per unit area of wall can then be written as $NKa/2$. Thus the total energy per unit area of the domain wall, including the contribution from the exchange energy (eqn 6.37) and the contribution from the anisotropy energy (eqn 6.41) is

$$\sigma_{\mathrm{BW}} = \mathsf{J} S^2 \frac{\pi^2}{Na^2} + \frac{NKa}{2}. \qquad (6.42)$$

This gives us the required behaviour since we have two terms, one proportional to $1/N$ (tending to unwind the wall and make it bigger) and the other proportional to $N$ (tending to tighten it and make it smaller). We find the equilibrium configuration using $dE_{\mathrm{BW}}/dN = 0$ which leads to a value of $N$ given by

$$N = \pi S \sqrt{2\mathsf{J}/Ka^3} \qquad (6.43)$$

so that the width of the Bloch wall is

$$\delta = Na = \pi S \sqrt{\frac{2\mathsf{J}}{Ka}}, \qquad (6.44)$$

where $a$ is the lattice spacing. Larger $\mathsf{J}$ makes the wall thicker, larger $K$ makes it smaller. The energy per unit area of the domain wall is

$$\sigma_{\mathrm{BW}} = \pi S \sqrt{\frac{2\mathsf{J}K}{a}}. \qquad (6.45)$$

In terms of the parameter $A$ defined in eqn 4.19 (for a cubic crystal) we have the domain wall thickness

$$\delta = \pi \sqrt{\frac{A}{K}}, \qquad (6.46)$$

and the energy per unit area as

$$\sigma_{\mathrm{BW}} = \pi \sqrt{AK}. \qquad (6.47)$$

### 6.7.4   Domain formation

Since it costs this energy to make a domain wall, one might wonder why more than one domain ever forms in the first place. The reason is that the formation of domains saves energy associated with dipolar fields. Because $\nabla \cdot \mathbf{H} = -\nabla \cdot \mathbf{M}$ (see Appendix B) then whenever $\mathbf{M}$ stops and starts, at the edges of a sample for example, the magnetic field diverges and this produces demagnetizing fields which fill space and which cost $B^2/2\mu_0$ Joules of energy per cubic metre. The energy associated with the demagnetizing field is called, variously, the demagnetization energy, magnetostatic energy or dipolar energy. It takes the value

$$-\frac{\mu_0}{2}\int_V \mathbf{M}\cdot\mathbf{H_d}\,d\tau, \tag{6.48}$$

where $\mathbf{H_d}$ is the demagnetizing field and the integral is taken over the volume of the sample (see Appendix D). For an ellipsoidally shaped sample magnetized along one of its principal axes, this energy reduces to

$$\frac{\mu_0}{2}NM^2V \tag{6.49}$$

where $N$ is the demagnetizing factor and $V$ is the sample volume.

This dipolar energy can be saved by breaking the sample into domains, but each domain created costs energy because of the cost of the domain walls. Therefore, in the same way as the size of a domain wall is a balance between the exchange and anisotropy energies, so the formation of domains is a balance between the cost of a demagnetizing field and the cost of a domain wall.

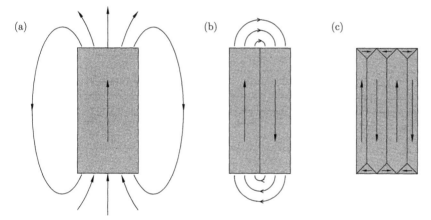

**Fig. 6.23** A sample which is (a) uniformly magnetized, (b) divided into two domains, and (c) with a simple closure domain structure.

Figure 6.23 shows three different choices for the domain structure for a ferromagnetic sample. The single domain structure in Fig. 6.23(a) has no domain walls but a large dipolar energy. The dipolar energy can be reduced by breaking the sample into two domains as shown in Fig. 6.23(b), albeit at the cost of introducing a domain wall. The so-called closure domain structure in Fig. 6.23(c) eliminates the dipolar energy but introduces a number of domain walls.

Dipolar energy can also determine the type of domain wall that can form. The Bloch wall is favoured in the bulk because it leads to a smaller dipolar energy. The Néel wall tends to be favoured in thin films (where there is a dipolar energy cost to rotating the spins out of the plane of the film).

## 6.7.5 Magnetization processes

Figure 6.24 shows a typical hysteresis loop expected for measuring the magnetization as a function of the applied magnetic field for a ferromagnetic sample. If the sample is magnetized to the saturation magnetization $M_s$ by an applied field, then when the applied field is reduced to zero the magnetization

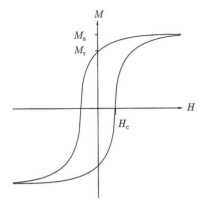

**Fig. 6.24** A hysteresis loop showing the saturation magnetization $M_s$, the remanent magnetization $M_r$ and the coercive field $H_c$.

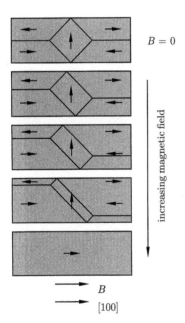

**Fig. 6.25** Effect of an applied field on the domain pattern on the surface of a single crystal iron whisker showing domain wall displacement, as the applied $B$ field increases from 0 up to a maximum value.

Heinrich Barkhausen (1881–1956)

John Kerr (1824–1907)

reduces to the **remanent magnetization** $M_r$. A magnetic field equal to the **coercive field** $H_c$ is needed to switch the magnetization into the opposite direction. The parameters $M_r$ and $H_c$ can be used to characterize a ferromagnet.

The length of a ferromagnet changes slightly when it is magnetized. This effect is known as **magnetostriction** and demonstrates that the magnetic and elastic behaviour of a material are connected (the connection is called **magnetoelastic coupling**). The crystal deforms because it can lower its anisotropy energy which can depend on the strain of the crystal. For example, in a cubic system it may be energetically favourable for the crystal to deform slightly from exactly cubic symmetry if doing so will save more anisotropy energy than it will cost in elastic energy.

When a demagnetized ferromagnet is magnetized, various processes occur. First is domain-wall motion; domains which are aligned favourably with respect to the applied magnetic field grow at the expense of domains which are unfavourably aligned. (This process is shown in Fig. 6.25.) At higher fields, domain rotation can occur in which the anisotropy energy can be outweighed and the magnetization can suddenly rotate away from the original direction of magnetization to the crystallographic easy axis which is nearest to the field direction. The final domain process at highest magnetic fields is coherent rotation of the domains to a direction aligned with the magnetic field, irrespective of the easy and hard axes.

The motion of domain walls through a magnetic material depends in detail upon the metallurgical properties of the material. Domain walls can be **pinned** by strains in the material, by surfaces and impurities because of the magnetoelastic coupling. Domain wall pinning therefore increases coercivity. The magnetization of a ferromagnet also changes by a series of discontinuous steps due to domain boundary motion, so that very small steps are sometimes seen on the magnetization curves. This is known as the **Barkhausen effect**. Low level acoustic emission also sometimes accompanies the magnetization process because of these sudden discontinous changes in magnetization and their coupling to the elastic modes of the system. This effect is known as **magnetoacoustic emission**.

### 6.7.6 Domain wall observation

Domains can be observed using a variety of techniques. In the **Bitter powder technique** a colloidal suspension of fine magnetic particles is placed on the surface of the sample. The magnetic particles collect near domain walls where there are strong local fields which attract them. These can be observed using optical microscopy. It is also possible to use polarized light reflecting from a polished magnetized sample to observe domains, a technique known as **Kerr microscopy**. The incident beam of light is polarized and its plane of polarization is rotated by reflection from the magnetized sample (this is the Kerr effect, see Section 8.8). The degree of rotation can be measured using a second polarizer (the analyser). If a beam of light is focussed to a spot and scanned across the surface of a magnetic sample, then for a particular setting of the analyser polarizer, some domains will appear bright and others dark. If the sample is transparent, the experiment can be performed in transmission (the Faraday effect).

Electron microscopy can also be effective as the electron beam is deflected by the Lorentz force due to the internal field. Boundaries between domains (across which this deflection can suddenly change in sign as the beam is scanned over) show up particularly well. This technique is called Lorentz microscopy.

In the technique of magnetic force microscopy[15] a tiny cantilever with a magnetized tip is scanned across a sample surface. The stray fields produced by a domain wall cause the cantilever to bend, and this bending (detected by, for example, a change in the resonant frequency of the cantilever) is used to image the surface.

[15] A magnetic force microscope is based on the well-known atomic force microscope.

### 6.7.7 Small magnetic particles

In many ferromagnetic samples the lowest energy state at zero applied field is the demagnetized state, so that the overall magnetic moment is zero. If the sample size is reduced, surface energies (such as the domain wall energies) become progressively more costly in comparison with volume energies (such as the demagnetizing energy).[16] Thus a critical dimension may be reached below which it is energetically favourable to remove the domain walls so that the sample consists of a single magnetized domain. It would thus behave like a small permanent magnet.

In the following example, we estimate the critical size for a single domain particle. The calculation will assume that this critical size is still larger than the domain wall width.

[16] This is because surface energies scale as the (sample size)$^2$ whereas volume energies scale as the (sample size)$^3$.

**Example 6.1**

Consider a ferromagnetic particle of radius $r$. The energy of the single domain state (Fig. 6.26(a)) is, using eqn D.12 and $N = \frac{1}{3}$ for a sphere, given by

The volume of a sphere is $V = \frac{4}{3}\pi r^3$.

$$E_{(a)} = \frac{1}{6}\mu_0 M^2 V = \frac{2}{9}\mu_0 \pi M^2 r^3. \qquad (6.50)$$

The state in Fig. 6.26(b), appropriate for cubic anisotropy, approximately removes all demagnetizing energy, but introduces 90° domain walls of area $2 \times \pi r^2$, leading to an energy cost

$$E_{(b)} = 2\pi r^2 \sigma_W^{90°}, \qquad (6.51)$$

where $\sigma_W^{90°}$ is the energy cost of a 90° domain wall per unit area. The state in Fig. 6.26(a) thus becomes more favourable than the state in Fig. 6.26(b) when $E_{(a)} < E_{(b)}$, i.e. when

$$r < \frac{9\sigma_W^{90°}}{\mu_0 M^2}. \qquad (6.52)$$

(a)

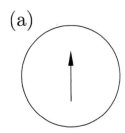

(b)

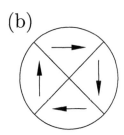

(c)

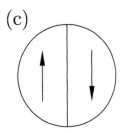

**Fig. 6.26** Three possible magnetization configurations for a small spherical ferromagnetic particle.

The state in Fig. 6.26(c), appropriate for uniaxial anisotropy, is harder to consider because neither domain is ellipsoidal in shape and there is still some demagnetizing energy. Roughly, though, the demagnetizing energy of this configuration is half that of Fig. 6.26(a) since the domains are half the size. The domain wall cost is then $\pi r^2 \sigma_{\mathrm{W}}^{180°}$ where $\sigma_{\mathrm{W}}^{180°}$ is the energy of a 180° domain wall per unit area. Thus

$$E_{(c)} = \frac{1}{9}\mu_0 \pi M^2 r^3 + \pi r^2 \sigma_{\mathrm{W}}^{180°}. \tag{6.53}$$

Hence the state in Fig. 6.26(a) becomes more favourable than the state in Fig. 6.26(c) when $E_{(a)} < E_{(c)}$, i.e. when

$$r < \frac{9\sigma_{\mathrm{W}}^{180°}}{\mu_0 M^2}. \tag{6.54}$$

If $\sigma_{\mathrm{W}}^{180°} \sim 10^{-2}$ Jm$^{-2}$ and $\mu_0 M \sim 1$ T, the critical radius is $\sim 10^{-7}$ m. Note, if this value came out to be smaller than the domain wall width, then the calculation would be nonsensical.

### 6.7.8   The Stoner–Wohlfarth model

It is possible to calculate the magnetization curve for a single-domain particle, via the Stoner–Wohlfarth model. Because the particle is a single magnetic domain, there are no domain walls and one need only consider coherent domain rotation.

We consider a single-domain magnetic particle in a magnetic field $H$ which is applied at an angle $\theta$ to the easy axis of its uniaxial anisotropy. If the magnetization of the particle then lies at an angle of $\phi$ to the magnetic field direction, the energy density of the system is

$$E = K \sin^2(\theta - \phi) - \mu_0 H M_s \cos \phi. \tag{6.55}$$

The energy can be minimized to find the direction of the magnetization at any given value of the applied magnetic field.

This is shown in Fig. 6.27 where the applied field is written in dimensionless units using the parameter $h$ given by

$$h = \frac{\mu_0 M_s H}{2K} \tag{6.56}$$

for two choices of the direction of the applied magnetic field. Analytic solutions are possible for this model for $\theta = 0$ and $\theta = \pi/2$, but in the figure it is shown how they can be solved numerically. In each case the trajectory of the magnetization direction which lowers the energy is shown by a curve across an energy surface (energy as a function of $\phi$ and $h$) and the corresponding hysteresis loops are also shown. The hysteresis loops are shown for a larger choice of values of $\theta$ in Fig. 6.28, together with the calculated result for a polycrystalline average of directions, as would be appropriate for an array of single-domain particles (ignoring any interactions between the particles). This model demonstrates how the anisotropies present in a system can lead to hysteresis, even in a system in which there are no irreversible effects associated with domain-wall pinning.

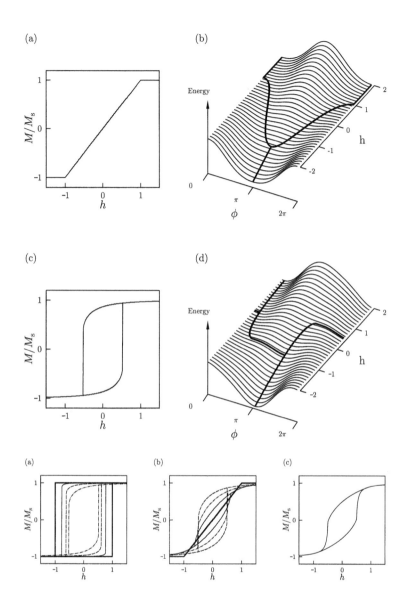

**Fig. 6.27** The hysteresis loops in the Stoner–Wohlfarth model for (a) $\theta = 90°$ and (c) $\theta = 30°$. These can be obtained by finding the minimum energy points on the energy surfaces as a function of $h$ and $\phi$, which are shown for (b) $\theta = 90°$ and (d) $\theta = 30°$.

**Fig. 6.28** The hysteresis loops in the Stoner–Wohlfarth model for (a) $\theta = 0°$ (bold), 5°, 15°, 30°, (b) $\theta = 45°$, 60°, 75° and 90° (bold). (c) A calculated hysteresis loop for a polycrystalline average.

## 6.7.9 Soft and hard materials

The energy dissipated (as heat) by a ferromagnet as it is taken around a circuit of its hysteresis loop is proportional to the area of that loop. If the area is small, the material is said to be magnetically **soft**. If the area is large, the material is said to be magnetically **hard**. Domain walls traverse a sample as the field is cycled and we can distinguish these two broad classes of ferromagnetic materials based on how easy it is for domains to move through a sample.

(1) Soft magnetic materials are easy to magnetize. Soft magnets are used in transformer coils, generators and motors. In these applications the magnetization must be reversed many times a second and it is important that the energy dissipated in each cycle is minimized. Soft materials have broad domain walls (small anisotropy energy $K$) which are thus easy

to move. This leads to small coercive fields. Low magnetostriction is often desirable so that internal strains will not induce a local anisotropy energy. An example is permalloy (a commercial Ni/Fe alloy with an additional ingredient) which has a coercive field $B_c \sim 2 \times 10^{-7}$ T.

(2) Hard magnetic materials are difficult to magnetize and thus difficult to demagnetize. Hard magnets are used as permanent magnets (e.g. in the back of loudspeakers, in motors, and of course on the front of your refrigerator!) and in magnetic tape recording (in powder form). In these applications, the magnetization needs to be preserved for as long as possible. The energy dissipated in a hysteresis loop cycle is as large as possible so that the magnetization will not occur spontaneously. Hard magnets have large hysteresis and narrow domain walls (large $K$) so that it is easy to have domain wall pinning. Large ion moments and large crystal fields are helpful for hard magnetic properties and suitable materials often involve rare earths, e.g. $Nd_2Fe_{14}B$ which has $T_C = 585$ K and coercive field $B_c = 1.2$ T.

An important application of hard magnetic materials is magnetic recording. The areal density of bits that can be recorded using magnetic materials has shot up dramatically over the last few years, reflecting a steady improvement in the understanding and fabrication of magnetic materials. The approximate figures are $\sim 0.1$ Mbits in$^{-2}$ in 1960, $\sim 10$ Mbits in$^{-2}$ in 1980 and $\sim 25$ Gbits in$^{-2}$ in 2000 (1 bit in$^{-2}$ is a bit of information, a '0' or '1', stored in a square-inch – this most modern of industries has yet to modernize its units!) As the areal density increases in the future, superparamagnetic fluctuations (i.e. thermal fluctuations of the magnetic particles, see Section 8.3) of the magnetic moments will become important.

Small magnetic particles of $Fe_2O_3$ were traditionally used on recording tapes and disks. Improvements in properties for high-fidelity audio recording and video tapes are based on high coercivity particles such as $Fe_2O_3$ doped with Co, $CrO_2$, metal particles (usually Fe), and barium ferrite ($BaO\cdot 6Fe_2O_3$, see Section 5.3). Hard drives typically consist of a large rigid disk with a thin film deposited by vacuum deposition or sputtering, usually of a Co/Cr alloy. The hard disks rotate rapidly and a reading-head moves across the surface of the disk (it can hover a fraction of a micron above the surface) and is able to access any part of it.

# Further reading

- The concept of broken symmetry is described in more detail in P. W. Anderson, *Basic notions of condensed matter physics*, Addison-Wesley 1984.

- The statistical mechanics of phase transitions is discussed in J. Yeomans, *Statistical mechanics of phase transitions*, OUP 1992.

- Also very helpful on critical phenomena is M. F. Collins, *Magnetic critical scattering*, OUP 1989.

- A very thorough description of phase transitions and

- broken symmetry phenomena, with particular emphasis on soft condensed matter, is P. M. Chaikin and T. C. Lubensky, *Principles of condensed matter physics*, CUP 1995.

- J. J. Binney, N. J. Dowrick, A. J. Fisher and M. E. J. Newman, *The theory of critical phenomena*, OUP 1992 contains an excellent account of renormalization group methods.

- Information on domains and magnetism may be found

in G. Bertoni, *Hysteresis in magnetism*, Academic Press 1998 and A. Hubert and R. Schäfer, *Magnetic domains*, Springer 1998.

## References

- K. Honda and S. Kaya, *Sci. Rep. Tohoku Univ.*, **15**, 721 (1926)

- S. Kaya, *Sci. Rep. Tohoku Univ.*, **17**, 639, 1157 (1928).

- Useful general references are D. Craik, *Magnetism: principles and applications*, Wiley 1995 and J. Crangle, *Solid state magnetism*, Edward Arnold 1991.

- T. G. Perring, G. Aeppli, S. M. Hayden, S. A. Carter, J. P. Remeika and S-W. Cheong, *Phys. Rev. Lett.*, **77**, 711 (1996).

- R. N. Sinclair and B. N. Brockhouse, *Phys. Rev.*, **120**, 1638 (1960).

# Exercises

(6.1) A one-dimensional ferromagnetic chain of $N$ spins is described by the Ising Hamiltonian

$$\hat{\mathcal{H}} = -2J \sum_{i=1}^{N-1} \hat{S}_i^z \hat{S}_{i+1}^z - 2J\hat{S}_N^z \hat{S}_1^z.$$

(The last term is used to give periodic boundary conditions.)

Show that the partition function $Z$ of this system can be obtained by introducing new operators

$$\zeta_i = 4S_i^z S_{i+1}^z$$

which have eigenvalues $+1$ or $-1$, and hence show that the partition function is $Z = (2\cosh(J/2k_BT))^N$.

Using these results obtain an expression for the heat capacity per spin of the chain as $N \to \infty$. Deduce the low and high temperature behaviour of the heat capacity and sketch the heat capacity as a function of temperature. Discuss the result.

(6.2) A uniaxial ferromagnet is described by the Hamiltonian

$$\hat{\mathcal{H}} = -\sum_{ij} J_{ij} \hat{\mathbf{S}}_i \cdot \hat{\mathbf{S}}_j - \sum_{ij} K_{ij} \hat{S}_i^z \hat{S}_j^z. \quad (6.57)$$

(a) Show that the state with all spins fully aligned along the $z$ axis is an eigenstate of the Hamiltonian.

(b) Obtain an expression for the spin wave spectrum as a function of wave vector $\mathbf{q}$.

(c) Simplify the expressions for the case where $J_{ij}$ and $K_{ij}$ are restricted to nearest neighbours, $J_0$ and $K_0$, and the ferromagnet is (i) a one-dimensional chain, (ii) a two-dimensional square lattice and (iii) a three-dimensional body–centred cubic material.

(6.3) Using the results of Exercise 6.2 for small wave vectors, deduce the temperature dependence at low temperatures

of the number of spin waves for the Heisenberg model ($K_0 = 0$) and the Ising model ($J_0 = 0$) for each of the structures (i), (ii) and (iii) of Exercise 6.2. Show that your results for the Ising model for case (i) agree with the results obtained in Exercise 6.1 and that there is no long range magnetic order above absolute zero for the Heisenberg model in one or two dimensions.

(6.4) Apply spin-wave theory to an antiferromagnet and show that the dispersion relation is

$$\hbar\omega = 4|J|S \sin(qa)|. \quad (6.58)$$

Hence show that for an antiferromagnet $\omega \propto |q|$ at small $q$, whereas for a ferromagnet $\omega \propto q^2$.

(6.5) The Landau theory of a ferromagnet in a magnetic field $H$ implies that the free energy is given by

$$F(M) = F_0 + a_0(T-T_C)M^2 + bM^4 - \mu_0 MH \quad (6.59)$$

where $a_0$ and $b$ are positive constants. Show that this implies that

$$M^2 = u + v\frac{H}{M} \quad (6.60)$$

where $u$ and $v$ are constants that you should determine. By sketching $M^2$ against $H/M$ for $T$ just above $T_C$, just below $T_C$, and exactly at $T_C$, show how this method can be used to determine $T_C$. This idea is the basis of the Arrott plot which is a plot of $M^2$ against $H/M$.

(6.6) Find analytic solutions to the Stoner–Wohlfarth model for $\theta = 0$ and $\theta = \pi/2$.

(6.7) To illustrate the ideas behind the Mermin–Wagner–Berenzinskii theorem, consider an excitation in two dimensions (see Fig. 6.29) in which the magnetization as a function of position $\mathbf{r} = (x, y)$ has constant magnitude $|\mathbf{M}|$ but varying direction specified by the vector $\mathbf{M}(\mathbf{r})$

given by

$$\mathbf{M}(\mathbf{r}) = \begin{cases} |\mathbf{M}|(\hat{\mathbf{y}}\cos(\pi r/R) + \hat{\mathbf{x}}\sin(\pi r/R)) & \text{if } r < R \\ -|\mathbf{M}|\hat{\mathbf{y}} & \text{if } r > R. \end{cases}$$
(6.61)

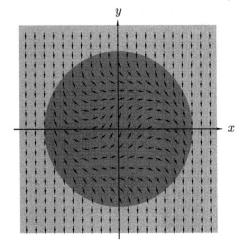

**Fig. 6.29** An excitation in a 2D magnet.

Explain why the exchange energy cost due to this excitation is proportional to

$$\int \left[ (\nabla M_x)^2 + (\nabla M_y)^2 \right] dx\, dy,$$
(6.62)

and show that it is a constant $(\pi^3 |\mathbf{M}|^2)$ and therefore independent of R. This implies that very large scale fluctuations can occur with no additional energy cost compared with a small-scale fluctuation, thereby destroying any long range order. This problem is based on an example presented in an article by A. S. Arrott and B. Heinrich, *Journal of Magnetism and Magnetic Materials* **93**, 571 (1991).

(6.8) Data are written onto a magnetic recording disk with an areal density of 25 Gbits in$^{-2}$. By using some educated guesses, convert this areal density into other units: (a) bits per cross-sectional area of a hydrogen atom, (b) copies of the complete plays by William Shakespeare (i.e. their information content) per postage stamp. Compare this density of information storage to that of the human genome.

(6.9) Consider an isotropic cubic ferromagnet (lattice parameter $a$) in a magnetic field, with Hamiltonian

$$\hat{\mathcal{H}} = -J \sum_{<ij>} \hat{\mathbf{S}}_i \cdot \hat{\mathbf{S}}_j + g\mu_B \sum_j \mathbf{B} \cdot \hat{\mathbf{S}}_j.$$
(6.63)

Show that it has a spin wave dispersion given by

$$\hbar\omega(q) = g\mu_B B + 2JSq^2 a^2$$
(6.64)

for $qa \ll 1$.

(6.10) Show that the Hamiltonian

$$\hat{\mathcal{H}} = -\sum_{\mathbf{n},\delta} J(\delta) \hat{\mathbf{S}}_\mathbf{n} \cdot \hat{\mathbf{S}}_{\mathbf{n}+\delta}$$
(6.65)

has a spin-wave dispersion given by

$$\hbar\omega(\mathbf{q}) = \hbar\omega(0) + S \sum_\delta J(\delta)(\mathbf{q} \cdot \delta)^2$$
(6.66)

for $|\mathbf{q} \cdot \delta_{\max}| \ll 1$ where $\delta_{\max}$ is the largest vector $\delta$ for which $J(\delta)$ gives any appreciable contribution.

(6.11) The state $|q\rangle$ in eqn 6.23 represents a delocalized single flipped spin in a background of unflipped spins. In three dimensions it becomes

$$|\mathbf{q}\rangle = \frac{1}{\sqrt{N}} \sum_\mathbf{j} e^{i\mathbf{q}\cdot\mathbf{R}_\mathbf{j}} |\mathbf{j}\rangle.$$
(6.67)

The probability that the flipped spin is at site $i$ is

$$P = |\langle \mathbf{q}|\mathbf{i}\rangle|^2.$$
(6.68)

Show that $P = 1/N$ and hence that the lowered spin is distributed with equal probability.

The operator for the transverse spin correlation function is

$$\hat{S}_\mathbf{j}^x \hat{S}_{\mathbf{j'}}^x + \hat{S}_\mathbf{j}^y \hat{S}_{\mathbf{j'}}^y.$$
(6.69)

Show that

$$\langle \mathbf{q}| \hat{S}_\mathbf{j}^x \hat{S}_{\mathbf{j'}}^x + \hat{S}_\mathbf{j}^y \hat{S}_{\mathbf{j'}}^y |\mathbf{q}\rangle = \frac{2S}{N} \cos(\mathbf{q} \cdot (\mathbf{R}_\mathbf{j} - \mathbf{R}_{\mathbf{j'}}))$$
(6.70)

(for $j \neq j'$) and interpret this with reference to Fig. 6.13.

(6.12) Our calculation for the domain wall width of a Bloch wall made the assumption that the angle between adjacent spins in the wall was $\pi/N$ everywhere throughout the $N$ spins of the wall. This was clearly an approximation! Here is an improved calculation.

Let the spin as a function of $z$, a coordinate running perpendicular to the plane of the wall, be described by an angle $\theta(z)$. We assume that $\theta(z) \to 0$ as $z \to -\infty$ and $\theta(z) \to \pi$ as $z \to +\infty$, corresponding to a 180° wall. The energy is given by

$$E = \int_{-\infty}^{\infty} \left[ A\left(\frac{\partial\theta}{\partial z}\right)^2 + f(\theta) \right] dz,$$
(6.71)

where $A$ is the continuum exchange constant and $f(\theta)$ is a function which describes the anisotropy. Show that the energy is minimized when

$$\frac{\partial f}{\partial \theta} - 2A \frac{\partial^2 f}{\partial z^2} = 0.$$
(6.72)

Multiply this equation by $\partial\theta/\partial z$ and integrate with respect to $z$ to show that

$$f(\theta) = A\left(\frac{\partial f}{\partial z}\right)^2. \qquad (6.73)$$

Now suppose that $f(\theta)$ is given by

$$f(\theta) = K\sin^2\theta, \qquad (6.74)$$

where $K$ is an anisotropy constant. Show that this implies that

$$\cos\theta(z) = -\tanh\left(\frac{\pi z}{\delta}\right), \qquad (6.75)$$

where $\delta = \pi\sqrt{A/K}$. Sketch $\theta(z)$ and argue that $\delta$ is a suitable measure of the domain wall width. Show further that

$$E = 4\sqrt{AK} \qquad (6.76)$$

in this case.

(6.13) Consider the closure domain structure in a ferromagnetic film as shown in Fig. 6.30. The easy axis is vertical and the hard axis is horizontal. The energy of the long 180° walls per unit area is $\sigma_W$ and the anisotropy energy density is $K$. Ignoring the contribution from the 90° walls

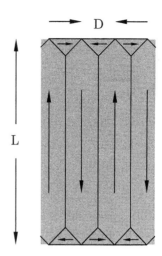

**Fig. 6.30** A closure domain.

(i.e. assume $L \gg D$), show that

$$D \approx \sqrt{\frac{2\sigma_W L}{K}}. \qquad (6.77)$$

Assuming that $\sigma_W = 2\times 10^{-3}$ Jm$^{-2}$ and that $K = 4\times 10^4$ Jm$^{-3}$ and $L = 4$ mm, estimate $D$.

# 7 Magnetism in metals

In this chapter the magnetic properties of metals are considered. In previous chapters we have concentrated on interacting but localized magnetic moments. The conduction electrons in metals are delocalized and can wander freely through the sample; they are known as *itinerant* electrons. In some cases the magnetic moments in metals are associated with the conduction electrons, in other cases the magnetic moments remain localized. In both cases paramagnetic and diamagnetic behaviour can occur. Ferromagnetism is possible under certain conditions. Most of the discussion in this chapter will be centred around the free electron model, which is introduced in the following section. The free electron model is a crude approximation to most real situations, but it is simple to consider and will allow the discussion to proceed a long way. Subsequent sections contain derivations of the magnetic properties of the electron gas which include Pauli paramagnetism, Landau diamagnetism, the origin of RKKY interactions, instabilities of the electron gas such as spin-density wave formation, and the Kondo effect which occurs when localized moments interact with the electron gas.

## 7.1 The free electron model

We begin our discussion of the magnetism of itinerant electrons by reviewing the **free electron model**. In this model, the periodic potential due to the lattice is ignored, and the electrons fill states up to the **Fermi wave vector** $k_F$. Points in $k$-space are separated by $2\pi/L$ (see Fig. 7.1(a)) where $V = L^3$ is the volume of the sample, so that the number of states between $k$ and $k + dk$ is equal to $4\pi k^2 \, dk$, the volume of a spherical shell of radius $k$ and width $dk$ (see Fig. 7.1(b)) divided by $(2\pi/L)^3$, the volume occupied by one point in $k$-space. Each state is doubly occupied, by an electron with spin-up and an electron with spin-down, so there is an additional factor of two. Hence the density of states $g(k) \, dk$ can be written as

$$g(k) \, dk = \frac{2}{(2\pi/L)^3} \times 4\pi k^2 \, dk, \tag{7.1}$$

where the factor of 2 takes care of the two spin-states of the electrons. Hence

$$g(k) \, dk = \frac{V k^2 \, dk}{\pi^2}. \tag{7.2}$$

If the material has $N$ electrons, then at absolute zero $(T = 0)$ these electrons will fill up the states up to a maximum wave vector of $k_F$. Hence

$$N = \int_0^{k_F} g(k) \, dk = \frac{V k_F^3}{3\pi^2}, \tag{7.3}$$

so that

$$k_F^3 = 3\pi^2 n \qquad (7.4)$$

where $n = N/V$ is the number of electrons per unit volume. The Fermi energy $E_F$ is defined by

$$E_F = \frac{\hbar^2 k_F^2}{2m_e}. \qquad (7.5)$$

The density of states as a function of energy $E \propto k^2$ is proportional to $E^{1/2}$, i.e.

$$g(E) = \frac{dn}{dE} \propto E^{1/2} \qquad (7.6)$$

so that

$$n = \int_0^{E_F} g(E)\, dE \propto E_F^{3/2} \qquad (7.7)$$

and hence

$$\frac{dn}{n} = \frac{3}{2} \frac{dE_F}{E_F}. \qquad (7.8)$$

The density of states at the Fermi energy is therefore given by[1]

$$g(E_F) = \left(\frac{dn}{dE}\right)_{E=E_F} = \frac{3}{2} \frac{n}{E_F}. \qquad (7.9)$$

We note in passing that another useful expression for $g(E_F)$ can be obtained by combining eqn 7.4, eqn 7.5 and eqn 7.9 to yield

$$g(E_F) = \frac{m_e k_F}{(\pi \hbar)^2}, \qquad (7.10)$$

which shows that $g(E_F) \propto m_e$. Many important properties depend on the density of states at the Fermi energy and therefore it is useful to know that it is proportional to the electron's mass. In many systems of interest, the electron's mass is enhanced above its free space value due to the effect of the band structure or interactions.

In the free electron model we ignore the periodic potential due to the lattice. However if it is included as a perturbation (the **nearly free electron model**) it turns out that it has very little effect *except* when the wave vector of the electron is close to a reciprocal lattice vector. At such points in $k$-space **energy gaps** appear in the dispersion relation (see Fig. 7.2).

So far, everything has been treated at $T = 0$. When $T > 0$, the density of states $g(E)$ is unchanged but the occupancy of each state is governed by the Fermi function $f(E)$ which is given by

$$f(E) = \frac{1}{e^{(E-\mu)/k_B T} + 1}, \qquad (7.11)$$

where $\mu$ is the chemical potential which is temperature dependent. This function is plotted in Fig. 7.3. At $T = 0$, $f(E)$ is a step function, taking the value 1 for $E < \mu$ and 0 for $E > \mu$. The step is smoothed out as the temperature $T$ increases. When the Fermi function is close to a step function, as is the usual case for most metals at pretty much all temperatures below their melting temperature, the electrons are said to be in the **degenerate limit**. The Fermi

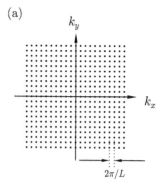

(a)

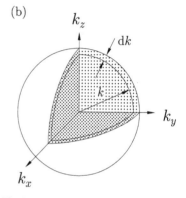

(b)

**Fig. 7.1** (a) Electron states are separated by $2\pi/L$. Each state can be doubly occupied and occupies a volume $(2\pi/L)^3$. (b) The density of states can be calculated by considering the volume in $k$-space between states with wave vector $k$ and states with wave vector $k + dk$, namely $4\pi k^2\, dk$.

[1]Equation 7.9 could also be derived using eqns 7.2, 7.4 and 7.5 directly.

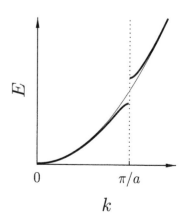

**Fig. 7.2** The energy gap at the Brillouin zone boundary.

(a)    $f(E)$

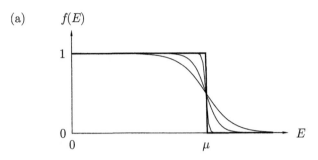

(b)    $g(E)$

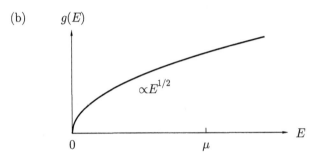

(c)    $f(E) g(E)$

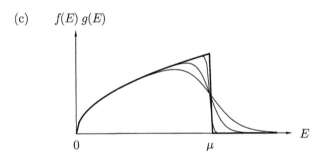

**Fig. 7.3** (a) The Fermi function $f(E)$ defined by eqn 7.11. The thick line is for $T = 0$. The step function is smoothed out as the temperature increases. The temperatures shown are $T = 0$, $T = 0.01\mu/k_B$, $T = 0.05\mu/k_B$ and $T = 0.1\mu/k_B$. (b) The density of states $g(E)$ for a free electron gas is proportional to $E^{1/2}$. (c) $f(E)g(E)$ for the same temperatures as in (a).

function is a consequence of the Pauli exclusion principle, that each electron must have a unique set of quantum numbers and no two electrons can sit in the same state. For $(E - \mu) \gg k_B T$ the Fermi function approaches the Maxwell–Boltzmann form $e^{-(E-\mu)/k_B T}$ which is known as the **non-degenerate limit**.

The Fermi energy is the energy of the highest occupied level at $T = 0$ and is determined by the equation

$$\int_0^{E_F} f(E)g(E)\,dE = n. \tag{7.12}$$

The function $f(E)g(E)$ is shown in Fig. 7.3(c). At $T = 0$ we easily find that the Fermi energy precisely equals the chemical potential: $E_F = \mu$. For $T > 0$, a tedious calculation gives

$$\mu = E_F \left[ 1 - \frac{\pi^2}{12} \left( \frac{k_B T}{E_F} \right)^2 + O\left( \left( \frac{k_B T}{E_F} \right)^4 \right) \right] \tag{7.13}$$

but this means that equating $E_F$ and $\mu$ is good to 0.01% for typical metals even at room temperature, although it is worthwhile keeping in the back of one's mind that the two quantities are not the same. The **Fermi surface** is the set of points in $k$-space whose energy is equal to the chemical potential. If the chemical potential lies in a gap, then the material is a semiconductor or an insulator and there will be no Fermi surface. Thus a metal is a material with a Fermi surface.

## 7.2 Pauli paramagnetism

Each $k$-state in a metal can be doubly occupied because of the two possible spin states of the electron. Each electron in a metal is therefore either spin-up or spin-down. When a magnetic field is applied, the energy of the electron is raised or lowered depending on its spin. This gives rise to a paramagnetic susceptibility of the electron gas and is known as Pauli paramagnetism.

### 7.2.1 Elementary derivation

Initially, we neglect the orbital contribution and take $g = 2$. We also neglect smearing of the Fermi surface due to finite temperature. As shown in Fig. 7.4, in an applied magnetic field, the electron band is spin-split into two spin subbands separated by $g\mu_B B = 2\mu_B B$. We will assume that $g\mu_B B$ is a very small energy so that the splitting of the energy bands is very small. The number of extra electrons per unit volume with spin-up is $n_\uparrow = \frac{1}{2}g(E_F)\mu_B B$. This is also the number per unit volume of the deficit of electrons with spin-down, $n_\downarrow = \frac{1}{2}g(E_F)\mu_B B$. Thus the magnetization is given by

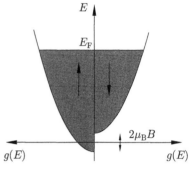

**Fig. 7.4** Density of states showing splitting of energy bands in a field $B$. The splitting is shown greatly exaggerated.

$$M = \mu_B(n_\uparrow - n_\downarrow) = g(E_F)\mu_B^2 B \qquad (7.14)$$

and the magnetic susceptibility $\chi_P$ (the subscript 'P' denoting the *Pauli* susceptibility) by

$$\chi_P = \frac{M}{H} \approx \frac{\mu_0 M}{B} = \mu_0\mu_B^2 g(E_F) \qquad (7.15)$$

$$= \frac{3n\mu_0\mu_B^2}{2E_F} \qquad (7.16)$$

where the final equality is obtained using eqn 7.9. Because $\chi_P \ll 1$ we are justified in writing $\chi_P \approx \mu_0 M/B$ (see Section 1.1.4).

Our expression for Pauli paramagnetism is temperature independent, although admittedly this is because we started out by ignoring the smearing of the Fermi surface due to finite temperature. However, if temperature is included, it makes only a rather small correction (see Exercise 7.1). Pauli paramagnetism is a weak effect, much smaller than the paramagnetism observed in insulators at most temperatures due to Curie's law. This is because in paramagnetic insulators at least one electron on every magnetic atom in the material contributes, but in a metal, it is only those electrons close to the Fermi surface which play a rôle. The small size of the paramagnetic susceptibility of most metals was something of a puzzle until Pauli pointed out that it was a consequence of the fact that electrons obeyed Fermi Dirac, rather than classical, statistics.

### 7.2.2   Crossover to localized behaviour

The effect of the Fermi Dirac statistics and the crossover between Pauli paramagnetism and localized moment behaviour can be illustrated by rederiving eqn 7.15. One should strictly write the number of electrons per unit volume of each spin state as

$$n_\uparrow = \frac{1}{2} \int_0^\infty g(E + \mu_B B) f(E) \, dE$$

$$n_\downarrow = \frac{1}{2} \int_0^\infty g(E - \mu_B B) f(E) \, dE. \tag{7.17}$$

Thus for small $B$, the magnetization is $M = \mu_B(n_\uparrow - n_\downarrow)$ and hence

$$M \approx \mu_B^2 B \int_0^\infty \frac{dg}{dE} f(E) \, dE$$

$$= \mu_B^2 B \left( \left[ g(E) f(E) \right]_0^\infty - \int_0^\infty \frac{df}{dE} g(E) \, dE \right) \tag{7.18}$$

where the second line is obtained by performing the integration of the first line by parts. Using the fact that $g(0) = 0$ and $f(\infty) = 0$ (see Fig. 7.3), the first term is zero. Hence

$$M \approx \mu_B^2 B \int_0^\infty \left( -\frac{df}{dE} \right) g(E) \, dE. \tag{7.19}$$

In the degenerate limit at $T = 0$, $-df/dE$, the differential of a step function, is a delta function at $E_F$, i.e.

$$-\frac{df}{dE} = \delta(E - E_F), \tag{7.20}$$

so that we recover $M = \mu_B^2 B g(E_F)$ and hence $\chi = \mu_0 \mu_B^2 g(E_F)$ in agreement with eqn 7.15.

In the non-degenerate limit, $f(E) \approx e^{-(E-\mu)/k_B T}$ so that

$$-\frac{df}{dE} = \frac{f}{k_B T} \tag{7.21}$$

and the magnetization is

$$M = \frac{\mu_B^2 B}{k_B T} \int_0^\infty f(E) g(E) \, dE = \frac{n \mu_B^2 B}{k_B T}, \tag{7.22}$$

so that

$$\chi = n \mu_0 \mu_B / k_B T, \tag{7.23}$$

in agreement with eqn 2.28. The susceptibility is then equivalent to $n$ localized moments (of magnitude $\mu_B$) per unit volume. As stated above, for most metals $E_F$ is a few eV and so the degenerate limit holds well at all temperatures below the melting temperature. However, for a material in which the carrier concentration is low (for example in a doped semiconductor) and $E_F \propto n^{2/3}$ is much smaller, the non-degenerate limit may be reached and a Curie-like susceptibility obtained.

### 7.2.3 Experimental techniques

The expressions that we have derived for the Pauli paramagnetism of metals agree moderately well with experiment, but can be improved by correcting for the effect of electron–electron interactions. The spin susceptibility of a metal can be extracted from NMR measurements which are much more sensitive to the field due to the spin magnetic moment of the conduction electrons than to the fields which arise from the electron's orbital motion (which give rise to the diamagnetic effects considered in Section 7.6).

The effect of the contact interaction between the conduction electron spin and the nuclear spin leads to a small shift $\Delta\omega$, known as a **Knight shift**, in the nuclear resonance frequency $\omega$. It can be understood by imagining that individual conduction electrons hop on and off a given nucleus; the net hyperfine coupling which the nucleus experiences is the result of averaging over all the electron spin orientations. This net hyperfine coupling will be zero without an applied field because the average of the electronic spin orientations will vanish; the net hyperfine coupling will be non-zero in a non-zero static field because this will polarize the electron spins. The Knight shift, $K = \Delta\omega/\omega$, is therefore proportional to the conduction electron density at the nucleus (which expresses the dependence on the coupling strength) and also to the Pauli spin susceptibility (which expresses the extent to which an applied field polarizes the electrons).

Walter D. Knight (1919–2000)

The static average of the hyperfine interactions causes the Knight shift. Fluctuations about this average provide a mechanism for $T_1$ relaxation (known as **Korringa relaxation**). The dominant $T_1$ processes are flip–flop transitions of the electron and nuclear (or muon) spins, in which the difference in electron and nuclear Zeeman energies is taken up by a change in kinetic energy of the conduction electron. The exchange in energy between the nucleus and the conduction electrons is very small, so only electrons within $k_B T$ of the Fermi surface are able to participate since only these have empty states nearby into which they can make a transition. Thus for simple metals the spin–lattice relaxation rate $T_1^{-1}$ is proportional to temperature. The Knight shift, usually expressed in the dimensionless form $\Delta\omega/\omega$, and the Korringa relaxation rate $T_1^{-1}$ are usually connected by the equation

$$\frac{1}{T_1} \propto T \left( \frac{\Delta\omega}{\omega} \right)^2 , \qquad (7.24)$$

which is known as the **Korringa relation**.

## 7.3 Spontaneously spin-split bands

The magnetic moment per atom in iron is about 2.2 $\mu_B$ (see Table 5.1). This non-integral value is not possible to understand on the basis of localized moments on atoms. It is therefore strong evidence for **band ferromagnetism** (also known as **itinerant ferromagnetism**) in which the magnetization is due to spontaneously spin-split bands. In this section we will explore some models which can be used to understand how bands in some materials can become spontaneously spin-split.

In molecular field theory we say that all spins 'feel' an identical average exchange field $\lambda M$ produced by all their neighbours. In a metal, the molecular field can magnetize the electron gas because of the Pauli paramagnetism $\chi_P$. The resulting magnetization of the electron gas $M$ would in turn be responsible for the molecular field. This is a chicken-and-egg scenario (also known as *bootstrapping*); can this positive feedback mechanism lead to spontaneous ferromagnetism? Presumably yes, if $\lambda$ (expressing how much molecular field you get for a given $M$) and $\chi_P$ (expressing how much magnetization you get for a given molecular field) are both large enough.

It is desirable to make the above heuristic argument a little more rigorous! The question that we need to ask is: can the system as a whole save energy by becoming ferromagnetic?

Let us first imagine that in the absence of an applied magnetic field we take a small number of electrons at the Fermi surface from the spin-down band and place them in the spin-up band. Specifically we take spin-down electrons with energies from $E_F - \delta E$ up to $E_F$ and flip their spins, placing them in the spin-up band where they sit with energies from $E_F$ up to $E_F + \delta E$. This situation is illustrated in Fig. 7.5. The number of electrons moved is $g(E_F)\delta E/2$ and they increase in energy by $\delta E$. The total energy change is $g(E_F)\delta E/2 \times \delta E$. The total kinetic energy change $\Delta E_{K.E.}$ is therefore

$$\Delta E_{K.E.} = \frac{1}{2} g(E_F)(\delta E)^2. \tag{7.25}$$

This is an energy cost so this process looks unfavourable. However, the interaction of the magnetization with the molecular field gives an energy reduction which can outweigh this cost. The number density of up-spins is $n_\uparrow = \frac{1}{2}(n + g(E_F)\,\delta E)$ and the number density of down-spins is $n_\downarrow = \frac{1}{2}(n - g(E_F)\,\delta E)$. Hence the magnetization is $M = \mu_B(n_\uparrow - n_\downarrow)$, assuming each electron has a magnetic moment of 1 $\mu_B$. The molecular field energy is

$$\Delta E_{P.E.} = -\int_0^M \mu_0(\lambda M')\,dM' = -\frac{1}{2}\mu_0\lambda M^2 = -\frac{1}{2}\mu_0\mu_B^2\lambda(n_\uparrow - n_\downarrow)^2. \tag{7.26}$$

Writing $U = \mu_0\mu_B^2\lambda$, where $U$ is a measure of the Coulomb energy,[2] we have

$$\Delta E_{P.E.} = -\frac{1}{2}U(g(E_F)\delta E)^2. \tag{7.27}$$

Hence the total change of energy $\Delta E$ is

$$\Delta E = \Delta E_{K.E.} + \Delta E_{P.E.} = \frac{1}{2} g(E_F)(\delta E)^2(1 - Ug(E_F)). \tag{7.28}$$

Thus spontaneous ferromagnetism is possible if $\Delta E < 0$ which implies that

$$Ug(E_F) \geq 1 \tag{7.29}$$

which is known as the **Stoner criterion**. This condition for the ferromagnetic instability requires that the Coulomb effects are strong and also that the density of states at the Fermi energy is large. If there is spontaneous ferromagnetism, the spin-up and spin-down bands will be split by an energy $\Delta$, where $\Delta$ is the exchange splitting, in the absence of an applied magnetic field.

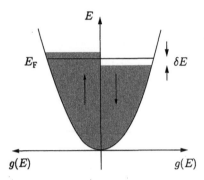

**Fig. 7.5** Density of states showing spontaneous splitting of energy bands without an applied magnetic field.

[2] The molecular field is due to exchange. Exchange is due to the Coulomb interaction. In more advanced treatments, the Coulomb energy is included directly as $Un_\uparrow n_\downarrow$ which yields the same result.

Edmund C. Stoner (1899–1968)

If the Stoner criterion is not satisfied, then spontaneous ferromagnetism will not occur. But the susceptibility may be altered. We can calculate this easily by including both the effects of an applied magnetic field and the interactions. The magnetization produced by an energy shift $\delta E$ is simply $M = \mu_B (N_\uparrow - N_\downarrow) = 2\mu_B g(E_F)\delta E$. Thus

$$\Delta E = \frac{1}{2}g(E_F)(\delta E)^2(1 - Ug(E_F)) - MB \tag{7.30}$$

$$= \frac{M^2}{2\mu_B^2 g(E_F)}(1 - Ug(E_F)) - MB \tag{7.31}$$

This is minimized when

$$\frac{M}{\mu_B^2 g(E_F)}(1 - Ug(E_F)) - B = 0 \tag{7.32}$$

so that the magnetic susceptibility $\chi$ is given by

$$\chi = \frac{M}{H} \approx \frac{\mu_0 M}{B} = \frac{\mu_0 \mu_B^2 g(E_F)}{1 - Ug(E_F)} = \frac{\chi_P}{1 - Ug(E_F)}. \tag{7.33}$$

This is larger than the value $\chi_P$ expected without the presence of Coulomb interactions by a factor $(1 - Ug(E_F))^{-1}$, a phenomenon known as **Stoner enhancement**. It is responsible for the enhanced Pauli susceptibility measured in the metals Pd and Pt which can both be thought of as systems on the verge of ferromagnetism; they have a large enough value of the parameter $Ug(E_F)$ to cause a significant enhancement of the magnetic susceptibility but not large enough (i.e. not sufficiently close to 1) to cause spontaneous ferromagnetism.

## 7.4 Spin-density functional theory

So far we have used free electron models or nearly free electron models in our discussion. It is possible to improve on this with more advanced methods, and one of these will be discussed in this section. In real systems one cannot ignore Coulomb interactions between electrons and the effect of exchange interactions on the motion of the electrons. The positions and motions of all the particles are correlated because the particles interact with each other and exert forces upon each other as they move. Thus the interactions lead to **correlations** appearing between particles. Such correlations can be very difficult to deal with theoretically, but a useful and successful approach is that of **density functional theory.**

In this theory it is recognized that the ground state energy of a many electron system can be written as a functional[3] of the electron density $n(\mathbf{r})$. The functional contains three contributions, a kinetic energy, a Coulomb energy due to the electrostatic interactions between the charged particles in the system, and a term called the **exchange-correlation energy** that captures all the many-body interactions. Rather than dealing with the wave function $\psi(\mathbf{r})$, in density functional theory one only has to consider the electron density $n(\mathbf{r}) = |\psi(\mathbf{r})|^2$, and this results in a considerable simplification. Minimizing the energy functional leads to an equation which can be used to find the ground state energy.

[3] A function is a rule which maps one number into another number. For example the function

$$f(x) = x^2$$

maps the number 2 into the number 4. A functional is a rule which maps an entire function into a number. For example, the functional

$$F[f] = \int_{-1}^{1} f(x)\,dx$$

maps the function $f(x) = x^2$ into the number 2/3.

It has been proved that minimizing the functional really does yield the exact ground state energy, even though one is using the electron density, not the wave function.

The problem is that the energy functional is not known in full! The kinetic energy and Coulomb energy can be written down but the exchange-correlation energy is unknown. However, one can use the value of the exchange-correlation energy from known results of many-electron interactions in a homogenous electron gas (an electron system of constant density). This is called the local density approximation (LDA) and it amounts to saying that at each point in space, $\mathbf{r}$, where the electron density is $n(\mathbf{r})$, an electron responds to the many-body interactions as if the electron density throughout all space (i.e. at positions $\mathbf{r}' \neq \mathbf{r}$) takes the value $n(\mathbf{r})$. The contributions to the energy from different positions in space are all added together (one integrates over all volume elements); these contributions are different for each volume element, depending on the local electron density. The LDA is exact for a perfect metal (which *does* have uniform electron density) but is not so good for systems with wildly spatially varying electron density.[4]

The extension of density functional theory to include the effects of spin polarization is called spin-density functional theory. In magnetic systems this is used together with the local spin-density approximation (often referred to as the LSDA) in which the exchange correlation potential depends not only on the local electron density but also on the local spin density (the difference between the electron density of spin-up and spin-down electrons). This technique can be used to perform realistic calculations of electronic band structure and obtain quantitative information concerning the spin density of real systems. In the next section, however, we retreat to the comparitive safety of the simple free electron model!

[4]The theory works very well for metals, but not so well for localized systems. This is because the Coulomb energy represents the response of a particular electron to the electron density, but the density is due to all electrons, including the particular electron. So electron self-interactions are unavoidably, but mistakenly, included. This is not so bad for a metal where any given electron is just one part of the vast conduction electron ocean, but disastrous in a localized system in which a particular state may be occupied by only one electron.

## 7.5 Landau levels

Pauli paramagnetism is an effect associated with the spin of the electrons. What about the orbital contribution of the electrons? This will be evaluated using the following argument. We assume that the free electron Hamiltonian is

$$\frac{\hat{\mathbf{p}}^2}{2m_e}\psi = -\frac{\hbar^2}{2m_e}\nabla^2\psi = E\psi, \tag{7.34}$$

and the wave functions are plane waves

$$\psi = \frac{1}{\sqrt{V}} e^{i\mathbf{k}\cdot\mathbf{r}}, \tag{7.35}$$

so that the energy $E$ is given by

$$E = \frac{\hbar^2 k_x^2}{2m_e} + \frac{\hbar^2 k_y^2}{2m_e} + \frac{\hbar^2 k_z^2}{2m_e}. \tag{7.36}$$

As before, with no magnetic field applied, the electron states are uniformly spaced in $k$-space, with each point separated by $2\pi/L$ (see Fig. 7.6(a)).

In the presence of a magnetic field $\mathbf{B} = (0, 0, B)$, the momentum operator $\hat{\mathbf{p}} = -i\hbar\nabla$ must be replaced by $-i\hbar\nabla + e\mathbf{A}$ where a useful choice for the

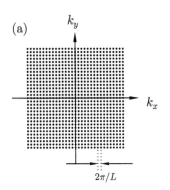

(a)

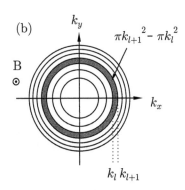

(b)

**Fig. 7.6** (a) With $B = 0$ the electron states are uniformly spaced in $k$-space, with each point separated by $2\pi/L$. (b) The application of a magnetic field results in only certain allowed energies. These Landau level states have a degeneracy which can be calculated by considering the area in $k$-space between the level with wave vector $k_{l+1}$ and the level with wave vector $k_l$.

magnetic vector potential is $\mathbf{A} = (0, Bx, 0)$. Hence eqn 7.34 becomes

$$\left(\frac{\hat{p}_x^2}{2m_e} + \frac{(\hat{p}_y + eB\hat{x})^2}{2m_e} + \frac{\hat{p}_z^2}{2m_e}\right)\psi = E\psi. \qquad (7.37)$$

Since this expression contains the operators $\hat{x}$, $\hat{p}_x$, $\hat{p}_y$ and $\hat{p}_z$, the wave function remains a plane wave in the $y$ and $z$ directions, so that $\hat{p}_y\psi = \hbar k_y\psi$ and $\hat{p}_z\psi = \hbar k_z\psi$. Thus eqn 7.37 can be rearranged (using separation of variables) to give

$$\left(\left[\frac{\hat{p}_x^2}{2m_e} + \frac{1}{2}m_e\omega_c^2(x - x_0)^2\right] + \frac{\hbar^2 k_z^2}{2m_e}\right)\psi = E\psi, \qquad (7.38)$$

where $x_0 = -\hbar k_y/eB$ and

$$\omega_c = \frac{eB}{m_e} \qquad (7.39)$$

is the **cyclotron frequency**. The term in square brackets in eqn 7.38 is the Hamiltonian for a one-dimensional harmonic oscillator. Hence the energy eigenvalues correspond to

$$E = (l + \tfrac{1}{2})\hbar\omega_c + \frac{\hbar^2 k_z^2}{2m_e} \qquad (7.40)$$

where $l$ is an integer. The energy eigenfunctions are a product of plane waves in the $y$ and $z$ directions and a one-dimensional harmonic oscillator wave function in the $x$ direction.

In the treatment presented here, the quantity $x_0$ is the 'centre' of the harmonic oscillator motion in the $x$ direction. This takes different values depending on the value of $k_y$.[5] A further useful way of attacking this problem is to choose a gauge which reflects the cylindrical symmetry inherent in this situation (this is done in Exercise 7.2) and this results in identical energy eigenvalues to those in eqn 7.40 but the eigenfunctions are a product of a plane wave in the $z$ direction and a function of $\sqrt{k_x^2 + k_y^2}$. In whichever gauge the problem is solved, the allowed states can be thought of as occurring on **Landau tubes** which lie parallel to $k_z$ (see Fig. 7.7). The application of a magnetic field causes the electrons to form **Landau levels** (as they are viewed when thinking of the states as a function of energy) and each Landau level can be labelled by the values of the quantum number $l$ (see Fig. 7.6(b)). The degeneracy of a Landau level can be calculated by noting that the area between one Landau level and the next in $k$-space (see Fig. 7.6(b)) is given by

$$\pi k_{l+1}^2 - \pi k_l^2 = \frac{2m_e\pi}{\hbar^2}\left((l + 1 + \tfrac{1}{2})\hbar\omega_c - (l + \tfrac{1}{2})\hbar\omega_c\right) = \frac{2m_e\pi\omega_c}{\hbar} \qquad (7.41)$$

so that since two electrons occupy each state, and one state occupies an area of $(2\pi/L)^2$ in the $k_x k_y$ plane, the degeneracy $p$ of one Landau level is

$$p = \frac{4m_e\pi\omega_c}{\hbar(2\pi/L)^2} = \frac{m_e L^2\omega_c}{\pi\hbar} \qquad (7.42)$$

Notice that the same energy eigenvalue would have been obtained if the magnetic vector potential was chosen to have been $\mathbf{A} = (-By, 0, 0)$ (which gives the same value for $\mathbf{B} = \nabla \times \mathbf{A} = (0, 0, B)$) but the energy eigenfunctions would have been in this case a product of plane waves in the $x$ and $z$ directions and a one-dimensional harmonic oscillator wave function in the $y$ direction. The 'choice of gauge' (i.e. the way you write down $\mathbf{A}$) doesn't affect the energy eigenvalues but does affect the set of wave functions that you derive.

[5]If you make the choice $\mathbf{A} = (-By, 0, 0)$ you obtain a parameter $y_0$ which is proportional to $k_x$ which is the centre of the harmonic oscillator motion in the $y$ direction. It turns out that the operators corresponding to $x_0$ and $y_0$ do not commute, so that $x_0$ and $y_0$ cannot take definite values simultaneously.

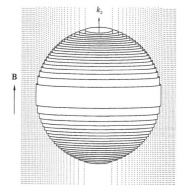

**Fig. 7.7** Landau tubes

## 7.6   Landau diamagnetism

Using the results from the previous section, we are now in a position to calculate the diamagnetic response of the electron gas. The idea is that as a magnetic field is applied, and the electron distribution breaks up into a series of Landau levels, the total energy of the system may change. This change of energy with field is equivalent to a magnetization of the system. The phenomenon is known as Landau diamagnetism.

The calculation is based on the observation that for different values of $k_z$ the highest occupied Landau level is different (see Fig. 7.7 which shows that the Fermi surface cuts different Landau tubes for different values of $k_z$). For a particular value of $k_z$, we define the energy $E_\perp$ associated with the electronic motion perpendicular to the field by the equation

$$E_\perp = E_F - \frac{\hbar^2 k_z^2}{2m_e}. \tag{7.43}$$

In the absence of a magnetic field electrons can fill up the states to a maximum value of $E_\perp$ for a particular choice of $k_z$, but the application of a magnetic field allows discrete Landau levels. We give the highest occupied Landau level the label $E_l$. The electrons in the $B = 0$ problem can be broken up into blocks whose average energy is the corresponding Landau level energy (except for the top block which is partially filled).

Let us define a parameter $x$ by

$$x = E_\perp - E_l \tag{7.44}$$

(see Fig. 7.8). The mean energy in the highest occupied block is

$$\frac{1}{2}\left[E_\perp + E_l - \frac{\hbar\omega_c}{2}\right] = E_\perp - \frac{x}{2} - \frac{\hbar\omega_c}{4} \tag{7.45}$$

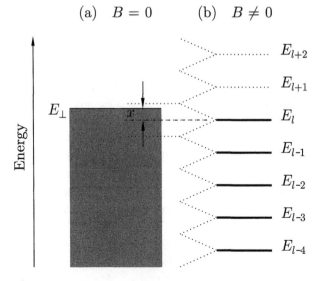

(a)   $B = 0$        (b)   $B \neq 0$

**Fig. 7.8** (a) In the absence of a magnetic field electrons can fill up the states to a maximum value of $E_\perp$ which is defined by eqn 7.43. The filled states at $T = 0$ are shown shaded. The 'highest occupied block' described in the text is labelled $E_l$. The parameter $x = E_\perp - E_l$ measures the energy between the middle of this block (dot-dash line) and the highest occupied state. (b) The magnetic field allows only discrete Landau levels with energy $\ldots E_{l-2}, E_{l-1}, E_l, E_{l+1} \ldots$. The occupied Landau levels at $T = 0$ are shown as heavy lines and the highest occupied Landau level is labelled as $E_l$.

and its occupancy is

$$\frac{p}{\hbar\omega_c}\left[E_\perp - (E_l - \frac{\hbar\omega_c}{2})\right] = \frac{p}{2}\left[1 + \frac{2x}{\hbar\omega_c}\right]. \tag{7.46}$$

Multiplying these together gives the energy of the top block. This has been evaluated for a particular choice of $k_z$ but it is necessary to average over all the electrons on the Fermi surface, and hence over all different values of $k_z$. This can be achieved by averaging over all possible values of $x$ assuming that $x$ is uniformly distributed between $-\hbar\omega_c/2$ and $\hbar\omega_c/2$. Hence

$$\langle x \rangle = \frac{1}{\hbar\omega_c}\int_{-\hbar\omega_c/2}^{\hbar\omega_c/2} x\, dx = 0, \tag{7.47}$$

and

$$\langle x^2 \rangle = \frac{1}{\hbar\omega_c}\int_{-\hbar\omega_c/2}^{\hbar\omega_c/2} x^2\, dx = \frac{\hbar\omega_c^2}{12}. \tag{7.48}$$

This leads to an average energy of the top block (the one labelled $E_l$) associated with motion perpendicular to the field equal to

$$\frac{p}{2}\left[E_\perp - \frac{\hbar\omega_c}{3}\right] \tag{7.49}$$

so that the total average energy is (using eqn 7.43)

$$\frac{p}{2}\left[E_F - \frac{\hbar\omega_c}{3}\right]. \tag{7.50}$$

With a non-zero magnetic field, the top block is full of $p$ electrons with energy $E_l$ if $0 < x < \hbar\omega_c/2$ and the block is totally empty if $-\hbar\omega_c/2 < x < 0$. (Note that this calculation is for a particular value of $k_z$ so that states are redistributed amongst those for different values of $k_z$.) Thus the average energy of the top block is

$$E = \begin{cases} p(E_\perp - x) & \text{if } x > 0 \\ 0 & \text{if } x < 0 \end{cases} \tag{7.51}$$

so that the mean energy associated with motion perpendicular to the field is

$$\frac{p}{2}\left[E_\perp - \frac{\hbar\omega_c}{4}\right] \tag{7.52}$$

and hence the total average energy (again using eqn 7.43) is

$$\frac{p}{2}\left[E_F - \frac{\hbar\omega_c}{4}\right]. \tag{7.53}$$

Including a factor $k_F L/\pi$ for the degeneracy of the states in the $k_z$ direction, the difference in total energy for $B \neq 0$ and $B = 0$ is

$$\Delta U = U(B \neq 0) - U(B = 0) \tag{7.54}$$

$$= \frac{p k_F L}{2\pi}\left[E_F - \frac{\hbar\omega_c}{4}\right] - \frac{p k_F L}{2\pi}\left[E_F - \frac{\hbar\omega_c}{3}\right] \tag{7.55}$$

$$= \frac{p k_F L \hbar\omega_c}{24\pi}. \tag{7.56}$$

Using the expression for the degeneracy $p$ (eqn 7.41) we have that

$$\Delta U = \frac{V k_F e^2 B^2}{24\pi^2 m}. \tag{7.57}$$

Hence the magnetic susceptibility is given by

$$\chi_L = -\frac{\mu_0}{V}\frac{\partial^2 \Delta U}{\partial B^2} \tag{7.58}$$

$$= -\frac{\mu_0 k_F e^2}{12\pi^2 m_e} \tag{7.59}$$

$$= -\frac{1}{3}\mu_0\left(\frac{e\hbar^2}{2m_e}\right)^2\left[\frac{m_e k_F}{\pi^2 \hbar^2}\right] \tag{7.60}$$

$$= -\frac{1}{3}\mu_0\mu_B^2 g(E_F). \tag{7.61}$$

This is the Landau diamagnetism and it is related to the Pauli paramagnetism $\chi_P$ by

$$\chi_L = -\frac{\chi_P}{3}. \tag{7.62}$$

We might therefore conclude that all metals will be paramagnetic since the (positive) Pauli paramagnetism is three times larger than the (negative) Landau diamagnetism. This would be rather rash, because we have assumed a free electron model and have therefore ignored band structure effects. If the electron has an effective mass $m^*$, then the density of states at the Fermi energy $g(E_F)$ will be enhanced by a factor $m^*/m_e$. This therefore enhances both the Pauli paramagnetism and the Landau diamagnetism by $m^*/m_e$. However, in eqn 7.61 we have used $\mu_B = e\hbar/2m_e$ to obtain the final result. A Bohr magneton is a fixed constant, independent of effective mass. This implies that the Landau diamagnetism must be further multiplied by $(m_e/m^*)^2$. The net result is that

$$\chi_L = -\left(\frac{m_e}{m^*}\right)^2 \frac{\chi_P}{3}, \tag{7.63}$$

and the total susceptibility of the metal will be

$$\chi = \chi_P\left[1 - \frac{1}{3}\left(\frac{m_e}{m^*}\right)^2\right]. \tag{7.64}$$

For most metals $m^* \sim m_e$ and they are therefore paramagnetic. If $m^* < m_e/\sqrt{3}$, the net susceptibility becomes negative. This explains the strong diamagnetism of bismuth, where $m^* \sim 0.01 m_e$.

A further effect which we have so far ignored is the diamagnetism of the bound electrons in the ion cores. This effect gets larger as the atomic number increases (because there are more electrons to contribute) but it is still usually much smaller than the effects associated with the conduction electrons which we have been discussing.

The electron motion on the Fermi surface can be described by the equation

$$\hbar\dot{\mathbf{k}} = -e\mathbf{v} \times \mathbf{B}, \tag{7.65}$$

where $\mathbf{k}$ is the electron wave vector on the Fermi surface, and $\mathbf{v} = \hbar^{-1}\partial E(\mathbf{K})/\partial\mathbf{k}$ is the electron velocity, and is always perpendicular to the

Fermi surface. Hence the electron motion is perpendicular to **B** but stays on the Fermi surface.[6] Diamagnetic orbital contributions to the magnetic susceptibility arise from closed orbits on the Fermi surface. As shown in Fig. 7.9, open orbits are also possible. In the simple free electron sphere, only closed orbits are possible. Both open and closed orbits are possible in real materials.

Interesting oscillatory changes in properties with magnetic field occur when the Landau levels break through the Fermi surface as the magnetic field is increased. Properties change because the density of states at the Fermi energy can oscillate as a function of magnetic field. This effect is maximized when a Landau tube crosses an extremal cross-section of the Fermi surface (for the Fermi sphere, this would be an equatorial cross-section). The oscillation of the density of states leads to an oscillatory magnetization and this is the de Haas–van Alphen effect and can be used to measure the cross-sectional area of the Fermi surface. There is an analogous effect on the electrical resistance known as the Shubnikov–de Haas effect. Both effects give rise to oscillations in properties as a function of $1/B$. The period of these oscillations gives the area $A_{\text{ext}}$ of the maximal and minimal cross-sectional area of the Fermi surface normal to the magnetic field according to

$$\Delta\left(\frac{1}{B}\right) = \frac{2\pi e}{\hbar A_{\text{ext}}}. \tag{7.66}$$

## 7.7 Magnetism of the electron gas

The Pauli paramagnetism effect describes the paramagnetic response of an electron gas to an applied magnetic field. But what if the applied magnetic field is spatially varying? The electron gas will respond to this perturbation according to the wave vector-dependent susceptibility, which we now calculate. (The diamagnetic response will be ignored for the moment.)

### 7.7.1 Paramagnetic response of the electron gas

A spatially varying magnetic field

$$\mathbf{H}(\mathbf{r}) = \mathbf{H_q} \cos \mathbf{q} \cdot \mathbf{r} \tag{7.67}$$

provides a perturbation $\hat{\mathcal{H}}'$ equal to

$$\hat{\mathcal{H}}' = \pm \frac{g\mu_0\mu_B}{2} |\mathbf{H_q}| \cos \mathbf{q} \cdot \mathbf{r} \tag{7.68}$$

where the $\pm$ refers to the electron spin. A plane wave state

$$\psi_{\mathbf{k}\pm}(\mathbf{r}) = \frac{1}{\sqrt{V}} e^{i\mathbf{k}\cdot\mathbf{r}} |\pm\rangle \tag{7.69}$$

is weakly perturbed into states $\psi_{(\mathbf{k}+\mathbf{q})\pm}(\mathbf{r})$ and $\psi_{(\mathbf{k}-\mathbf{q})\pm}(\mathbf{r})$. The amplitude of these states can be calculated using first-order perturbation theory (see Appendix C). The perturbed wave function is therefore

$$\psi_{\mathbf{k}\pm}(\mathbf{r}) = \frac{1}{\sqrt{V}} \left( e^{i\mathbf{k}\cdot\mathbf{r}} \pm \frac{g\mu_0\mu_B H_q}{4} \left[ \frac{e^{i(\mathbf{k}+\mathbf{q})\cdot\mathbf{r}}}{E_{\mathbf{k}+\mathbf{q}} - E_{\mathbf{k}}} + \frac{e^{i(\mathbf{k}-\mathbf{q})\cdot\mathbf{r}}}{E_{\mathbf{k}-\mathbf{q}} - E_{\mathbf{k}}} \right] \right) |\pm\rangle \tag{7.70}$$

[6]This is because in a time d$t$, the change in **k** is d**k** $\propto$ $(\partial E(\mathbf{K})/\partial \mathbf{k}) \times \mathbf{B}$ and hence is both normal to **B** and in the plane of the Fermi surface.

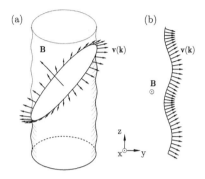

**Fig. 7.9** Examples of (a) a closed orbit and (b) an open orbit. The vectors on the orbits show the real-space velocity of the electron, given by $\mathbf{v} = \hbar^{-1}\partial E(\mathbf{K})/\partial\mathbf{k}$, which is a vector normal to the Fermi surface. With the magnetic field **B** oriented as in (a), a closed orbit results. With **B** along the $x$ direction, an open orbit on the same Fermi surface can result, as shown in (b).

P. M. van Alphen (1906–1967)

W. J. de Haas (1878–1960)

L. V. Shubnikov (1901–1945)

where $E_k = \hbar^2 k^2 / 2m_e$. Keeping only leading orders of $H_q$ this implies that

$$|\psi_{k\pm}(\mathbf{r})|^2 = \frac{1}{V}\left(1 \pm \frac{g\mu_0\mu_B H_q m_e}{\hbar^2}\right.$$
$$\left. \times \left[\frac{1}{(\mathbf{k}+\mathbf{q})^2 - k^2} + \frac{1}{(\mathbf{k}-\mathbf{q})^2 - k^2}\right]\cos\mathbf{q}\cdot\mathbf{r}\right). \quad (7.71)$$

The magnetization $M(\mathbf{r})$ due to this perturbed wave function is then

$$M(\mathbf{r}) = \frac{g\mu_0\mu_B}{2}(|\psi_{k+}(\mathbf{r})|^2 - |\psi_{k-}(\mathbf{r})|^2)$$
$$= \frac{g^2\mu_0\mu_B^2 m_e H_q \cos\mathbf{q}\cdot\mathbf{r}}{\hbar^2 V}\left[\frac{1}{(\mathbf{k}+\mathbf{q})^2 - k^2} + \frac{1}{(\mathbf{k}-\mathbf{q})^2 - k^2}\right]$$
$$(7.72)$$

To calculate the magnetic response of the electron gas as a whole, it is necessary to sum the expression in eqn 7.72 over all the electrons in the Fermi sphere. Hence the spatially varying magnetization of the electron gas $M(\mathbf{r})$ becomes

$$M(\mathbf{r}) = M_q \cos\mathbf{q}\cdot\mathbf{r} \quad (7.73)$$

where $M_q$ is given by

$$M_q = \frac{g^2\mu_0\mu_B^2 m_e H_q}{\hbar^2 V}\int_{|k|<k_F} g(\mathbf{k})d^3k\left[\frac{1}{(\mathbf{k}+\mathbf{q})^2 - k^2} + \frac{1}{(\mathbf{k}-\mathbf{q})^2 - k^2}\right]$$
$$= \frac{k_F m_e g^2\mu_0\mu_B^2 H_q}{\pi^2\hbar^2}\left[1 + \frac{4k_F^2 - q^2}{4k_F q}\log\left|\frac{q + 2k_F}{q - 2k_F}\right|\right] \quad (7.74)$$

where the density of states $g(|k|) = Vk^2/\pi^2$ and use has been made of the integral

$$\int_{|k|<k_F}\frac{d^3k}{(\mathbf{k}+\mathbf{q})^2 - k^2} = \int_0^{k_F} 2\pi k^2\, dk \int_0^\pi \frac{\sin\theta\, d\theta}{q(q - 2k\cos\theta)}$$
$$= \int_0^{k_F}\frac{\pi k\, dk}{q}\log\frac{q + 2k}{|q - 2k|} \quad (7.75)$$
$$= \frac{\pi k_F}{2}\left[1 + \frac{4k_F^2 - q^2}{4k_F q}\log\left|\frac{q + 2k_F}{q - 2k_F}\right|\right]. \quad (7.76)$$

Using the equation

$$\chi_q = M_q/H_q \quad (7.77)$$

which we shall justify below, and the expression for $g(E_F)$ in eqn 7.10, the wave vector-dependent susceptibility $\chi_q$ is then given by

$$\chi_q = \chi_P\, f\left(\frac{q}{2k_F}\right) \quad (7.78)$$

where the function $f(x)$ is given by

$$f(x) = \frac{1}{2}\left(1 + \frac{1 - x^2}{2x}\log\left|\frac{x + 1}{x - 1}\right|\right). \quad (7.79)$$

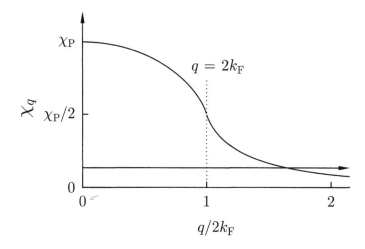

$\chi_P$

$q = 2k_F$

$\chi_q$  $\chi_P/2$

0

0  1  2

$q/2k_F$

**Fig. 7.10** The $q$-dependent paramagnetic susceptibility. At $q = 0$ it takes the value of the Pauli paramagnetic susceptibility.

$\chi_q$ is plotted in Fig. 7.10. At $q = 0$, $f(0) = 1$ so that $\chi_{q=0} = \chi_P$ and so this result happily reduces to the expression for Pauli paramagnetism derived in Section 7.2. The kink in the curve which appears at $q = 2k_F$ is due to the existence of the Fermi surface (the diameter of the Fermi sphere is $2k_F$). This kink becomes progressively more severe as the dimensionality is reduced down from three dimensions, as will be discussed later in the chapter.

In general the $q$-dependent magnetic field $\mathbf{H_q}$ can be defined by a Fourier transform of $\mathbf{H(r)}$ by

$$\mathbf{H_q} = \int d^3\mathbf{r}\, \mathbf{H(r)}e^{-i\mathbf{q}\cdot\mathbf{r}}, \qquad \mathbf{H(r)} = \frac{1}{(2\pi)^3}\int d^3\mathbf{q}\, \mathbf{H_q}e^{i\mathbf{q}\cdot\mathbf{r}}. \qquad (7.80)$$

In the same way the $q$-dependent magnetization $\mathbf{M_q}$ is defined by

$$\mathbf{M_q} = \int d^3\mathbf{r}\, \mathbf{M(r)}e^{-i\mathbf{q}\cdot\mathbf{r}}, \qquad \mathbf{M(r)} = \frac{1}{(2\pi)^3}\int d^3\mathbf{q}\, \mathbf{M_q}e^{i\mathbf{q}\cdot\mathbf{r}}, \qquad (7.81)$$

and the $q$-dependent susceptibility $\chi_q$ is similarly

$$\chi_q = \int d^3\mathbf{r}\, \chi(\mathbf{r})e^{-i\mathbf{q}\cdot\mathbf{r}}, \qquad \chi(\mathbf{r}) = \frac{1}{(2\pi)^3}\int d^3\mathbf{q}\, \chi_q e^{i\mathbf{q}\cdot\mathbf{r}}, \qquad (7.82)$$

where for each of eqns 7.80, 7.81 and 7.82 the inverse transformation is also shown. The $q$-dependent susceptibility provides a means of calculating the magnetization of the electron gas. The magnetization $\mathbf{M(r)}$ at any point $\mathbf{r}$ in the electron gas is related to the magnetic field $\mathbf{H(r)}$ by

$$\mathbf{M(r)} = \int d^3\mathbf{r'}\, \chi(\mathbf{r} - \mathbf{r'})\mathbf{H(r')} \qquad (7.83)$$

which is a convolution. Thus $\mathbf{M(r)}$ responds not just to the field at $\mathbf{r}$, but to a weighted average over nearby values. The convolution theorem implies that the relationship between the corresponding $q$-dependent quantities is

$$\mathbf{M_q} = \chi_q \mathbf{H_q}, \qquad (7.84)$$

which is the same as eqn 7.77 which we used above.

### 7.7.2 Diamagnetic response of the electron gas

The $q$-dependent diamagnetic susceptibility can also be calculated, although this is considerably more complicated. Only the result is quoted here, and it is

$$\chi_{\mathbf{q}} = \chi_{\mathrm{L}} \frac{3k_{\mathrm{F}}^2}{2q^2} \left[ 1 + \frac{q^2}{4k_{\mathrm{F}}^2} - \frac{k_{\mathrm{F}}}{q} \left( 1 - \frac{q^2}{4k_{\mathrm{F}}^2} \right)^2 \log \left| \frac{q + 2k_{\mathrm{F}}}{q - 2k_{\mathrm{F}}} \right| \right]. \tag{7.85}$$

The diamagnetic susceptibility $\chi_{\mathbf{q}}$ is plotted in Fig. 7.11. At $q = 0$, the expression reduces to $\chi_{\mathbf{q}=0} = \chi_{\mathrm{L}} = -\chi_{\mathrm{P}}/3$. Expression 7.85 also has a singularity at $q = 2k_{\mathrm{F}}$ because of the existence of the Fermi surface.

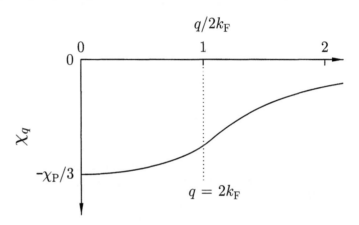

**Fig. 7.11** The $q$-dependent diamagnetic susceptibility. At $q = 0$ it takes the value of the Landau diamagnetic susceptibility.

### 7.7.3 The RKKY interaction

The $q$-dependent paramagnetic susceptibility has consequences for the real space behaviour of a system with a delta-function perturbation. A delta-function perturbation $\mathbf{H}(\mathbf{r}) = \delta(\mathbf{r})\mathbf{H}$ can be decomposed into a sum of all possible spatial frequencies

$$\mathbf{H}(\mathbf{r}) = \delta(\mathbf{r})\mathbf{H} = \frac{1}{(2\pi)^3} \int \mathbf{H}_{\mathbf{q}} e^{i\mathbf{q}\cdot\mathbf{r}} \mathrm{d}^3 q \tag{7.86}$$

where $\mathbf{H}_{\mathbf{q}} = \mathbf{H}$. Thus a delta function can be considered as a sum over all frequencies, comprising both very long wavelength and short wavelength oscillations. If $\chi_{\mathbf{q}} = \chi$ is independent of $q$, then $\mathbf{M}_{\mathbf{q}} = \chi \mathbf{H}_{\mathbf{q}}$ and

$$\mathbf{M} = \frac{1}{(2\pi)^3} \int \mathbf{M}_{\mathbf{q}} e^{i\mathbf{q}\cdot\mathbf{r}} \mathrm{d}^3 q = \chi \delta(\mathbf{r})\mathbf{H}, \tag{7.87}$$

in other words the magnetization is also a delta function and the electron gas responds completely to the perturbation, shadowing its behaviour perfectly. (In this case $\chi(\mathbf{r} - \mathbf{r}') = \chi \delta(\mathbf{r} - \mathbf{r}')$.)

However the susceptibility of the electron gas is a $q$-dependent. Because of the presence of the Fermi surface, the electron gas therefore becomes much less responsive for wave vectors above $q = 2k_{\mathrm{F}}$. Thus although it can follow the long wavelength components of the delta-function perturbation, the short wavelengths (that is wavelengths below about $\pi/k_{\mathrm{F}}$) get attenuated because the electron gas is less able to respond to fields with such large spatial frequencies. This produces a spatial 'ringing' effect in the magnetization which is analogous to the oscillations observed in the diffraction pattern from a slit. The resulting magnetization has spatial oscillations.

This effect can be calculated as follows. The real-space susceptibility is given by

$$\chi(\mathbf{r}) = \frac{1}{(2\pi)^3} \int d^3\mathbf{q}\, \chi_{\mathbf{q}} e^{i\mathbf{q}\cdot\mathbf{r}}$$

$$= \frac{1}{(2\pi)^3} \int d^3\mathbf{q}\, \frac{\chi_P}{2} \left( 1 + \frac{4k_F^2 - q^2}{4k_F q} \log\left| \frac{q + 2k_F}{q - 2k_F} \right| \right) e^{i\mathbf{q}\cdot\mathbf{r}}$$

$$= \frac{2k_F^3 \chi_P}{\pi} F(2k_F r) \tag{7.88}$$

where $r = |\mathbf{r}|$ and the function $F(x)$ is given by

$$F(x) = \frac{-x\cos x + \sin x}{x^4} \tag{7.89}$$

and is plotted in Fig. 7.12. This is known as the RKKY interaction (which was presented earlier in Section 4.2.4). The integral in eqn 7.88 is far from trivial and is examined in more detail in Exercise 7.5. The susceptibility (and hence the magnetization due to a delta function perturbation) is proportional to $\cos(2k_F r)/r^3$ at large distances ($r \gg k_F^{-1}$) and is oscillatory. At small $r$ the susceptibility diverges, which is a consequence of the delta function perturbation which we have been assuming. The RKKY interaction was first proposed in order to understand the effect of a localized nuclear moment on the electron gas; the nuclear moment is pretty much as close as one can get to a delta function perturbation, but is not infinitely localized, but spread over the (albeit tiny) nuclear volume. The RKKY interaction provides an important mechanism for magnetic coupling between localized electronic moments in metals; one moment produces an oscillatory magnetization of the electron gas which can interact with a second moment. The interaction can be ferromagnetic or antiferromagnetic depending on the distance between the two moments.

A similar phenomenon is observed in the non-magnetic case, i.e. the response of the electron gas to a delta-function charge. This is of interest in the study of localized charges in metallic alloys. Oscillations in the charge density are produced by a similar mechanism and are known as Friedel oscillations.

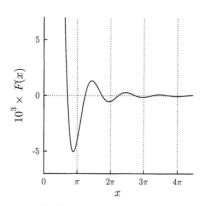

**Fig. 7.12** The function $F(x)$ of eqn 7.89 which describes the real space magnetization produced by a delta function field.

# 7.8   Excitations in the electron gas

The electron gas also contains a rich spectrum of excitations. These can be treated elegantly by calculating $\chi(\mathbf{q}, \omega)$, the $q$- and $\omega$-dependent susceptibility, but we shall not proceed with this approach in this book.

In an itinerant ferromagnet in which the bands are spin-split in the absence of a magnetic field, there are two types of excitations. First there are the familiar spin-wave excitations which follow a conventional dispersion relation. Our treatment of spin-waves in Section 6.6 was based on the Heisenberg model and is therefore applicable to localized moment systems, but it turns out that spin waves can also be derived for metallic systems. Second, there is also a continuum of electron-hole excitations in which an electron transfers from a filled state in one of the spin-split bands to an empty state in the other spin-split band. These latter excitations are known as **Stoner excitations**. In a Stoner excitation an electron with wave vector $\mathbf{k}+\mathbf{q}$ and spin down is excited to a state with wave vector $\mathbf{k}$ and spin up. The energy of the excitation is given by

$$\hbar\omega = E_{\mathbf{k}+\mathbf{q}} - E_{\mathbf{k}} + \Delta, \qquad (7.90)$$

where $E_{\mathbf{k}} = \hbar^2 k^2/2m_e$ and $\Delta$ is the exchange splitting, the energy cost to flip a spin. These results are illustrated in Fig. 7.13.

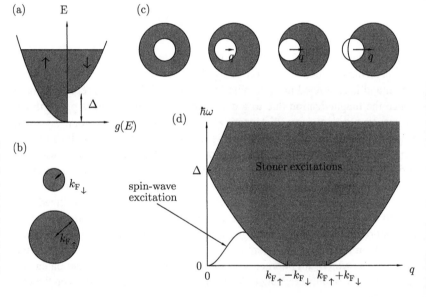

**Fig. 7.13** Excitations in an electron gas. (a) The spin-split bands are separated by $\Delta$, the exchange splitting. (b) The Fermi surfaces for spin-up and spin-down electrons. (c) A Stoner excitation can be created if there is a filled $\uparrow$ state with wave vector $\mathbf{k}$ and an $\downarrow$ empty state with wave vector $\mathbf{k} + \mathbf{q}$. The shaded area shows the possible choices of $\mathbf{k}$ for different values of $\mathbf{q}$. (d) The dispersion relation (a graph of $\omega$ against $q$) shows a spin-wave branch and a continuum of Stoner excitations.

In paramagnetic metals the two spin-split bands are not spin-split in the absence of a magnetic field. The spin-wave excitations have a short lifetime and thus are very heavily damped (they are in the overdamped limit, if considered as damped harmonic oscillators). They are known as **paramagnons** and, like conventional spin-wave excitations, they can be studied by inelastic neutron scattering.

In this book we have been working with two entirely distinct pictures: insulating materials in which the magnetism is associated with localized moments on atoms and metals in which the magnetism is associated with entirely delocalized moments. Many real materials are somewhere in between, with magnetism associated with spin density fluctuations intermediate between the localized moment and band ferromagnet regimes. A local moment fluctuation is localized in real space while a spin fluctuation in a weakly ferromagnetic metal may be regarded as being localized in $q$-space.

## 7.9   Spin-density waves

The Stoner criterion demonstrates that the electron gas is unstable against spontaneous ferromagnetism if $Ug(E_F)$ becomes large enough (see Section 7.3) and that the magnetic susceptibility can be enhanced by a factor $(1 - Ug(E_F)^{-1})$. It turns out that the $q$-dependent susceptibility, which we will write as $\chi_{\mathbf{q}}^0 = \chi_P f(q/2k_F)$ without the presence of Coulomb interactions, is also enhanced by the Coulomb interactions and becomes

$$\chi_{\mathbf{q}} = \frac{\chi_P f(q/2k_F)}{1 - Ug(E_F) f(q/2k_F)} = \frac{\chi_{\mathbf{q}}^0}{1 - \alpha\chi_{\mathbf{q}}^0} \tag{7.91}$$

where $\alpha = U/\mu_0\mu_B^2$. It may occur that $\chi_{\mathbf{q}}^0$ has a maximum at some value of $\mathbf{q}$ not equal to zero. In this case the interactions, parametrized by $\alpha$, can make the susceptibility diverge at this value of $\mathbf{q}$ if $\alpha\chi_{\mathbf{q}}^0$ reaches unity. Hence an oscillatory static magnetization could spontaneously develop in the sample. If $\mathbf{q} = 0$ this would correspond to ferromagnetic order; if $|\mathbf{q}| = \pi/a$ then antiferromagnetic order could develop. In general, one would expect spiral structures or **spin-density wave** structures with wave vector $\mathbf{q}$.

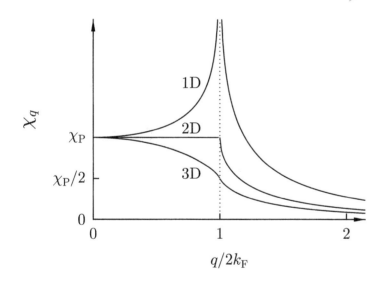

**Fig. 7.14** The $q$-dependent paramagnetic susceptibility in one, two and three dimensions. The response function diverges at $q = 2k_F$ in one-dimension. At $q = 0$ it takes the value of the Pauli paramagnetic susceptibility in all dimensions.

We have evaluated the $q$-dependent susceptibility for an electron gas in three dimensions, but the calculation can be repeated (see Exercise 7.6) for two dimensions and one dimension. The results are shown in Fig. 7.14 and demonstrate that as you lower the dimension, the knee in the curve for three dimensions at $q = 2k_F$ becomes a kink in two dimensions and a singularity in one-dimension. The peak in one-dimension is known as a **Kohn anomaly** and shows that the one-dimensional electron gas is unstable to the formation of spin-density waves with wave vector $2k_F$. This instability is associated with the fact that a periodic modulation of the magnetization with wavelength $2\pi/q$ and wave vector $q$ opens up a gap in the energy dispersion at wave vectors

Walter Kohn (1923–)

Rudolf E. Peierls (1907–1995)

±$q/2$ (any periodic potential can do this). If $q/2 = k_F$ the gap is at the Fermi surface and this can lead to a lowering of the total energy. This opportunity to lower the electronic energy can therefore itself drive the formation of the spin-density wave.

There is an analogous instability for the formation of charge-density waves with wave vector $2k_F$. Both effects are connected with the well-known **Peierls instability** in which the dimerization of a one-dimensional chain of atoms can occur spontaneously because the electronic energy saved by opening up a gap in the energy spectrum near the Fermi surface can outweigh the elastic energy cost of the dimerization. This is illustrated in Fig. 7.15.

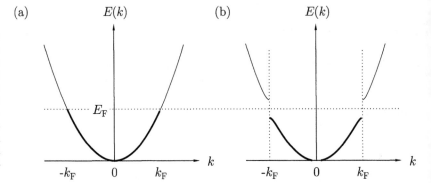

**Fig. 7.15** (a) Electrons in a band fill up to the Fermi energy $E_F$. (b) The energy can be lowered by a periodic potential of wave vector $q = 2k_F$ which lowers the total electronic energy by producing a gap at the Fermi surface. The gapping of the Fermi surface (i.e. going from (a) to (b)) is accompanied by a metal–insulator transition. The corresponding real space distortion is known as a **Peierls distortion**. A one-dimensional electron gas is unstable with respect to this Peierls instability.

In three dimensions a spin-density wave cannot produce an energy gap at all points on the Fermi surface. Sometimes the Fermi surface is such that translation of one part of the Fermi surface by a vector **q** can place it on top of another part of the Fermi surface. This phenomenon is known as **nesting**. A spin-density wave, with wave vector **q**, can produce energy gaps along the region of the Fermi surface for which this nesting is possible. In real metals it very often happens that two pieces of Fermi surface are approximately translated from one another in $k$-space by a fixed wave vector **q**. This can give rise to a peak in the susceptibility and a resulting instability. The formation of the density wave is said to **nest** the Fermi surface.

If the nesting wave vector **q** turns out to be $\pi/a$ where $a$ is the spacing between atoms, the spin-density wave is commensurate with the lattice and antiferromagnetic order results (Fig. 7.16(a)). However it is much more usual

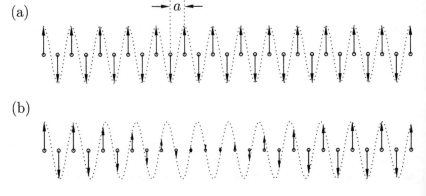

**Fig. 7.16** (a) Commensurate spin-density wave, with wave vector $q = \pi/a$. (b) Incommensurate spin-density wave.

for the nesting wave vector $\mathbf{q}$, which is equal to $2k_F$, not to be a simple multiple of $\pi/a$ so that the spin-density wave is incommensurate (Fig. 7.16(b)).

Chromium has a body-centred cubic structure, with the side of the cube equal to $a$, and develops a spin-density wave structure below $T_N = 310$ K. The wave vector $\mathbf{q} \approx 0.96(2\pi/a)$ (thus the material is 'almost' an antiferromagnet) varies slowly with temperature and is in the [100] direction. The polarization is longitudinal for $T < 115$ K and transverse for $T > 115$ K. The electrical resistivity of Cr increases just below $T_N$ due to the energy gaps introduced into parts of the Fermi surface by the spin-density wave which reduce the number of carriers. Spin density waves are also found in a number of organic metals.

## 7.10  The Kondo effect

In very dilute alloys of magnetic ions in a non-magnetic host, the magnetic moments of the magnetic ions can be considered as independent if we ignore the RKKY coupling. The only remaining interaction is between the magnetic moment and the conduction electron spins. At high temperatures the magnetic moments behave like free, paramagnetic moments but below a characteristic temperature, known as the **Kondo temperature** $T_K$, the interaction between the magnetic moment and the conduction electrons leads to the impurity spin becoming non-magnetic. What is happening is that the conduction electrons begin to form a cloud of opposite spin-polarization around the impurity spin resulting in a quasi-bound state. This process of magnetic screening of a magnetic impurity by the conduction electrons is known as the **Kondo effect** and has two profound experimental consequences. First, the magnetization falls below its free-moment value and the susceptibility therefore falls below the value expected from Curie's law. Second, because the impurity moment is strongly interacting with the conduction electrons, the electron scattering cross-section of the moment is strongly enhanced and there is a new term in the resistivity which is proportional to $J \ln T$ where $J < 0$ is the exchange coupling between the local moment and the conduction electrons.

The resistivity of most materials decreases with decreasing temperature. This is because the number of phonons decreases as the material cools, and the resistivity is strongly determined by the number of phonons. Thus it is usually dominated by a term proportional to $T$ at high temperatures and $T^5$ at lower temperatures. There is also a constant term, determined by the (temperature-independent) concentration of impurities which becomes important at the lowest temperatures.

The new $J \ln T \propto - \ln T$ (because $J < 0$) term from the Kondo effect kicks in at low temperature and leads to the appearance of a resistivity minimum (see Fig. 7.17), below which the resistivity increases with decreasing temperature.

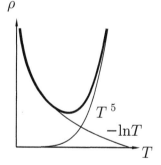

**Fig. 7.17** The temperature dependence of the resistivity shows a minimum. The resistivity is the sum of a term proportional to $T^5$ and a term given by $J \ln T \propto - \ln T$.

## 7.11  The Hubbard model

Metallic behaviour occurs because electrons save kinetic energy by being delocalized over the whole of a crystal. However, in some systems the on-site Coulomb energy (the cost of putting two electrons on the same lattice site) is so strong that electrons cannot move freely through the crystal; the high energetic cost of double occupancy of a site means that metallic behaviour breaks down. Electrons can no longer be treated as free particles but electron correlations, the features associated with the way electrons have to avoid each other if the Coulomb effects are large, become very significant. The competition between metallic behaviour (parametrized by the transfer integrals in the tight-binding model, or the size of the electronic bandwidth) and Coulomb energy (parametrized by $U$, sometimes known as the Hubbard-U) is expressed by the Hubbard model.

Each site in a material can be doubly occupied, with one electron having spin-up and one having spin-down. The Pauli exclusion principle precludes double occupancy of a site by electrons with the same spin. In the half-filled limit, where there are just enough electrons to have one electron per lattice site, the on-site Coulomb energy can be minimized if one electron sits on each site and doesn't move. This is the ground state if $U = \infty$. If $U$ is large but not infinite, each site remains singly occupied but it is possible for an electron briefly to visit its neighbour's site if their spins are antiparallel. This saves a bit of kinetic energy and therefore there is an energetic saving for the electrons on neighbouring sites to be antiparallel, hence the system will be an antiferromagnet. However an electron cannot get on to its next-nearest neighbour's site because that will have the same spin and a site cannot be occupied by two electrons of the same spin by the Pauli exclusion principle. Therefore the system is also an insulator. Such an antiferromagnetic insulator, with a large $U$ and one electron per site, is known as a Mott insulator. When $U$ is smaller than the electronic bandwidth $W$, metallic behaviour is realized. The crossover occurs when $U \sim W$ where there is a metal–insulator transition.

Nevill F. Mott (1905–1996)

## 7.12   Neutron stars

We conclude this chapter with a rather extreme example of magnetic condensed matter: the neutron star. Formed during the rapid collapse of a star in a supernova explosion, neutron stars consist of highly dense nuclear matter. Conservation of angular momentum during the stellar collapse means that any initial rotational rate of the precursor star is greatly increased (as the star collapses, its moment of inertia plummets and so its angular rotation rate shoots up). The radius of a neutron star is typically $\sim 10$ km and the rotation rate can be up to $10^3$ s$^{-1}$. The collapse also is associated with flux compression, so that the magnetic field of neutron stars is enormous, up to $\sim 10^8$ T. For comparison, the field at the surface of the Sun is $\sim 10^{-4}$ T and at the surface of the Earth is $\sim 5 \times 10^{-5}$ T. The magnetic field is most intense near the poles, but the magnetic field axis of a neutron star, like the Earth's, is at an angle to the rotation axis, so that it precesses as the neutron star rotates. The resulting changing magnetic field leads to a large electric field by Faraday's law and this accelerates electrons along the magnetic axis. The electrons are accelerated along curved trajectories and therefore emit electromagnetic radiation. The result is a tightly concentrated beam of radiation that revolves as the neutron star spins. When the Earth happens to be in the way, we receive regular pulses of radiation. Hence neutron stars were first called *pulsars*, i.e. pulsating stars.

We can make some (very rough) estimates of the behaviour of neutrons inside the neutron star by using a 'free neutron' model and treating them non-relativistically. Neutrons, like electrons, are fermions and must obey the Pauli exclusion principle. Therefore, as a first-guess model of neutron stars, we can consider the filling up of $k$-states with neutrons in a box the size of a neutron star following the method of Section 7.1. The neutron star interior contains mainly neutrons at a density about 12 orders of magnitude above that of conduction electrons in a metal. The neutron mass is of course about 3 orders of magnitude higher than the electron mass. Since the Fermi energy is proportional to $n^{2/3}/m$, it should be about 5 orders of magnitude higher in a neutron star than in a metal. Thus the Fermi temperature[7] of the neutrons in a neutron star should be about $10^9$ K which is well above the temperature of the neutron star (up to $\sim 10^6$ K) showing that the neutrons in the neutron star are in the degenerate limit. The surface of a neutron star is thought to be solid, but the matter in the interior may possibly be in a superfluid state. This is analogous to the pairing of helium atoms that is found in superfluid helium, but in neutron stars it may be pairs of neutrons, and also pairs of protons, that condense. Neutrons and protons have a pairing energy of 1–2 MeV, which is equivalent to a temperature of $\sim 10^{10}$ K, much higher than the temperature of the neutron star, so that pairing can be expected to occur.

[7] By comparison, for the case of electrons in a metal we expect $E_F \sim 1$ eV, so that $T_F \sim 10^4$ K

# Further reading

- Landau diamagnetism is discussed in A. Dupré, *American Journal of Physics*, **49**, 34 (1981), and the treatment in this chapter has been adapted from that presented in this article.

- The band theory of metals and the de Haas–van Alphen effect is clearly described in J. Singleton, *Electronic properties of solids*, OUP 2001.

- A useful treatment of the physics of Landau levels is in Chapter 6 of J. H. Davies, *The physics of low-dimensional semiconductors*, CUP 1998.

- An advanced treatment of the magnetism of band electrons is in K. Yosida, *Theory of magnetism*, Springer 1996.

- Spin density functional theory is described in J. Kübler, *Theory of itinerant electron magnetism*, OUP 2000.

- Useful information on fluctuations in metallic magnets may be found in T. Moriya, *Spin fluctuations in itinerant electron magnetism*, Springer 1985.

- The Kondo effect is treated in A. C. Hewson, *The Kondo problem to heavy fermions*, CUP 1993.

- The physics of neutron stars and related astrophysical objects are described in S. L. Shapiro and S. A. Teukolsky, *Black holes, white dwarfs and neutron stars*, Wiley 1983.

# Exercises

(7.1) Show that Pauli paramagnetism leads to a magnetic susceptibility that can be written in a Curie-like form

$$\chi_P = \frac{n\mu_0\mu_B^2}{k_B T_0} \qquad (7.92)$$

where the constant $T_0$ is given by

$$T_0 = \frac{2T_F}{3} \qquad (7.93)$$

where $T_F = E_F/k_B$ is the Fermi temperature. Show further that if $T \ll T_F$ the temperature-dependent correction to $\chi_P$ is

$$\chi_P(T) = \frac{\mu_0\mu_B^2}{3k_B T_0}\left(1 - \frac{\pi^2 T^2}{12 T_F^2}\right) \qquad (7.94)$$

and estimate the size of this correction for a typical metal at room temperature.

(7.2) (a) Show that the paramagnetic susceptibility of a non-degenerate electron gas containing $N$ electrons is identical to that of $N$ independent localized electrons whose orbital motion is quenched.

(b) Consider a semiconductor with $3 \times 10^{22}$ electrons per cubic metre in its conduction band and an effective mass $m^* = 0.1m_e$. Estimate the temperature below which the magnetic susceptibility is independent of temperature. Below this temperature, calculate the magnitude of the Pauli paramagnetism and the Landau diamagnetism.

(7.3) In the presence of a magnetic field $\mathbf{B} = (0, 0, B)$, the momentum operator $\hat{\mathbf{p}} = -i\hbar\nabla$ must be replaced by $-i\hbar\nabla + e\mathbf{A}$ leading to the energy eigenvalues in eqn 7.40. In Section 7.5 the magnetic vector potential was chosen to be $\mathbf{A} = (0, Bx, 0)$. Show that the same results could have been obtained if $\mathbf{A} = (-By, 0, 0)$ was used instead. The problem can be solved in cylindrical polars where $A_\phi = \frac{1}{2}Br$, $A_z = A_r = 0$. Show that this choice yields the same value for $\mathbf{B}$ and for the energy eigenvalues.

(7.4) Show that the Fourier transform of the Fermi sphere is related to a function related to the RKKY function in eqn 7.89, namely that

$$\int_{|\mathbf{k}|<k_F} d^3k\, e^{i\mathbf{k}\cdot\mathbf{r}} = \int_0^{k_F} 2\pi k^2\, dk \int_0^\pi e^{ikr\cos\theta} \sin\theta\, d\theta$$
$$= \frac{4\pi}{r^3}[\sin k_F r - k_F r \cos k_F r] \quad (7.95)$$

(7.5) The integral in eqn 7.88 is a bit complicated to evaluate and this problem is designed to take it a stage at a time. First show that

$$\int d^3q\, e^{i\mathbf{q}\cdot\mathbf{r}}\varphi(\mathbf{q}) = \frac{2\pi}{ir}\int_{-\infty}^\infty dq\, q e^{iqr}\varphi(q), \quad (7.96)$$

where $\varphi(q) = \varphi(-q)$ is a function of the variable $q$. Hence show that

$$\frac{1}{(2\pi)^3}\int d^3q\, \frac{\chi_P}{2}\left(1 + \frac{4k_F^2 - q^2}{4k_F q}\log\left|\frac{q + 2k_F}{q - 2k_F}\right|\right)e^{i\mathbf{q}\cdot\mathbf{r}}$$
$$\qquad (7.97)$$
$$= \frac{\chi_P k_F^2}{i\pi^2 r}\int_{-\infty}^\infty dx\, x e^{iyx} f(x),$$
$$\qquad (7.98)$$

where $x = q/2k_F$, $y = 2k_F r$ and where $f(x)$ is the function defined in eqn 7.79. To do this integral it is necessary to resort to a trick of contour integration.

Consider the two integrals

$$I_1 = \frac{1}{ir}\int_{-\infty}^\infty dx\, \frac{xe^{iyx}}{2}\left(1 + \frac{1 - x^2}{2x}\log\left|\frac{x + 1}{x - 1}\right|\right)$$
$$\qquad (7.99)$$

$$I_2 = \frac{1}{ir}\int_{-\infty}^\infty dx\, \frac{xe^{iyx}}{2}\left(1 + \frac{1 - x^2}{2x}\log\frac{x + 1}{x - 1}\right).$$
$$\qquad (7.100)$$

The two integrands are equal when $|x| > 1$. Show that $I_2$ is zero by completing the contour with an infinite semicircle in the upper half of the $x$ plane and using Jordan's lemma. Hence, using this result and a contour for $I_1$ that is displaced infinitesimally above the real axis from $x = -1$ to $x = 1$, show that

$$I_1 = \frac{1}{ir}\int_{-1}^1 dx\, \frac{e^{iyx}(1 - x^2)}{4}\left(\log\left|\frac{x + 1}{x - 1}\right| - \log\frac{x + 1}{x - 1}\right)$$
$$= \frac{\pi}{r}\int_{-1}^1 dx\, \frac{e^{iyx}(1 - x^2)}{4}$$
$$= \frac{\pi}{r}\left[\frac{\sin y - y\cos y}{y^3}\right]$$
$$= 2\pi k_F F(2k_F r), \quad (7.101)$$

and hence verify eqn 7.88.

(7.6) (a) Show that the density of the states at the Fermi energy levels in one, two and three dimensions are given by

$$\text{one-dimension:} \quad g(E_F) = \frac{n}{2E_F} = \frac{2m}{\hbar^2\pi k_F}$$

$$\text{two-dimensions:} \quad g(E_F) = \frac{n}{E_F} = \frac{m}{\hbar^2\pi} \qquad (7.102)$$

$$\text{three-dimensions:} \quad g(E_F) = \frac{3n}{2E_F} = \frac{mk_F}{\hbar^2\pi^2}$$

(b) Show that the $q$-dependent susceptibility of the electron gas in one dimension is given by

$$\chi_q = \chi_P \frac{k_F}{q}\log\left|\frac{q + 2k_F}{q - 2k_F}\right| \quad (7.103)$$

and in two-dimensions is given by

$$\chi_q = \begin{cases} \chi_P & q < 2k_F \\ \chi_P\left(1 - \sqrt{1 - (4k_F^2/q^2)}\right) & q > 2k_F. \end{cases}$$
$$\qquad (7.104)$$

These results are plotted in Fig. 7.14.

(7.7) Show that the continuum of Stoner excitations described in Section 7.8 and pictured in Fig. 7.13 are bounded by the two parabolae

$$\hbar\omega = \Delta + \frac{\hbar^2 q}{2m_e}(q \pm 2k_{F\uparrow}) \quad (7.105)$$

and that the $\omega = 0$ excitations are contained within the range of $q$ given by

$$k_{F\uparrow} - k_{F\downarrow} \le q \le k_{F\uparrow} + k_{F\downarrow}, \qquad (7.106)$$

as shown in Fig. 7.13 (Hint: use the fact that $\hbar^2 k_{F\uparrow}^2 / 2m_e = \hbar^2 k_{F\downarrow}^2 / 2m_e + \Delta$.)

(7.8) Consider the splitting of energy bands which is shown in Fig. 7.5, but now in the case in which $\delta E$ is not an infinitesimal quantity. The Fermi energy has now become spin-dependent. If the system of electrons has $n$ electrons per unit volume, show that $n_+$, the number of up electrons per unit volume, and show that $n_-$, the number of down electrons per unit volume, are given by

$$n_\pm = \int_0^{E_{F\pm}} g_\pm(E) \, dE \qquad (7.107)$$

$$= \int_0^{E_{F\pm}} \frac{3}{4} \frac{n}{E_F^{3/2}} E^{1/2} \, dE \qquad (7.108)$$

$$= \frac{n}{2} \left( \frac{E_{F\pm}}{E_F} \right)^{3/2}, \qquad (7.109)$$

where $E_F$ is the Fermi energy in the absence of exchange splitting.

Using $n_+ + n_- = n$ and defining $x = (n_+ - n_-)/n$, show that

$$\frac{E_{F\pm}}{E_F} = (1 \pm x)^{2/3}. \qquad (7.110)$$

The kinetic energy is given by

$$T = \sum_{\sigma = \pm} \int_0^{E_{F\sigma}} E g(E) \, dE. \qquad (7.111)$$

Show that

$$T = \frac{3}{10} n E_F \left[ (1+x)^{5/3} + (1-x)^{5/3} \right]. \qquad (7.112)$$

The interaction energy can be written

$$V = -\int_0^M \mu_0 (\lambda M') \, dM' = -\frac{1}{2} \mu_0 \lambda M^2 = -\frac{1}{2} U n^2 x^2 \qquad (7.113)$$

where $\lambda$, $U = \mu_0 \mu_B^2 \lambda$ and $M = (n_+ - n_-)^2$ are the molecular field parameter, Coulomb energy and magnetization respectively. The total energy is therefore

$$E = T + V. \qquad (7.114)$$

Show that the condition $dE/dx = 0$ leads to stable solutions when

$$\frac{(1+x)^{2/3} + (1-x)^{2/3}}{x} = \frac{4U g(E_F)}{3}. \qquad (7.115)$$

(Hint: use $g(E_F) = 3n/2E_F$.) Interpret this result graphically.

Explain why the non-magnetic state is unstable when $d^2 E / dx^2 < 0$ at $x = 0$. Show that this is equivalent to

$$U g(E_F) > 0, \qquad (7.116)$$

which is the Stoner criterion.

# 8 Competing interactions and low dimensionality

In this chapter we will examine some of the ways in which competing interactions and low dimensionality can lead to some extremely subtle, complex, and sometimes even useful magnetic behaviour in solids. Competing interactions and low dimensionality both occur naturally in some systems; for example some crystals grow in such a way that the magnetic moments couple strongly in chains or occupy sites on frustrated lattices. However, these features can be introduced artificially, for example, by fabricating a ferromagnetic multilayer or using electron-beam lithography to produce ferromagnetic wires. Many of these topics are currently under active study in research groups throughout the world. This chapter presents an account of some of these developments, many of which contain mysteries which are not yet fully unravelled. Because many of the topics are new and complex, the discussion will be somewhat oversimplified in some cases; the interested reader can consult the references in the further reading at the end of the chapter for more details. Nevertheless I hope to give some flavour of a few of the problems which are, at the time of writing, providing enormous challenges to condensed matter physicists.

We begin with frustration, which can lead to a number of different ground states including spin glasses, and then continue to the effects of low dimensionality.

## 8.1 Frustration

In some lattices it is not possible to satisfy all the interactions in the system to find the ground state. Often this leads to there being no single unique ground state but a variety of similar low energy states of the system in which the unhappiness (by which I mean non-minimization of the energy) is shared around as much as possible. In this case the system is said to show frustration. As an example, consider the lattices in Fig. 8.1 in which only nearest-neighbour antiferromagnetic interactions operate. On the square lattice (Fig. 8.1(a)) it is easily possible to satisfy the requirement that nearest-neighbour spins must be antiparallel. However on a triangular lattice things are not so simple. As shown, if two adjacent spins are placed antiparallel, one is faced with a dilemma for the third spin. Whichever choice is made, one of the two neighbours will not have their energy minimized. The system therefore cannot achieve a state that entirely satisfies its microscopic constraints, but does possess a multiplicity of equally unsatisfied states. As a result the frustrated system shows metastability, hysteresis effects (dependence on the sample's magnetic or thermal history), and time-dependent relaxation towards equilibrium.

(a)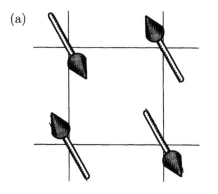

(b)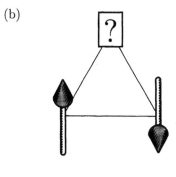

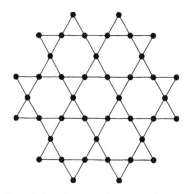

**Fig. 8.1** Antiferromagnetic nearest-neighbour interactions on the (a) square lattice and (b) triangular lattice. The triangular lattice shows frustration, because it is not possible to orient the spin on the third site to satisfy the requirement of antiferromagnetic nearest-neighbour interactions with the other two spins.

Hence in some systems the geometry of the lattice can frustrate the ordering of the spins. In two dimensions, Heisenberg spins on triangular and corner sharing kagomé lattices (see Fig. 8.2) show this effect, while in three dimensions the most well studied systems have a pyrochlore structure, in which the magnetic ions occupy a lattice of corner sharing tetrahedra. These systems are believed not to order but to display a classical ground state with macroscopic degeneracy, sometimes described as **cooperative paramagnetism** in which only short-range correlations between spins are found for all temperatures. Some of these systems are also believed to possess a dispersionless spin-wave branch, known as a **zero mode**, which strongly affects the low temperature thermodynamic behaviour and leads to the persistence of spin fluctuations down to zero temperature.

**Fig. 8.2** A section of the kagomé lattice.

## 8.2   Spin glasses

Spin glasses were described earlier in Section 5.5. In that section a spin glass was defined as a random, mixed-interacting magnetic system characterized by a random, yet cooperative, freezing of spins at a well defined temperature $T_f$ below which a highly irreversible, metastable frozen state appears without the usual magnetic long range ordering. We now examine the different parts of this definition in much more detail, beginning with the randomness which can be introduced in several ways.

First is **site-randomness**, which can be achieved in an alloy. A commonly studied spin glass is $Cu_{1-x}Mn_x$ with $x \ll 1$ in which the substitution of small amounts of Mn into the Cu matrix occurs completely randomly with no short-range ordering. This directly leads to a random distance between magnetic Mn ions in the non-magnetic Cu matrix (see Fig. 5.1(c)). Random distances between magnetic ions can also be engineered by taking an intermetallic such as $GdAl_2$ and making it amorphous (melting it and then cooling very rapidly from the melt) so that the crystalline lattice is destroyed.

Another possibility is **bond-randomness** in which the nearest neighbour interactions vary between $+J$ and $-J$. This can be achieved by fixing the magnetic ions in a regular crystalline array but modulating the indirect exchange interactions between the magnetic ions. This is performed in $Rb_2Cu_{1-x}Co_xF_4$ for which both Cu and Co have effective spins of $\frac{1}{2}$ (the crystal field on the Co splits the $S = \frac{3}{2}$ levels in such a way that only the lowest $S = \frac{1}{2}$ doublet is

See Example 5.1 on page 100

populated at low temperatures). The crystal field also causes the spins to point either parallel or antiparallel to the $z$ axis (a uniaxial single ion anisotropy, see Section 3.2.2) which is said to give them Ising character. However the size and sign of the superexchange interaction between magnetic ions depends crucially on whether the coupling is between Co and Co, or Co and Cu, or Cu and Cu, and which orbital on the Cu is occupied. The net result is that the bonds between ions can have different values of the exchange $J$ and these bonds are randomly distributed throughout the sample.

The randomness inherent in a spin glass is important, but equally important is the presence of competing interactions. The distribution of distances between moments in a random-site spin glass leads to competing interactions because the interactions are of RKKY-type and therefore their sign (ferromagnetic or antiferromagnetic) depends on the distance between the spins. Another contributing feature is the magnetic anisotropy, due to single-ion anisotropy or Dzyaloshinsky–Moriya interactions. In an amorphous material these anisotropies vary from site to site so that so-called amorphous magnets often possess random anisotropy, so that there can be a locally varying 'easy-axis' for the magnetization.

Competing interactions are automatically present in a random-bond spin glass, because different bonds 'pull' the system in different ways. These competing interactions lead to frustration, so that as in systems such as the kagomé lattices, there is a multidegenerate ground state. Spin glasses share this multidegenerate ground state but also show a new effect, a cooperative freezing transition, which will now be described.

At high temperature the behaviour of all magnetic systems is dominated by thermal fluctuations, so that in a spin glass all the spins are independent. As a spin glass is cooled from high temperature, the independent spins slow down and build up into locally correlated units, known as clusters. The spins which are not in clusters take part in interactions between clusters. As the temperature cools to $T_f$ the fluctuations in the clusters progressively slow down. The interactions between spins become more long range so that each spin becomes more aware of spins in a progressively growing region around it. At $T_f$ the system finds one of its many ground states and freezes. This process is not fully understood but it appears to be a cooperative phase transition. However this is not a phase transition to a magnetically ordered state; there are no magnetic Bragg peaks found in scattering experiments as would be the case if the system showed long range magnetic order. Below $T_f$ the ground state appears to be 'glassy', possessing metastability and slow relaxation behaviour. There is a divergence between the field-cooled and zero-field cooled magnetic susceptibility below $T_f$, reflecting this metastability (see Fig. 8.3).

One of the signatures of spin glass behaviour is a sharp peak close to $T_f$ in the real part of the a.c. susceptibility, $\chi(\omega)$. In this technique the magnetic susceptibility is measured using a very small alternating magnetic field of frequency $\omega$, sometimes with a constant (d.c.) magnetic field also applied. The position of the peak varies slightly with $\omega$. The imaginary part of $\chi(\omega)$, related to the absorption, shows a sudden onset near $T_f$. The dynamics of the fluctuations associated with the freezing process can be studied using a.c. susceptibility, neutron spin echo (a method of measuring time correlation functions on relatively long time scales) and $\mu$SR. These techniques show

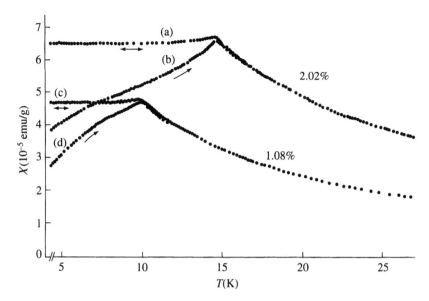

**Fig. 8.3** The static susceptibility of $Cu_{1-x}Mn_x$ for $x = 0.0108$ and $x = 0.0202$. After zero-field cooling, $\chi$ was measured in a small field of 0.59 mT for increasing temperature ((b) and (d)). If the samples were field-cooled in 0.59 mT $\chi$ was reversible ((a) and (c)) (Adapted from Nagata *et al.* 1997).

that there is a large distribution of relaxation times for the clusters which form above $T_f$. As the sample is cooled, these clusters grow and fluctuate at a rate dependent on the size and nature of each cluster. A spin outside the clusters will fluctuate rapidly but as soon as it joins a growing cluster (something of a 'hostile takeover'!) it is forced to slow down and keep step with its new host. At $T_f$ the random anisotropy becomes important and the system freezes into a random orientation in which the clusters are fixed. A wide range of relaxation times is also observed below $T_f$, showing that some free spins or small superparamagnetic (see below) clusters are still there. The co-operative spin freezing in spin glasses is still not yet fully understood.

The $Cu_{1-x}Mn_x$ spin glass is usually studied when $x$ is a few percent or less. In a concentrated spin glass the behaviour is very different because the greater number of magnetic nearest neighbours, which can be enhanced by chemical clustering, leads to the presence of large, ordered magnetic clusters which then dominate the magnetic behaviour. The system then becomes known as a **cluster glass** (sometimes known as a **mictomagnet**) which has some characteristics of a spin glass but long range magnetic order is always in the background waiting to spring (the technical term is that the long range magnetic order is **incipient**). Sometimes the high temperature state can actually be an ordered ferromagnet, rather than a paramagnet, and the system actually freezes to a low temperature non-magnetic disordered state. Such a system is called a **re-entrant spin glass** and the behaviour is believed to be connected with either temperature-dependent random anisotropy or some sort of anisotropy associated with the freezing. In very high concentrations of magnetic ions, the system approaches the **percolation limit** at which it is possible to find a path of nearest-neighbour links through the whole sample along which each ion is magnetic; this results in the sample having long range magnetic order. The transitions between these different states can be seen in the phase diagram of $Au_{1-x}Fe_x$ shown in Fig. 8.4.

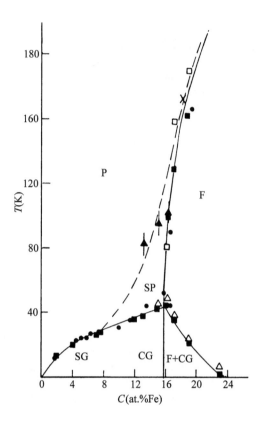

**Fig. 8.4** The phase diagram of $Au_{1-x}Fe_x$. P = paramagnetic, SP = superparamagnetic, SG = spin glass, CG = cluster glass, F = ferromagnetic. After Coles *et al.* 1978.

## 8.3   Superparamagnetism

If ferromagnetic particles are small enough, they will be single-domain (see Section 6.7.7), because the energy cost of domain wall formation does not outweigh any saving in demagnetizing energy. The magnetization of a small, single-domain ferromagnetic particle is often constrained to lie parallel or antiparallel to a particular direction. This can be due to magnetocrystalline anisotropy, or shape anisotropy (associated with the demagnetizing energy and the shape of the particle), or a variety of other reasons. However, we will assume that the energy density of the particle contains a term $K \sin^2 \theta$ where $\theta$ is the angle between the magnetization and this particular direction and $K$ is a constant which quantifies the energy density associated with this anisotropy. Thus the energy is minimized when $\theta = 0$ or $\pi$ (see Fig. 8.5). A particle of volume $V$ needs an activation energy of $\Delta E = KV$ to flip its magnetization from $\theta = 0$ to $\pi$ or from $\pi$ to 0. For very small particles, such that $KV$ is small compared to $k_B T$, the magnetization can be easily flipped in this way by thermal fluctuations.

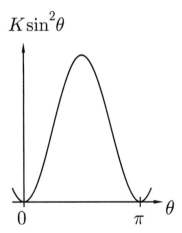

**Fig. 8.5** The energy density of a magnetic particle contains a term $K \sin^2 \theta$. The energy is minimized when $\theta = 0$ or $\pi$.

Consider now a distribution of these small ferromagnetic particles in a non-magnetic matrix and assume that the particles are separated far enough apart so that interparticle interactions can be neglected. For $k_B T \gg KV$ the system will therefore behave like a paramagnet, albeit one in which the independent moments are not atomic moments but large groups of moments, each group inside a ferromagnetic particle. The system is therefore called a **superparamagnet**. At high temperatures the moments on the particles are able to fluctuate rapidly. The relaxation time $\tau$ of the moment on a particle is given by

$$\tau = \tau_0 \exp\left(\frac{KV}{k_B T}\right),\tag{8.1}$$

where $\tau_0$ is typically $10^{-9}$ s. The fluctuations therefore slow down ($\tau$ increases) as the sample is cooled (see Fig. 8.6) and the system appears static when $\tau$ becomes much longer than the measuring time $t$ of the particular laboratory experimental technique which you are using. If we define 'much longer' to be $\tau > \alpha t$ where $\alpha = 100$, then below the **blocking temperature** $T_B$, given by $T_B = KV/k_B \ln(\alpha t/\tau_0)$, each magnetic particle appears to be locked into one of its two minima.

The experimental measuring time $t$ is in the range $10^{-12}$–$10^{-10}$ s for inelastic neutron scattering, $10^{-10}$–$10^{-7}$ s for Mössbauer (the decay processes in the Mössbauer transition) are $\sim 10^{-7}$ s, $10^{-10}$–$10^{-5}$ s for $\mu$SR (a measurable fraction of muons live for up to $\sim 10\tau_\mu$ where $\tau_\mu = 2.2\ \mu$s is the average muon lifetime) and a.c. susceptibility typically probes $10^{-1}$–$10^{-5}$ s. Because of the logarithmic dependence on $\alpha t/\tau_0$, any of $\alpha$, $t$ or $\tau_0$ can be changed by a few orders of magnitude with only a relatively small change to $T_B$. If the particles in the superparamagnetic system have a range of sizes then they will 'block' at different temperatures.

Superparamagnets have some similarities with spin glasses but are quite different in a number of respects: the interactions between the magnetic particles are not important in a superparamagnet, whereas they are vital in a spin glass; also, the spin glass shows a cooperative phase transition while the superparamagnet shows a gradual blocking of the superparamagnetic particles. Superparamagnetism is technologically important since many important recording materials are particulate. This is crucial because magnetic tape is expected to store information for years, not $\mu$s, which puts a limit on the minimum size of the particles. In studying rock magnetism it is sometimes necessary to consider the exponential decay of magnetism over millions of years. Thus the stability of particles against superparamagnetic fluctuations depends on the timescale which you are using; a particle may be stable in a Mössbauer experiment but fluctuate in an a.c. susceptibility; it may be stable for a century, but decay on geological time scales.

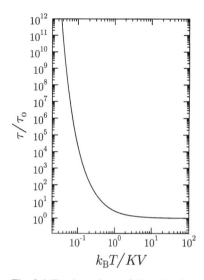

**Fig. 8.6** The dependence of the relaxation time $\tau$ as a function of temperature $T$ (scaled by $k_B/KV$) according to eqn 8.1. The fluctuations therefore slow down ($\tau$ increases) as the temperature is reduced.

## 8.4 One-dimensional magnets

No long range order is possible in one-dimension for $T > 0$, so it might be thought that one-dimensional magnets were rather tedious and uninteresting. In fact nothing could be further from the truth! The one-dimensionality implies the possibility of complex excitations which are still far from being completely understood.

### 8.4.1  Spin chains

A one-dimensional ($d = 1$) line of spins is known as a **spin chain**. The individual spins can be constrained to lie parallel or antiparallel to a particular direction (Ising spins), or may be free to point anywhere in a fixed plane (XY spins), or free to point in any direction (Heisenberg spins). Spin chains can be approximately realized in crystals, if the crystal structure is such so as to keep the chains reasonably far apart. The single ion anisotropy due to the crystal field may lead to the magnetic moments behaving as Ising spins ($D = 1$), XY spins ($D = 2$), Heisenberg spins ($D = 3$), or somewhere in between. A commonly studied family of systems is based on crystals of the type $ABX_3$ where A is a non-magnetic cation of a single charge, B is a doubly charged magnetic cation and X is a halide anion. This leads to a simple hexagonal lattice with transition metal ions forming chains along the $c$ direction. For example $CsCoCl_3$ behaves almost as a one-dimensional Ising spin chain since the anisotropy constrains the spins along a particular direction; $KCuF_3$ behaves like a one-dimensional Heisenberg spin chain, as do a number of Cu salts with organic ligands.

Very often these systems show three-dimensional long range order at very low temperatures because there will always be some small interchain interaction which can couple the chains together. $CsCoCl_3$ shows three-dimensional long range magnetic order below 21 K because of this interchain coupling. Nevertheless, there is a wide region of temperature above the crossover to a three-dimensional region, where the magnetic behaviour is that of a one-dimensional system.

The spin quantum number for each spin on the chain depends on the atom. For chains with $Cu^{2+}(3d^9)$ the spin quantum number is $S = \frac{1}{2}$, with $Mn^{2+}$ ($3d^5$) $S = \frac{5}{2}$ and for $Co^{2+}$ ($3d^7$) in $CsCoCl_3$ the ground state has an effective spin of $S = \frac{1}{2}$ (the $S = \frac{3}{2}$ free-ion ground state is split by the crystal field leaving a ground state doublet and excited states). In the next section we will just consider chains with $S = \frac{1}{2}$ on each site.

### 8.4.2  Spinons

What makes these chains interesting is not their ordering (because unless the interchain interactions are strong enough, they do not show order), but their excitations. As shown in Section 6.6.1, in three-dimensional Heisenberg magnets the excitations are magnons, which are bosons. (In a metallic magnetic material there may also be Stoner excitations, see Section 7.8.) Each magnon is an excitation with $S = 1$ and so a single magnon can interact with a neutron in an inelastic scattering experiment.

In Ising spin chains, the excitations are associated with the creation of domain walls. There are no gapless excitations (i.e. excitations with vanishingly small energy in the long wavelength limit; 'massless Goldstone modes' to use the particle physics parlance) because even to create one domain wall costs a finite amount of energy (and in fact for an excitation you have to create the domain walls in pairs). Once the excitation is created it can move freely along the chain. The excitation energy has no wave vector dependence, but if the chain is not perfectly Ising-like (as very often happens in real systems) then there will be some modulation of the dispersion relation.

In Heisenberg spin chains, the excitations are known as **spinons**. These have spin-$\frac{1}{2}$ (in contrast to magnons which have spin-1) and are fermions. They have a dispersion relation which is given by

$$\hbar\omega = \pi|J\sin(qa)|, \tag{8.2}$$

where $J$ is the antiferromagnetic exchange coupling, and $a$ and $q$ are the lattice constant and wave vector, both measured along the chain direction. Equation 8.2 is the bold line in Fig. 8.7. This can be compared with the conventional spin wave dispersion relation in eqn 6.58 with $S = \frac{1}{2}$, but there is an additional factor of $\pi/2$. The excitations in both cases are gapless because when $q \to 0$ (long wavelength limit) $\omega \to 0$. A neutron scattering experiment involves a change of spin of one and so although this implies a creation or annihilation of a single magnon in a three-dimensional system, it implies a creation or annihilation of two spinons in a one-dimensional system. Neutron experiments therefore measure the momentum $q = q_1 + q_2$ and energy $\hbar\omega = \hbar\omega_1 + \hbar\omega_2$ associated with creation or annihilation of a pair of these spinons and so the experimental data show a continuum of excitations between eqn 8.2 and $\hbar\omega = 2\pi|J\sin(qa/2)|$ (see Fig. 8.7). Neutron scattering experiments have confirmed that these excitations do exist in some one-dimensional antiferromagnetic chains (see Fig. 8.8). (For a derivation of eqn 8.2, see des Cloizeaux and Pearson 1962.)

### 8.4.3 Haldane chains

The previous section described how the excitations in spin-$\frac{1}{2}$ antiferromagnetic Heisenberg spin chains are believed to be spinons which are gapless excitations. This result is believed to be true also for half-integer spin chains (i.e. chains of spins with $S = \frac{1}{2}, \frac{3}{2}, \frac{5}{2}, \ldots$). Haldane conjectured that something different would happen for integer spin chains (i.e. chains of spins with $S = 1, 2, 3, \ldots$), namely that there would be a gap in the excitation spectrum which occurs because of nonlinear quantum fluctuations in the ground state. A one-dimensional chain of integer spins is therefore known as a **Haldane chain** and the gap in the excitation spectrum is known as a **Haldane gap**. This fundamental difference between half-integer and integer spin chains is related to the difference between fermions and bosons under exchange; this different exchange symmetry has a topological origin and has a dramatic effect on the nature of the excitations.

Most tests of Haldane's conjecture have been carried out on materials with chains of $Ni^{2+}$ ($S = 1$) ions, including $CsNiCl_3$, $Ni(C_2H_8N_2)_2NO_2ClO_4$ and $Y_2BaNiO_5$. These $S = 1$ spin chains all seem to possess gaps in their excitation spectra as predicted. Half-integer antiferromagnetic chains are gapless, unless magnetoelastic coupling opens up a spin-Peierls gap, as discussed in the following section.

### 8.4.4 Spin-Peierls transition

Although spin-$\frac{1}{2}$ antiferromagnetic chains are gapless, they are susceptible to an analogous kind of instability that afflicts one-dimensional metals (see Section 7.9) which *can* open up a gap. This occurs at the **spin-Peierls transition**.

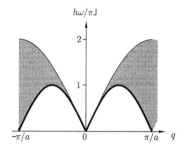

**Fig. 8.7** Dispersion relation for spinon excitations in a one-dimensional antiferromagnetic Heisenberg spin chain (bold line). The shaded region shows the continuum of excitations measured in a neutron scattering experiment.

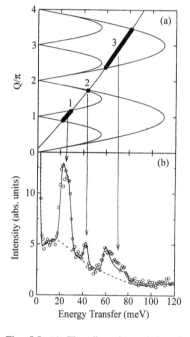

**Fig. 8.8** (a) The dispersion relation for spinons in $KCuF_3$. The experiment was performed using a time-of-flight technique so the line shows the scattering trajectory for neutrons with incident energy 148.9 meV and an incident momentum aligned $8°$ away from the $c^*$ direction in $KCuF_3$. Scattering results when the trajectory intersects with the continuum states. (b) Observed scattering at 20 K. The non-magnetic background is indicated by the dashed line. After Tennant *et al.* 1993.

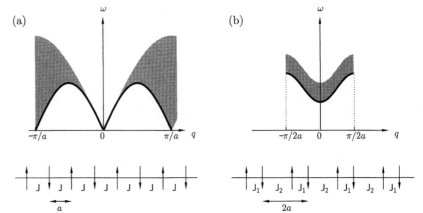

**Fig. 8.9** Schematic representation of the elementary excitations in (a) a uniform Heisenberg antiferromagnetic chain and (b) an alternating chain (for which the ground state is a singlet state at $q = 0$), and for which the unit cell is doubled. Adapted from Bray *et al.* 1983.

The driving force of this intrinsic lattice instability is the magnetoelastic coupling between the one-dimensional electronic structure and the three-dimensional lattice vibrations (phonons). This coupling arises because the exchange energy of the chains is a function of the separation between adjacent lattice sites. A distortion of the lattice influences the magnetic energy (see Fig. 8.9). The name spin-Peierls reflects the similarity with the Peierls distortion (discussed in Section 7.9).

Above the transition temperature $T_{SP}$, there is a uniform antiferromagnetic next-neighbour exchange in each chain; below $T_{SP}$ there is an elastic distortion resulting in dimerization, and hence two, unequal alternating exchange constants. The dimerization increases progressively as the temperature is lowered and reaches a maximum at zero temperature. The alternating chain possesses an energy gap between the singlet ground state and the lowest lying band of triplet excited states. The magnitude of the gap is related to the degree of dimerization and hence to the degree of lattice distortion, becoming zero for the uniform chain (zero dimerization) so that one returns to the gapless spinon case. Thus the magnetic susceptibility $\chi(T)$ shows a knee at $T_{SP}$, with a rather abrupt fall of $\chi$ below $T_{SP}$, corresponding to the opening of the gap (see Fig. 8.10). Whereas the normal Peierls distortion (the electronic analogue of the spin-Peierls transition, see Section 7.9) occurs at a temperature $T_P$ of the order of $k_B T_P \sim E_F \exp(-1/\lambda_{el-ph})$, where $\lambda_{el-ph}$ is the electron–phonon coupling constant, the spin-Peierls transition will occur at $k_B T_{SP} \sim |J| \exp(-1/\lambda_{sp-ph})$, where $J$ is the exchange interaction between adjacent spins and $\lambda_{sp-ph}$ is the spin–phonon coupling constant. Since $J \ll E_F$ (e.g. $J$ is typically 50 K, $E_F$ is typically 500–5000 K), $T_{SP}$ is always small in comparison with $T_P$.

There are only very few materials which show a spin-Peierls transition. This is because antiferromagnetic chains often become three-dimensionally ordered at low temperature due to interchain coupling. Only in very few materials is the spin–phonon coupling able to dominate the interchain spin–spin coupling and allow the formation of a spin-Peierls ground state. Examples of such materials include $CuGeO_3$ ($T_{SP} = 14$ K) and a number of organic systems such as $MEM(TCNQ)_2$ ($T_{SP} = 18$ K) and $TTF-CuS_4C_4(CF_3)_4$ ($T_{SP} = 12$ K).[1]

[1]MEM, TCNQ and TTF are organic molecules with lengthy chemical names.

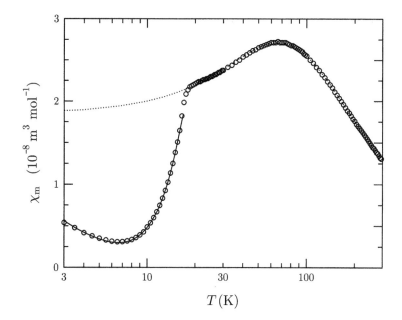

**Fig. 8.10** The molar magnetic susceptibility of the organic spin-Peierls material MEM(TCNQ)$_2$ which consists of stacks of the organic molecule MEM and stacks of the organic molecule TCNQ. At high temperature the susceptibility fits to a model appropriate for a uniform Heisenberg antiferromagnetic chain (dotted line), but on cooling the susceptibility drops rapidly at the spin-Peierls transition as a gap in the excitation spectrum opens up. The rise at very low temperatures (as $T$ is lowered) is due to the Curie-like ($\sim T^{-1}$) susceptibility from defects. After Lovett *et al.* 2000.

## 8.4.5 Spin ladders

Before considering two-dimensional magnets, we can consider a system which is somewhere in between a one-dimensional magnet and a two-dimensional magnet. Consider two parallel spin chains with bonds between them such that the interchain coupling is of comparable strength to the intrachain coupling. Such a system is known as a two-leg **spin ladder** (see Fig. 8.11(a)). It is also possible to have three-leg (see Fig. 8.11(b)), four-leg, and in fact $n$-leg, spin ladders. Promising experimental systems include SrCu$_2$O$_3$ and La$_{1-x}$Sr$_x$CuO$_{2.5}$ (both two-leg spin-$\frac{1}{2}$), and Sr$_2$Cu$_3$O$_5$ (three-leg spin-$\frac{1}{2}$). In fact there is a general system Sr$_{n-1}$Cu$_{n+1}$O$_{2n}$ with $n$ odd which consists of $(n+1)/2$-leg spin-$\frac{1}{2}$ because its structure has strips of a CuO$_2$ square lattice which have $(n+1)/2$ Cu$^{2+}$ ions across their width.

The spin-$\frac{1}{2}$ two-leg ladder is known to have a finite gap in its excitation spectrum, which is easy to see in the 'strong-rung' limit in which the rung

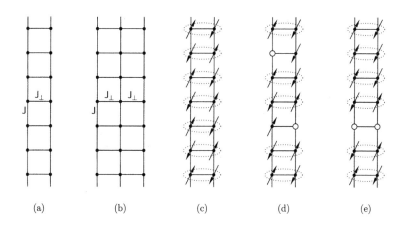

**Fig. 8.11** (a) A two-leg ladder; (b) A three-leg ladder. (c) In a two-leg the ground state consists of spin singlets on each rung of the ladder and hence the two-leg ladder has a gap in its excitation spectrum. (d) Doping holes on to the ladder breaks up the singlets (e) but this energy cost can be minimized if the holes pair up.

coupling $J_\perp$ is larger than the coupling $J$ along the legs. In this case the ground state consists simply of spin singlets along each rung of the ladder (see Fig. 8.11(c)). To create an excitation, you must promote a rung-singlet into a rung triplet (costing an energy $J_\perp$), hence the energy gap. If $J_\perp = 0$ the system is two isolated spin-$\frac{1}{2}$ chains which do not have a gap in their excitation spectrum. However it is believed that a gap appears as soon as $J_\perp$ is non-zero, no matter how small it is. For the $n$-leg ladder, the situation is identical if $n$ is even. However if $n$ is odd, then on a given rung the spins will pair up into singlets leaving one left over. At large $J_\perp$ the system can be mapped into a spin-$\frac{1}{2}$ chain which is gapless. So an even-leg ladder has a gap in its excitation spectrum, while an odd-leg ladder is gapless. This appears to be supported by the results of neutron, $\mu$SR and transport experiments.

Doping the spin ladder with holes breaks up singlets in a spin ladder (see Fig. 8.11(d)). For two-leg ladders there is an energetic advantage for the holes to pair up since they can then share a common rung, reducing the number of 'damaged' singlets from two to one (Fig. 8.11(e)). Hence it is favourable for holes to pair on two-leg ladders and this shows how it might be possible to engineer superconductivity in two-leg spin ladders. Superconductivity ($T_c \sim$ 14 K) has in fact been discovered in $Sr_{14-x}Ca_xCu_{24}O_{21}$ (sometimes called [14-24-41] or the 'phone number' compound) for $x = 13.6$ at 5 GPa. The interest in spin ladders derives from the fact they can have a gap in their excitation spectrum, they can become superconducting, and yet are simple well defined systems which theorists (who like working in one-dimension) can try to model. Hence these magnetic systems may shed light on the problem of high-$T_c$ superconductivity (see Section 8.5 below).

## 8.5    Two-dimensional magnets

Two-dimensional magnetism is often studied in systems with formula $A_2BX_4$, where as before A is a non-magnetic cation of a single charge, B is a doubly charged magnetic cation and X is a halide anion. The crystal structure is tetragonal and the magnetic ions sit on a square lattice in two-dimensions. A typical material is $K_2NiF_4$ and very often these systems are said to have the $K_2NiF_4$ structure. The Mermin–Wagner–Berezinskii theorem demonstrates that in dimensions $d \leq 2$, thermal fluctuations prohibit the existence of long range magnetic order at non-zero temperature in an isotropic system. However it says nothing about the $T = 0$ ground state. In one-dimension, it turns out that for the spin-$\frac{1}{2}$ Heisenberg antiferromagnet no long range order exists even at $T = 0$. The case for the two-dimensional Heisenberg antiferromagnet is not so clear cut, and there is currently evidence that this does show long range magnetic order at $T = 0$. This system is of great interest because two-dimensional Heisenberg models appear to be important in understanding the high-$T_c$ cuprate superconductors. These systems contain planes of copper and oxygen (see Fig. 8.12) which appear to be crucial for the superconducting properties.

For example, pure $La_2CuO_4$ is an antiferromagnetic insulator and has the $K_2NiF_4$ structure. The $Cu^{2+}$ ions ($3d^9$) have one hole in the d band and thus have spin-$\frac{1}{2}$. They sit on a square lattice, each $Cu^{2+}$ ion separated from its

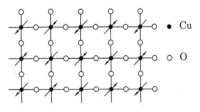

• Cu

o O

**Fig. 8.12**    Copper–oxygen planes in $La_2CuO_4$.

neighbour by oxygen ions which mediate an antiferromagnetic superexchange interaction between the $Cu^{2+}$ spins (see Fig. 8.12). With one electron per site you would expect the system to be metallic, but the holes are localized because of the correlations. Doping the material with extra holes (by replacing some $La^{3+}$ ions with $Sr^{2+}$) leads to high temperature superconductivity at around 40 K for the optimally doped samples (about 20 % of the $La^{3+}$ replaced by $Sr^{2+}$). You can make isostructural materials by replacing $Cu^{2+}$ with $Ni^{2+}$ (which has $S = 1$) but these do not become superconducting, even when doped. There is therefore something special about $Cu^{2+}$. It is particularly surprising that these magnetic ions should be helpful in promoting superconductivity because magnetism normally destroys superconductivity. This is because local magnetic fields due to magnetic ions act as pair-breakers, splitting up the Cooper pairs that are responsible for the supercurrents. The magnetic properties of the two-dimensional $S = \frac{1}{2}$ square lattice Heisenberg antiferromagnet are therefore probably pertinent to the problem of high-$T_c$ superconductivity, although this contention is unproven since, at the time of writing, no definitive theory of high-$T_c$ superconductivity is universally believed.

The two-dimensional $S = \frac{1}{2}$ square lattice Heisenberg antiferromagnet is therefore of considerable interest and has attracted a number of theoretical and experimental investigations. As the temperature is reduced, the system does not order but the correlated regions of short range order increase in size. This size is known as the spin–spin correlation length, $\xi$, and it diverges exponentially at low temperature leading to true long range order at $T = 0$. The behaviour of $\xi$ as a function of temperature can be calculated by mapping the two-dimensional $S = \frac{1}{2}$ square lattice Heisenberg antiferromagnet on to a continuum model known as the two-dimensional quantum nonlinear sigma model. This procedure is outside the scope of this book but the experimental results, which have given much support to this approach, are shown in Fig. 8.13. The results are obtained by using inelastic neutron scattering from $Sr_2CuO_2Cl_2$ which is an extremely good approximation to the ideal two-dimensional $S = \frac{1}{2}$ square lattice Heisenberg antiferromagnet. It has a larger interplane spacing than $La_2CuO_4$ and remains tetragonal down to low temperatures. Interplane interactions lead to long range three-dimensional antiferromagnetic order at $T_N = 256.5$ K, but well above this temperature the spin fluctuations are dominated by the two-dimensional fluctuations. Three-dimensional order leads to sharp Bragg peaks in the inelastic neutron scattering data, but short range order with a correlation length $\xi$ leads to broad peaks with width proportional to $\xi^{-1}$ (for long range order $\xi \to \infty$ and the peaks become very sharp). The experimental data show that the peaks narrow as the material is cooled and the data for different incident energies are shown in Fig. 8.13 with a calculated line based on the predictions of the two-dimensional quantum nonlinear sigma model. There are no fitting parameters in this theory apart from J which has been obtained from Raman scattering measurements; the agreement is therefore extremely impressive and demonstrates that this system is beginning to be understood in some detail.

Nevertheless the application of these results to the problem of high-$T_c$ superconductivity remains controversial. Recent tantalizing results have shown that in some of these systems the doped holes are not randomly arranged but segregate into stripes, the arrangement of the stripes depending rather subtlely

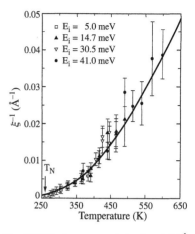

**Fig. 8.13** The inverse correlation length $\xi^{-1}$ in $Sr_2CuO_2Cl_2$ derived from inelastic neutron scattering. The data are for different values of the initial neutron energy $E_i$ and the theoretical curve is based on the predictions of the two-dimensional quantum nonlinear sigma model. After Greven *et al.* 1994.

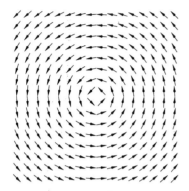

**Fig. 8.14** A vortex state in the XY model.

on the doping level. This intriguing cooperative behaviour appears to be part of the puzzle, but whether it is the crucial piece of the jigsaw or another red herring remains to be seen.

The situation dealt with so far in this section has been Heisenberg spins which have dimensionality $D = 3$, i.e. even though they lie on a $d = 2$ dimensional lattice, the spins can point in any three-dimensional direction. If the magnetic moments are constrained to lie in a plane, we have the two-dimensional XY model ($D = 2$). In this case there is no long range order at $T \neq 0$, but nevertheless there is an interesting phase transition which can occur. The spin–spin correlation function only decays away algebraically at low temperatures, not exponentially as they do at high temperatures. There is a transition temperature between these two regimes which is associated with the thermal stability of vortices.

The energy to create a vortex in an XY system, such as the one drawn in Fig. 8.14 is given approximately by $\pi J \ln(R/a)$ where $R$ is the size of the system and $a$ is a lattice constant. This energy cost diverges as $R \to \infty$ so it would seem that vortices are very costly to produce. However, the centre of the vortex could be at any one of the $(R/a)^2$ sites of the system, so the entropy of the vortex is $S = k_B \ln[(R/a)^2]$. Hence the free energy of a vortex is

$$F = (\pi J - 2k_B T) \ln(R/a) \qquad (8.3)$$

which becomes negative above a temperature $T_{KT}$ given by

$$T_{KT} = \frac{\pi J}{2k_B}, \qquad (8.4)$$

at the **Kosterlitz–Thouless transition** where vortices can be spontaneously thermally produced. This is a special phase transition because it occurs between two completely disordered phases, one at low temperature and one at high temperature. It is accompanied by a change of rigidity; the low temperature state has an elastic rigidity and the high temperature state does not. The transition can therefore also be thought of as a vortex unbinding transition since any vortices in the system below $T_{KT}$ are strongly bound. It is called a **topological phase transition** because there is no symmetry breaking (in contrast with a ferromagnetic–paramagnetic transition).

If the spins are constrained to lie in a line, we have a two-dimensional Ising model ($d = 2$, $D = 1$) which does have a magnetic phase transition. Materials such as $K_2CoF_4$ and $Rb_2CoF_4$ are approximate realizations of the two-dimensional Ising model and have been studied in detail using neutron scattering and other techniques.

## 8.6 Quantum phase transitions

The phase transitions which we considered in chapter 6 are all driven by temperature. In such phase transitions, it is the thermal fluctuations (whose energy scale is controlled by $k_B T$) which destroy the order as the sample is warmed through its transition. However, if one has a transition which is controlled by some other variable (such as pressure, magnetic field, or doping level) then at some critical value of this variable one can have a transition which can, in principle, occur at absolute zero. Such a zero temperature phase transition is called a **quantum phase transition** and the point at which it occurs is a **quantum critical point**. The relevant fluctuations are no longer thermal but the quantum mechanical fluctuations determined by Heisenberg's uncertainty principle. Quantum fluctuations have quite different properties from thermal fluctuations, and require completely different theories to describe them. A quantum system is described by a complex-valued wave function and this introduces features which are not present in classical systems.

In a classical phase transition driven by temperature, the correlation length and correlation time diverge as you approach the transition. The order parameter therefore fluctuates more slowly and over an increasing scale as the transition is approached. Therefore there is some frequency, say $\omega^*$, associated with the thermal fluctuations, which tends to zero at the critical temperature $T_c$. As long as $k_B T_c \gg \hbar\omega^*$ close to the phase transition, the critical fluctuations will behave classically. Hence for a quantum phase transition, where $T_c = 0$, quantum fluctuations cannot be ignored. Systems with quantum critical points are expected to show unusual dynamics which is controlled by quantum fluctuations and can be probed using such techniques as NMR and neutron scattering.

An example of a quantum phase transition is the Ising magnet $LiHoF_4$ in which the ferromagnetic order can be destroyed at absolute zero by applying a magnetic field perpendicular to the easy-axis of the Ising spins (see Fig. 8.15). This counterintuitive behaviour occurs because the magnetic field facilitates quantum tunnelling between the up and down spin states. Above a critical magnetic field the quantum fluctuations are sufficient to destroy the ferromagnetic order. The paramagnetic state is different from the usual high temperature paramagnet in which the spins constantly, but classically, fluctuate between the up and down states. In $LiHoF_4$ there is a unique phase-coherent wave function for the ground state which is a quantum superposition of up and down spin states.

There are a number of materials which show quantum critical points involving the proximity of magnetic and superconducting regions of the phase diagram. These include **heavy-fermion**[2] compounds such as $CePd_2Si_2$ (see Fig. 8.16(a)) which is an antiferromagnet at ambient pressure but upon application of hydrostatic pressure[3] can be made to superconduct. The phase transition at 20 kbar and $T = 0$ is a quantum critical point.

A second example is a family of **organic superconductors** based upon the organic molecule ET.[4] One can make very pure crystals of charge transfer salts of ET in which a pair of ET molecules jointly donate an electron to an inorganic anion, such as $Cu(NCS)_2^-$. The crystal structure consists of alternating layers of ET molecules and layers of the inorganic anion. The conductivity is high in the plane of the organic layers because of the molecular overlaps of the ET molecules, but is low between the planes. Organic materials are much softer than inorganic materials and so hydrostatic pressure has a much larger effect than in $CePd_2Si_2$. Also, by making changes to the anion, for example by making it bulkier, one can apply negative pressure by chemical means. Figure 8.16(b) shows that one obtains a similar phase diagram using a combination of chemical pressure and applied pressure. Again, an antiferromagnetic phase lies close to a superconducting phase.

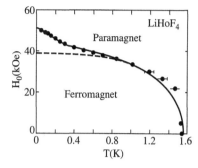

**Fig. 8.15** The phase diagram of $LiHoF_4$. The dashed line is a mean-field theory including only electronic spin degrees of freedom, while the solid line also incorporates the nuclear hyperfine interaction. After Bitko *et al.* 1996.

[2]Heavy-fermion compounds have very large effective masses, hence the name. They are often compounds of Ce or U.

[3]Hydrostatic pressure is pressure applied isotropically, the same way in all directions.

[4]ET is short for BEDT-TTF, which is in turn short for bis-ethylenedithiotetrathiafulvalene.

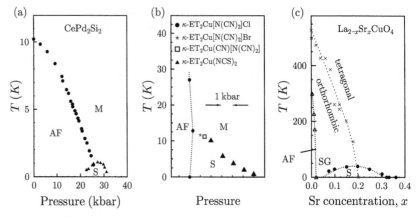

**Fig. 8.16** (a) Temperature–pressure phase diagram of the heavy-fermion superconductor $CePd_2Si_2$ (after Mathur *et al.* 1998); AF = antiferromagnetism, S = superconducivitity, M = metal. (b) Temperature–pressure phase diagram of the organic superconductors $\kappa$-$(ET)_2Cu(NCS)_2$, $\kappa$-$(ET)_2Cu[N(CN)_2]Br$, $\kappa$-$(ET)_2Cu(CN)[N(CN)_2]$ and $\kappa$-$(ET)_2Cu[N(CN)_2]Cl$. The pressure axis includes the effect of 'chemical pressure' caused by chemically varying the unit cell size (see text) as well as conventionally applied hydrostatic pressure (adapted from Kanoda 1997, Lubczynski *et al.* 1996). (c) Phase diagram of the cuprate superconductors $La_{2-x}Sr_xCuO_4$. (adapted from Kastner *et al.* 1998); SG = spin glass.

A third example is one we have already described, namely the antiferromagnet $La_2CuO_4$ which becomes superconducting when it is doped with Sr. Here the variable which controls the properties is the level of Sr doping, $x$. At a critical value of $x$ the system begins to switch from antiferromagnetic to superconducting.

This proximity between magnetic[5] and superconducting regions of the phase diagram is evidence that the spin fluctuations, rather than phonons,[6] are important in mediating the superconductivity in some of these materials. This story is however, at the time of writing, still far from being wrapped up.

## 8.7   Thin films and multilayers

So far we have considered two-dimensional magnets which occur in crystals. But it is also possible to directly engineer such systems by growing a very thin film of magnetic material using surface–science techniques. By using molecular–beam epitaxy, a thin magnetic film can be grown on to a flat substrate with a thickness of a single atomic layer. The growth is carried out under ultra-high vacuum conditions so that contamination and oxidation is reduced as far as possible. The lattice constant and symmetry of the substrate are chosen so as to closely match the material being grown. Not only films but mulitlayers can be grown, so that sandwich structures can be prepared and magnetic coupling studied through non-magnetic spacer layers, as will be considered in a later section.

The magnetic properties of a monolayer film are expected to be greatly different from the bulk material from which it is grown. An atom at the surface of a film has a smaller number of nearest neighbours in comparison with an atom in the bulk. The reduction in nearest neighbours reduces the electronic

[5]In $UGe_2$ the superconducting region of the phase diagram is adjacent to a ferromagnetic region (Saxena *et al.* 2000).

[6]In the BCS theory of superconductivity, the attractive pairing interaction between electrons is mediated by phonons. This appears to be correct for conventional superconductors, but is not the right description of superconductivity in heavy-fermion, organic and high temperature superconductors.

bandwidth and hence increases the density of states at the Fermi energy level, $g(E_F)$. This increases the chance of the Stoner criterion (see Section 7.3) being satisfied and hence increases the propensity to magnetic order. Magnetic thin films are therefore expected to have enhanced magnetic moments. In addition, a monolayer of Pd or V (normally non-magnetic metals) is predicted to be ferromagnetic because of this enhancement.

Another trick which can be achieved with these techniques is to grow metastable phases of magnetic elements by appropriately choosing a substrate on which to grow the film. Thin film growth of magnetic materials on crystalline substrates allows the forces present at the interface to drive the film into a different crystallographic phase. This alternative phase may already be known as a high temperature or high pressure phase, or alternatively as a phase which is completely unknown in the bulk. At room temperature Ni is normally face-centred cubic (fcc), Fe is normally body-centred cubic (bcc) and Co is usually hexagonally close-packed (hcp). However, by using particular substrates that lattice-match to each other, metastable crystallographic phases can be grown. For example metastable phases of fcc Co can be stabilized on fcc Cu surfaces and bcc Co can be stabilized on GaAs(110) surfaces. Since the energies associated with a change in crystal structure ($\sim 0.1$ eV/atom) are of the same order of magnitude as those associated with a change of magnetic structure (e.g. ferromagnetic to antiferromagnetic), often the magnetic properties of thin films dramatically depend on the growth conditions, as well as on the structure of the substrate.

The change in symmetry of atoms at the surface of a thin film also has an impact on the magnetic anisotropy and the easy direction of magnetization. To lowest order, the anisotropy energy of a ferromagnetic layer may be written as

$$E_{an} = K \sin^2 \theta \qquad (8.5)$$

where $K$ is an effective anisotropy constant which is the sum of three terms,

$$K = \frac{2K_s}{t} + K_v - \mu_0 M^2, \qquad (8.6)$$

and $\theta$ is the angle between the magnetization and the surface normal. The first term in eqn 8.6 represents the surface anisotropy. $K_s$ is the surface or interface anisotropy constant and $t$ is the thickness of the layer. The factor 2 appears because each layer has in general two faces. The second term is the volume anisotropy and may be due to lattice strains, or may appear because we have a uniaxial single crystal with its axis perpendicular to the plane of the film; $K_v$ is the volume anisotropy constant. The third term is the shape anisotropy (see Section 6.7.2) of the film. It is a dipolar contribution and is calculated by assuming a uniform distribution of magnetic poles on the plane surfaces, and can therefore in practice be significantly reduced from this calculated value by the presence of interface roughness.

In thick films, the dipolar term dominates and the magnetization lies in the plane of the film. In thin films, the surface or interface term may dominate (because of the $t^{-1}$ factor) and the spontaneous magnetization may become perpendicular to the plane of the film. This perpendicular anisotropy can be particularly useful in recording applications.[7] $K_s$ arises because the surface states have lower symmetry which enhances the magnetocrystalline anisotropy

[7]In recording applications, small regions on the film store bits of information. The dipolar field experienced by one of these small regions due to its neighbours is lower if the magnetization is perpendicular to the plane. Therefore the bits can, in principle, be stored at a higher density. However, this explanation glosses over a large number of practical issues which can sometimes lead manufacturers to prefer systems in which the magnetization lies in the plane.

at the surface. It is affected by the roughness of the interface and also by the lattice mismatch at each interface. The preparation of systems with large perpendicular magnetization is of great importance in the development of magnetic recording technology.

A number of experimental techniques have been used to study the magnetization of thin films. Mössbauer techniques have already been described (see Section 3.2.3) and magneto-optic methods will be introduced in the next section. Polarized neutron reflection (PNR) is a technique which involves spin-polarized neutrons reflected at grazing incidence from a magnetic multilayer. The reflectivity is different for spin-up and spin-down neutrons and can be used to infer the depth-dependent magnetization (direction and magnitude). In magnetic force microscopy (MFM), a magnetic moment on a cantilever is scanned across the surface of a sample. There is a force on the tip whenever there is a gradient in the magnetic field above the sample, so the technique is sensitive to domain walls (see Section 6.7.6). Another microscopy technique is SEMPA (Scanning electron microscopy with polarization analysis) which is a conventional scanning electron microscope which studies the ejection of secondary electrons (these retain their spin polarization and give information on the magnetization of the surface). Other techniques include Kerr microscopy (in which a polarized laser beam is scanned across a surface to image the magnetization) and scanning SQUID microscopy.

## 8.8    Magneto-optics

Michael Faraday (1791–1867)

Magneto-optical effects were first studied by Michael Faraday in 1845 who showed that when polarized light passed through a piece of glass that was placed in a magnetic field, the light emerged with its plane of polarization rotated. This effect is now known as the Faraday effect. A related phenomenon was found by John Kerr in 1877; while Faraday's effect occurs in transmission, the magneto-optic Kerr effect occurs in reflection. John Kerr reflected light from the polished pole of an electromagnet and noticed that its plane of polarization was rotated.

John Kerr (1824–1907)

Both effects are related to the spin–orbit interaction, and a rigorous treatment of both effects requires the use of perturbation theory. Nevertheless the general principles can be understood rather more simply as follows. Right-handed and left-handed circularly polarized light cause charges in a material to rotate in opposite senses, and each polarization therefore produces a contribution to the orbital angular momentum having an opposite sign. A magnetic field gives rise to a spin-polarization along the magnetic field direction and the spin–orbit interaction then leads to an energy contribution for the two circular polarizations having the same magnitude but with opposite sign. This leads to right-handed and left-handed polarizations having different refractive indices in the material. If plane polarized light is incident on a magnetic material, it should be considered as a sum of right-handed and left-handed circularly polarized beams which propagate through the material at different speeds. When they emerge, these two beams recombine but the phase-lag between them implies that the emerging beam has a rotated plane of polarization.

In an isotropic medium the dielectric tensor becomes

$$\epsilon = \epsilon \begin{pmatrix} 1 & iQ_z & -iQ_y \\ -iQ_z & 1 & iQ_x \\ iQ_y & -iQ_x & 1 \end{pmatrix}, \qquad (8.7)$$

where $\mathbf{Q} = (Q_x, Q_y, Q_z)$ is known as the **Voigt vector** which is aligned with the magnetic field and has a magnitude which depends on the material. It is the off-diagonal terms in the dielectric tensor which contribute to magneto-optic effects. The tensor leads to the two circularly polarized normal modes which have dielectric constants $\epsilon_\pm = \epsilon(1 \pm \mathbf{Q} \cdot \hat{\mathbf{k}})$ where $\hat{\mathbf{k}}$ is the direction of propagation of the light. The circular modes travel with different velocities and attenuate differently. The emerging waves therefore combine to yield a rotated axis of polarization and also an ellipticity (the different attenuations lead to the polarization of the light that emerges being slightly elliptical).

Woldemar Voigt (1850–1919)

Studies of metallic magnets use the Kerr effect rather than the Faraday effect because metals absorb light and so cannot be studied in transmission, unless a sample is very thin. The Kerr effect is particularly useful for thin magnetic films and surfaces. In this case the technique is called **SMOKE** (which stands for the surface magneto-optic Kerr effect). The technique can be used during film growth inside an ultra-high vacuum (UHV) chamber by passing a laser beam through a window of the chamber such that it reflects from the sample and emerges from the chamber through another window. The change of polarization of the light can be monitored using crossed polarizers.

The Kerr effect can be performed in various geometries (see Fig. 8.17): in the **polar Kerr effect** the magnetization direction $\mathbf{M}$ is perpendicular to the plane of the film. If $\mathbf{M}$ is in the plane of the film, then one can measure the **longitudinal Kerr effect** if $\mathbf{M}$ is in the scattering plane of the light or the **transverse Kerr effect** if $\mathbf{M}$ is perpendicular to the plane of incidence. Because it can be performed during film growth, SMOKE is useful in studies of surface magnetism, magnetism of films as a function of film thickness, and surface anisotropies (for example it can be used to show at what thickness of film the magnetization changes from in-plane to out-of-plane).

Magneto-optical effects are extremely useful in magnetic recording since one can scan a polarized laser beam across a disk very rapidly and measure the magnetic information encoded on the disk by the changes in polarization of the reflected light. Although magneto-optical effects are used to read the disk, data are written by using another laser to locally heat a tiny region of the disk above $T_C$ and applying a magnetic field to the disk; only the heated region switches – the information on the rest of the disk is left unaltered. The laser beam then moves on to the next region to write the next bit. At the turn of the 21st century, magneto-optical recording disks have started to be of enormous importance, providing a cheap and portable method of storage of relatively large amounts of data.

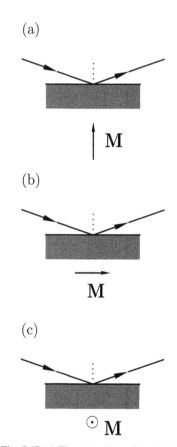

**Fig. 8.17** (a) The polar Kerr effect; (b) the longitudinal Kerr effect; (c) the transverse Kerr effect.

## 8.9 Magnetoresistance

The change in resistance, $R$, of a material under an applied magnetic field $H$ is known as **magnetoresistance**. The magnetoresistance $\Delta\rho/\rho$ is usually defined by

$$\frac{\Delta\rho}{\rho} = \frac{R(H) - R(0)}{R(0)}. \tag{8.8}$$

As we shall see, many interesting magnetoresistances are large and negative, and you can make the magnetoresistance of your new 'wonder magnetic material' appear deceptively larger by defining the magnetoresistance as $(R(H) - R(0))/R(H)$, i.e. by dividing by the smaller number. This trick has been used more than once in the research literature!

William Thomson (Lord Kelvin) (1824–1907)

It is a technologically useful quantity because magnetoresistive sensors are extensively used in applications (e.g. for measuring the magnetic field from the magnetic strip on a credit card). A free electron gas shows no magnetoresistance, even with an anisotropic effective mass; magnetoresistance only appears in models with more than one carrier, and its high field behaviour depends on the topology of the Fermi surface. The effect was first discovered in 1856 by Lord Kelvin (then William Thomson) who was examining the resistance of an iron sample. He found a 0.2 % increase in the resistance of the iron when a magnetic field was applied longitudinally, and a 0.4 % decrease when the field was applied in the transverse direction.

### 8.9.1   Magnetoresistance of ferromagnets

The observation of negative magnetoresistance in ferromagnets is a very puzzling one. When a metal carries a current, the displacement of electrons to different parts of the Fermi surface is such that scattering is minimized; the electrons find the path of least dissipation to cross the sample. Hence, if electrons are forced to take a different path, because of the presence of an applied magnetic field for example, they would take a path which leads to more scattering. Thus, in general, a *positive* magnetoresistance is expected. A negative magnetoresistance can sometimes be observed at low temperatures in samples which are thin compared to the mean free path. If a magnetic field is applied in the plane and perpendicular to the current direction, the electron paths describe orbits with smaller diameters and therefore surface scattering is reduced. However, in ferromagnets, the explanation of the negative magnetoresistance must be entirely different.

A very important insight into this problem was provided by Mott who considered the transport properties of Ni (see Fig. 8.18 for a schematic diagram of the spin-split bands), in which only a few eV are needed to change the configuration from $(3d^8 4s^2)$ to $(3d^9 4s^1)$ or $(3d^{10})$. In general, Ni is considered to be $(3d^{9.4} 4s^{0.6})$. The d band is very narrow (which is a necessary condition for ferromagnetism in transition metals so that $g(E_F)$ is large and the Stoner criterion is satisfied), and hence $m_d^* \gg m_e$. As the s bands are nearly free, $m_s^* \sim m_e$. Hence, the conductivity $\sigma$, which is given by

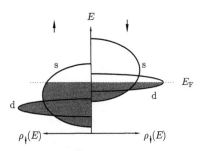

**Fig. 8.18** The spin-split bands in a ferromagnet

$$\sigma = \frac{n_s e^2 \tau_s}{m_s^*} + \frac{n_d e^2 \tau_d}{m_d^*}, \tag{8.9}$$

is dominated by the first term and conduction is mainly due to the s electrons. In eqn 8.9, $n_s$ and $n_d$ are the number of electrons in the s and d bands respectively, and the scattering times are $\tau_s \sim \tau_d$.

The transition probability is mainly due to s→d transitions. At low temperatures, $T \ll T_c$, all the unoccupied d states are antiparallel, so only half of the s electrons can make transitions. For $T > T_c$, all the s electrons can make transitions, and so there is more scattering. Hence, a decrease in the resistivity below $T_c$ is expected. Now, consider applying a magnetic field to

Ni. A magnetic field may increase the spin polarization and allow fewer s→d transitions; therefore a negative magnetoresistance is observed.

Most elastic collision processes are such that the electron conserves its spin. These collisions are characterized by a relatively short relaxation time $\tau$. However, even a weak spin–orbit coupling can allow a weak spin-flip scattering which has a much longer relaxation time, $\tau_{sf}$. In the absence of any external forces, a perturbation created in the equilibrium distribution of spin-up and spin-down electrons, will decay first into a uniform distribution in each spin in a time $\tau$, and after that it will reach the equilibrium distribution in a time $\tau_{sf}$. For $T \ll T_c$, spin-flip scattering is not expected to be dominant, and a ferromagnet can be well approximated by the **two-current model** in which the ↑ and ↓ electron currents are considered independently. This has been particularly successful in describing the properties of alloys in which a small quantity of one transition metal (the impurity) is dissolved in another transition metal (the host). The scattering due to certain transition-metal impurities is strongly spin-dependent. This is due to the combined effects of the spin-splitting of the host d band, the spin-splitting of the impurity d levels and the different hybridization between the host and impurity states for the spin ↑ and spin ↓ directions. For example, Cr impurities in Fe scatter the spin ↑ electrons much more strongly, resulting in a ratio of the resistivities for each spin-state of $\rho_\uparrow/\rho_\downarrow \sim 6$. At high temperatures, the spin-flip scattering due to collisions with spin waves leads to **spin-mixing**, i.e. the blurring of the distinction between the two spin channels. This concept should be kept in mind since it will return in the discussion of the giant magnetoresistance in sandwich structures.

## 8.9.2  Anisotropic magnetoresistance

In ferromagnets, the measured magnetoresistance depends on the orientation of the magnetization with respect to the direction of the electric current in the material. This effect is known as **anisotropic magnetoresistance**. Its origin is connected with the spin–orbit interaction and its influence on s-d scattering. The symmetry of atomic wave functions is lowered by the spin–orbit interaction which also mixes states. The crystal axes determine the direction of **L** and the magnetization determines the direction of **S**, so that the mixing of states leads to anisotropic scattering. Using symmetry arguments, it is possible to predict the general form of the dependence of the magnetoresistance on the direction of the magnetization and the current density in particular crystalline or polycrystalline materials.

## 8.9.3  Giant magnetoresistance

Anisotropic magnetoresistive effects are rather small, and so something of a revolution occurred in the late 1980s with the discovery of a very large effect (given the name **giant magnetoresistance**, or GMR for short) in Fe/Cr/Fe multilayers. A large negative magnetoresistance of more than 50% was discovered at high magnetic field at low temperatures (see Fig. 8.19). The effect is associated with samples with magnetic Fe layers which were antiferro-magnetically coupled. It was found that the coupling between magnetic layers through a spacer layer oscillates in sign as the spacer thickness increases. For

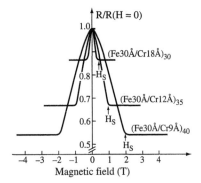

**Fig. 8.19** Giant magnetoresistance in Fe/Cr/Fe multilayers (Baibich *et al.* 1988).

certain thicknesses it is ferromagnetic, and then is antiferromagnetic at larger thicknesses, and then returns to antiferromagnetic. It appears that the period of the oscillation is of the order of ten atomic spacings (usually between 9 and 18 Å) and its value is mainly determined by the spacer metal, not by the ferromagnetic metal. Good matching between the spacer and ferromagnetic lattices favours large couplings. In the best cases, like Co/Ru superlattices, the amplitude of the oscillation decays with spacer thickness $t$ as $1/t^2$.

The origin of these oscillations is connected with the RKKY interaction $J(r) \sim \cos(2k_F r)/r^3$ between two localized spins separated by a distance $r$ in a bulk metal. When summed over all the spins on the interfaces, the coupling becomes $J \sim \cos(2k_F t)/t^2$ where $t$ is the separation of the two ferromagnetic layers. The oscillations in the coupling can be directly related to the topology of the Fermi surface of the spacer layer. If the ferromagnetic metal is assumed to have a full majority spin d band, then in the case of ferromagnetic coupling, minority spin holes will be confined in the spacer layer, but in the case of antiferromagnetic coupling, there will be no confinement. The energy difference between these different couplings, with the number of particles conserved, is then entirely determined by the size quantization of the energy of minority spin holes in the spacer layer (which has thickness $t$). This arises because the energy levels in the spacer layer are *discrete*; therefore the density of states consists of a series of steps, the width of each being proportional to $1/t^2$. As $t$ increases, the step width decreases, and one of the steps must eventually cross the Fermi level. The formalism which is used to calculate this effect can be understood as a one-dimensional analogue of the de Haas–van Alphen effect; in the case considered here, the size quantization is due to one-dimensional (rather than two-dimensional) confinement in a direction perpendicular to the layer planes

The type of magnetic coupling in a sandwich structure can directly influence the observed magnetotransport behaviour since this is very sensitive to the alignment of the magnetic layers, with the GMR effects being largest for antiferromagnetic coupling. The first explanation for GMR was given in terms of the two-current model (see above) which separately considers the individual currents of ↑ and ↓ electrons (↑ means parallel to the majority spin band). In this discussion, I will initially assume that the mean free path $\lambda$ is much greater than the Cr interlayer thickness $t_{Cr}$. Imagine an Fe/Cr/Fe structure in which the magnetization in each of the two Fe layers are aligned antiparallel in zero applied field, and suppose further that $\rho_\downarrow \ll \rho_\uparrow$. Then there are two cases to consider:

(1) $H > H_s$ (where $H_s$ is the saturation field). Here the magnetic moments in the Fe layers are aligned (see Fig. 8.20(a)) so that the resistivity $\rho$ is given by

$$\frac{1}{\rho} = \frac{1}{\rho_\uparrow} + \frac{1}{\rho_\downarrow} \quad \Rightarrow \quad \rho \sim \rho_\downarrow. \tag{8.10}$$

There is an effective *short circuit* by the less scattered electrons.

(2) $H = 0$. The magnetic moments in the two iron layers are now antiparallel (see Fig. 8.20(b)). In this case the electrons are alternately ↑ and ↓ in each of the layers with respect to the local magnetization, and the spin ↑ and ↓ channels are effectively 'mixed' (cf. the spin mixing due

to spin waves in alloys at high temperature, considered above), so that $\rho_\uparrow \to \rho_{av}$ and $\rho_\downarrow \to \rho_{av}$ where $\rho_{av} = (\rho_\uparrow + \rho_\downarrow)/2$ so that the total resistivity $\rho$ is given by

$$\rho = \frac{\rho_\uparrow + \rho_\downarrow}{4} \gg \rho_\downarrow. \qquad (8.11)$$

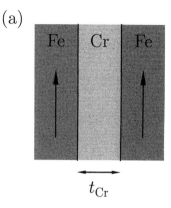

This again predicts the negative magnetoresistance which is observed. The large effect is due to the inequality of $\rho_\uparrow$ and $\rho_\downarrow$, which can be very large, and the ease in which the 'spin-mixing' can be switched on and off, simply by the application of the magnetic field.

For thicker Cr layers, the spin-dependent scattering at the interface affects the electron distribution function near the interface within a layer whose thickness is of the order of the mean free path $\lambda$. If $t_{Cr} \gg \lambda$, the magnetoresistance is expected to fall roughly as $\exp(-t_{Cr}/\lambda)$. However, there will be no giant magnetoresistance effect if there is no antiferromagnetic coupling, and since this appears periodically in $t_{Cr}$, the magnetoresistance is also expected to oscillate with increasing interlayer thickness, as well as die away exponentially.

Magnetic coupling through multilayers can be measured using many of the experimental techniques that we described in Section 8.7 on magnetic thin films. Useful additions to these are **ferromagnetic resonance** (the analogue of NMR or ESR for ferromagnets) and **Brillouin light scattering** techniques (which study excitations using the inelastic scattering of light). These can both be used to study spin waves and magnetostatic modes in magnetic thin films. The spins in the multilayers are coupled together through exchange, dipolar and anisotropic interactions. Hence the spin-wave excitations, which are the eigenmodes of this magnetic system, have a dispersion relation which can depend quite sensitively on the exchange coupling and anisotropies and magnetoelastic effects.

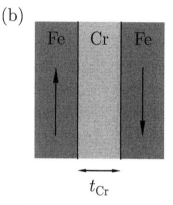

**Fig. 8.20** (a) The magnetic moments in the Fe layers are parallel when $H > H_s$. (a) The magnetic moments in the Fe layers are antiparallel when $H = 0$ if the thickness of the chromium layer, $t_{Cr}$, is chosen correctly.

The preparation of multilayers is not the only way to achieve GMR. It has also been observed in heterogeneous Cu–Co alloy films. The relative orientation of the magnetic moments inside the Co-rich grains inside the Cu-rich matrix determines the magnetoresistance and this can be varied by an applied field. GMR is however not observed in *homogeneous* alloys which do not possess isolated, large, magnetically-rich grains of the appropriate size. Alloys are easier to prepare than multilayers and therefore this offers exciting prospects for sensor applications.

### 8.9.4 Exchange anisotropy

**Exchange anisotropy** (or unidirectional anisotropy) is the interfacial exchange that can be observed between a ferromagnet and an antiferromagnet. If the Curie temperature $T_C$ of a ferromagnet is greater than the Néel temperature $T_N$ of an antiferromagnet, then by depositing one over the other (for an appropriate choice of the ferromagnet and antiferromagnet, and for a sufficiently thick ferromagnetic layer), and cooling them in an applied field through the Néel temperature, the measured magnetic hysteresis loop observed at $T \ll T_N$ appears to be shifted as if another magnetic field was present in addition to the applied magnetic field. It appears that it is energetically

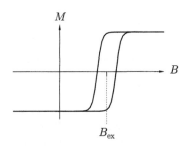

**Fig. 8.21** A hysteresis loop shifted by an exchange anisotropy. Without the exchange anisotropy the loop would be centred at zero magnetic field.

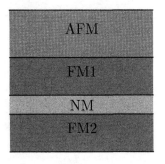

**Fig. 8.22** A spin valve. The antiferromagnet (AFM) 'clamps' the magnetic moment in the upper ferromagnetic layer (FM1) along a particular direction. The magnetic moment in the lower magnetic layer (FM2) is free to rotate and can be aligned with an applied magnetic field. The angle between the magnetic moments in the two ferromagnetic layers controls the resistance of the device. In zero applied magnetic field, the relative alignment of the layers is controlled by the thickness of the non-magnetic spacer layer (NM).

favourable for the ferromagnetic film to be magnetized one way (in the direction in which it was cooled) rather than the other. It is sometimes said to be **exchange biased** by the antiferromagnetic layer. The effect of the exchange biasing is to produce a unidirectional exchange field $\mathbf{B}_{ex}$ which can act in competition with the applied field $\mathbf{B}$ (the total energy of the magnetic layer is given by $-\mathbf{M} \cdot (\mathbf{B} + \mathbf{B}_{ex})$). If $\mathbf{B}$ and $\mathbf{B}_{ex}$ are in the same direction, the effect is simply to shift the hysteresis loop (see Fig. 8.21). If they are at right angles, a hard axis response is obtained. In both cases, there is a *unique* angle of the magnetization which minimizes the energy at each value of the magnetic field.

This technique can be used to add an 'exchange bias' field to magnetoresistive sensor heads in order to bias them into their linear regions. It is also used to constrain the direction of the magnetization in one soft ferromagnetic layer in a sandwich structure. The magnetization in the other layer can then be rotated by an applied field, allowing magnetoresistance and coupling energies to be measured in a carefully controlled way. This is also the basis of the **spin-valve** (see Fig. 8.22) which is a giant magnetoresistive sensor consisting of two magnetic layers with a non-magnetic spacer, with one of the magnetic layers adjacent to an antiferromagnetic layer. The resistance across the sandwich structure is then sensitive to the value of the applied magnetic field along a particular direction.

### 8.9.5   Colossal magnetoresistance

The transport properties of Mn oxides have recently generated enormous interest. $LaMnO_3$ contains Mn in the $Mn^{3+}$ state which is a Jahn–Teller ion. $LaMnO_3$ shows A-type antiferromagnetic ordering (see Section 5.2.4). If a fraction $x$ of the trivalent $La^{3+}$ ions are replaced by divalent $Sr^{2+}$, $Ca^{2+}$ or $Ba^{2+}$ ions, holes are introduced on the Mn sites. This results in a fraction $1 - x$ of the Mn ions remaining as $Mn^{3+}$ ($3d^4$, $t_{2g}^3 e_g^1$) and a fraction $x$ becoming $Mn^{4+}$ ($3d^4$, $t_{2g}^3 e_g^0$). When $x = 0.175$ the Jahn–Teller distortion vanishes and the system becomes ferromagnetic with a Curie temperature around room temperature. Above $T_C$ the material is insulating and non-magnetic, but below $T_C$ it is metallic and ferromagnetic. Particularly near $T_C$ the material shows an extremely large magnetoresistive effect (see Fig. 8.23) which has been called **colossal magnetoresistance** (abbreviated to CMR), the term 'giant' having been already taken! The origin of the CMR is connected with the presence of a metal–insulator transition.

The origin of the behaviour described above is partly connected with the phenomenon of double exchange (see Section 4.2.5). In a $Mn^{3+}$ ion, the $t_{2g}$ electrons are tightly bound to the ion but the $e_g$ electron is itinerant. Because of the double exchange interaction, the hopping of $e_g$ electrons between Mn sites is only permitted if the two Mn core spins are aligned (in fact the hopping probability is proportional to $|\cos(\theta/2)|$ where $\theta$ is the angle between the two Mn core spins). The magnetic field aligns the core spins and therefore increases the conductivity, especially near $T_C$.

The situation is actually more complicated because the carriers interact with phonons because of the Jahn–Teller effect. The strong electron–phonon coupling in these systems implies that the carriers are actually **polarons** above $T_C$, i.e. electrons accompanied by a large lattice distortion. These polarons are

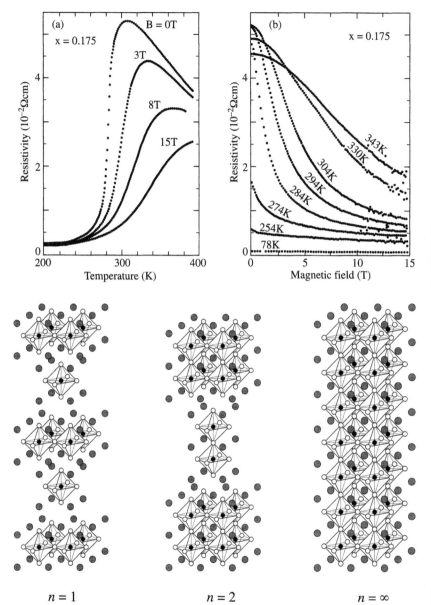

**Fig. 8.23** Colossal magnetoresistance in $La_{1-x}Sr_xMnO_3$ for $x = 0.175$. (a) Temperature dependence of the resistivity. (b) Isothermal magnetoresistance. After Tokura *et al.* 1994.

**Fig. 8.24** The Ruddlesden–Popper phases of the manganites.

magnetic and are self-trapped in the lattice. The transition to the magnetic state can be regarded as an unbinding of the trapped polarons. There are other signatures of the electron–phonon couplings, including magnetic-field dependent structural transitions and charge ordering. The charge-ordered ground state can compete with ferromagnetism and is enhanced near commensurate values of the doping (e.g. at $x = \frac{1}{2}$ where there is one hole for every two Mn sites). These effects are not fully understood and are under active current study.

The manganese perovskites are just one type of oxide material which shows colossal magnetoresistance. Perovskites are a member of a large family of crystals known as **Ruddlesden–Popper phases** (see Fig. 8.24). These phases have a general formula $X_{n+1}Mn_nO_{3n+1}$ (where X is a lanthanide, strontium, or mixture) and can be thought of as stacks of perovskite blocks[8] $n$ layers thick with each block separated by a rock-salt like $(Sr,Ln)_2O_2$ layer which tends to decouple the blocks electrically and magnetically. The perovskite compounds are realized when $n = \infty$, whereas the case $n = 1$ is equivalent to the $K_2NiF_4$ structure, adopted by the high-$T_c$ cuprates. Manganese compounds can be made which adopt a range of structures (e.g. $n = 1, 2, 3$ and $\infty$ have all been prepared at the time of writing). In the perovskites $n = \infty$, each MnO$_6$ octahedron is surrounded by six others. In the $n = 2$ phases, this 'coordination number' drops to five, and it is four for the $n = 1$ phases. The reduction in the number of nearest-neighbours is expected to produce an anisotropic reduction in the width of the energy bands which are derived (largely) from the Mn 3d orbitals and this modifies the electrical conductivity and magnetic behaviour of these materials. These materials can thus be carefully controlled by using different dopants, different doping levels, and adjusting the dimensionality.

### 8.9.6  Hall effect

A magnetic field applied normal to the current direction in a conducting material produces a transverse force on the conduction electrons in the film. This force on the conduction electrons gives rise to a transverse **Hall voltage**. This is known as the **ordinary Hall effect**[9] and is proportional to the applied magnetic field **B**, because the Lorentz force on conduction electrons is $\mathbf{F} = e(\mathbf{E}+\mathbf{v}\times\mathbf{B})$. In ferromagnets an additional effect occurs, known as the **extraordinary Hall effect** which depends on the magnetization. Note that it is *not* just a Lorentz force due to **M** rather than **B**, because this is already included in the ordinary Hall effect since $\mathbf{B} = \mu_0(\mathbf{H}+\mathbf{M})$. Empirically the transverse resistivity $\rho_H$ is given

$$\rho_H = R_o B + \mu_0 R_e M, \tag{8.12}$$

where $R_o$ is the ordinary Hall coefficient, and $R_e$ is the extraordinary Hall coefficient (sometimes the spontaneous Hall coefficient). The ordinary Hall coefficient $R_o$ tends to be fairly temperature independent, whereas $R_e$ is usually very temperature dependent. The effect is not only seen in ferromagnets, but also in strong antiferromagnets or paramagnets. The effect only requires the presence of localized moments. (For example, it is seen in Tb in both its ferromagnetic and paramagnetic states.) If you measure the Hall resistivity as a function of field, a straight line graph results but there is a sudden change of gradient when the saturation field is reached (see Fig. 8.25). This is because at low magnetic field both $B$ and $M$ increase, so both ordinary and extraordinary Hall effects are seen and the gradient is due to $R_o$ and $R_e$. Above the saturation field $M$ can no longer show a further increase and the gradient results from $R_o$ alone.

The mechanism for the extraordinary Hall effect is associated with the spin–orbit interaction between the conduction electrons and localized moments. The carriers have an orbital angular momentum about the scattering centre whose sign depends on whether they pass the centre on one side or the other. This orbital angular momentum **L** couples to the spin angular momentum **S** of the scattering centre because of the spin–orbit interaction which is proportional to $\mathbf{L} \cdot \mathbf{S}$. The electrons thus find it energetically favourable to pass to one side

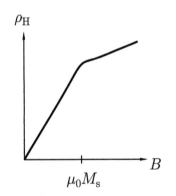

$\rho_H$

$B$

$\mu_0 M_s$

**Fig. 8.25** The Hall resistivity in a ferromagnet.

rather than the other. This results in asymmetric scattering and hence to a transverse current which is responsible for the extraordinary Hall effect.

# 8.10   Organic and molecular magnets

Magnetic materials have been conventionally prepared from substances which are essentially the products of inorganic chemistry. However there has been recent interest in *organic* magnetic materials which are based upon organic chemistry. The tunability, resulting from the rich structure of carbon chemistry which allows many small adjustments to be made to each molecule, means that in principle materials can be tailor-made to exhibit desired properties. This is only 'in principle' however, since the structure–property relations can turn out to be remarkably intricate. The impetus to study organic materials in condensed matter physics stems particularly from the recent remarkable discoveries of three types of new material: first, conducting polymers which can have electrical conductivities as high as conventional metals such as copper and which have been used in the fabrication of polymer transistors and light-emitting-diode devices; second, superconducting charge transfer salts which can exhibit superconducting transition temperatures as high as 15 K; third, carbon 'bucky-balls' $C_{60}$, a new allotrope of carbon which when appropriately doped may exhibit superconductivity or even unusual magnetic properties.

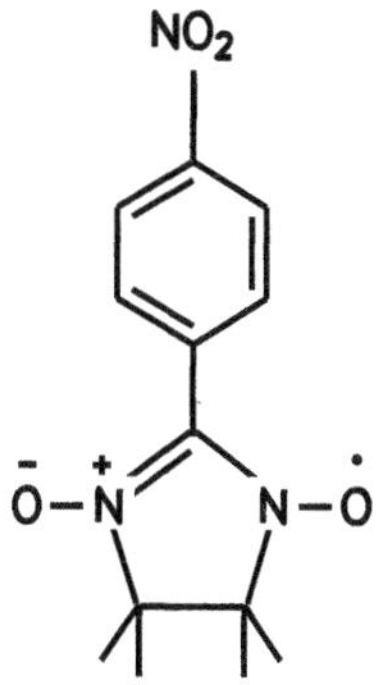

Ferromagnets are rather rare even among the elements and are exclusively found in the d or f block. To achieve a magnetic moment on an organic material, it is natural to search for an organic free radical which has unpaired spins. Many organic radicals exist, but few are stable enough to be assembled into crystalline structures. Moreover, even when that is possible, *aligning* these spins ferromagnetically is usually impossible. Thus the discovery of ferromagnetism, albeit at rather low temperatures, in certain nitronyl nitroxide molecular crystals in the early 1990s was particularly remarkable. The first material of this sort to be found was *para*-nitrophenyl nitronyl nitroxide (*p*-NPNN) (see Fig. 8.26) which shows ferromagnetism up to $T_C \sim 0.65$ K only in one of its crystal phases. Nitronyl nitroxides contain only the elements C, H, N, and O and are therefore fully **organic magnets**. On each nitronyl nitroxide molecule there is an unpaired spin associated with the two N–O groups. Small chemical changes to the rest of the molecule lead to significant changes in crystal structure, thereby altering the intermolecular overlaps and thus the magnetic interactions between unpaired spins on neighbouring molecules. Thus, different compounds have greatly different magnetic ground states. These materials mostly have transition temperatures below 1 K.

Some of the most technologically promising materials are **molecular magnets** in which a transition metal ion provides the localized moment and organic bridges act as exchange pathways. Progress has also been achieved using materials with unpaired electrons on both the metal ions and the organic molecule. These have much higher transition temperatures than purely organic magnets, sometimes above room temperature. One of the promising aspects of these materials is their optical properties. Some of them are transparent and change colour when they become magnetic, often when this is associated with a spin transition. They therefore have a number of potential applications in cases

**Fig. 8.26** The molecular structure of the organic ferromagnet *para*-nitrophenyl nitronyl nitroxide (*p*-NPNN).

when it is necessary to see by eye that a magnetic change has taken place (e.g. on telephone cards or display applications).

## 8.11    Spin electronics

The ability to manipulate and amplify currents of different spin-type is the basis of the emerging field of **spin electronics** (also known as **spintronics**) which is an attempt to fabricate devices out of combinations of metallic ferromagnets, non-magnetic metals and semiconductors. The ideas at the basis of this field are all connected with the fact that spin-up and spin-down electrons have different mobilities in a ferromagnet. A crucial feature is the degree of spin polarization in a given ferromagnet. The **spin-polarization** $P$ is defined by

$$P = \left| \frac{n_\uparrow - n_\downarrow}{n_\uparrow + n_\downarrow} \right|, \tag{8.13}$$

and is less than unity for elemental metallic ferromagnets (0.44 for Fe, 0.34 for Co and 0.11 for Ni). However in some materials $P = 1$, implying that there is one spin-split band which is completely empty. Such materials are called **half-metallic magnets** (because the electrons of one spin are metallic and those of the other spin are insulating) and $CrO_2$, $Fe_3O_4$ and some of the manganites are believed to fall into this class. Such materials have a very long spin–diffusion length, and may be exciting materials for use in spin electronics applications, for example in **magnetic tunnel junctions** in which spin polarized electrons are used to tunnel through an insulating barrier.

A few different designs for a **spin transistor** have recently appeared which attempt to use the tunnelling of electrons of different spins from ferromagnets into semiconducting layers. Conventional electronics is based only on electronic charge and ignores electronic spin. There is much current optimism that the use of magnetic materials to control the spin of electronic currents could be a promising new development for electronic devices in this century.

## Further reading

- J. A. Mydosh, *Spin glasses: an experimental introduction*, Taylor and Francis 1993 and K. H. Fischer and J. A. Hertz, *Spin glasses*, CUP 1991, both contain useful accounts of theoretical and experimental work on spin glasses.

- E. Dagotto, *Reports on Progress in Physics* **62**, 1525 (1999) is a readable review on spin ladders.

- A. M. Tsvelik, *Quantum field theory in condensed matter physics*, CUP 1995, is an advanced text which contains a description of current theories used in describing low-dimensional magnetism.

- A. Auerbach, *Interacting electrons and quantum magnetism*, Springer-Verlag 1994, is a lively treatment of the quantum mechanics of low-dimensional magnets and the

Haldane gap. These topics are also reviewed in I. Affleck, *J. Phys.: Condensed Matter*, **1**, 3047 (1989).

- M. A. Kastner, R. J. Birgeneau, G. Shirane and Y. Endoh, *Rev. Mod. Phys.*, **70**, 897 (1998) contains a review of magnetic properties of single layer copper oxides and the high-$T_c$ problem.

- E. Manousakis, *Rev. Mod. Phys.*, **63**, 1 (1991) reviews the theory of the two-dimensional Heisenberg antiferromagnet on a square lattice.

- P. M. Chaikin and T. C. Lubensky, *Principles of condensed matter physics*, CUP 1995, is an excellent source of information concerning the statistical mechanics of systems in condensed matter.

- Quantum phase transitions are reviewd in S. L. Sondhi, S. M. Girvin, J. P. Carini and D. Shahar, *Rev. Mod. Phys.*, **69**, 315 (1997).

- A review of magnetic multilayers with an emphasis on spin-wave excitations is R. E. Camley and R. L. Stamps, *J. Phys.: Condensed Matter*, **5**, 3727 (1993).

- Organic magnetism is reviewed in O. Kahn, *Molecular magnetism*, VCH (1993).

- K. De'Bell, A. B. MacIsaac and J. P. Whitehead, *Rev. Mod. Phys.*, **72**, 225 (2000) contains an account of dipolar effects in thin film magnets.

- The current status of various topics in magnetism research may be discerned by reading the conference proceedings of major magnetism conferences which are published from time to time in the *Journal of Applied Physics*, the *Journal of Magnetism and Magnetic Materials* and *Physica B*. Recent views of various aspects of magnetism research are published in the *Journal of Magnetism and Magnetic Materials* volume 100 × *n* where *n* is an integer.

# References

- M. N. Baibich, J. M. Broto, A. Fert, F. Nguyen van Dau, F. Petroff, P. Etienne, G. Creuzet, A. Friederich, and J. Chazelas, *Phys. Rev. Lett.*, **61**, 2472 (1988).

- D. Bitko, T. F. Rosenbaum and G. Aeppli, *Phys. Rev. Lett.* **77**, 940 (1996).

- J. W. Bray, L. V. Interrante, I. S. Jacobs, J. C. Bonner, p.353 of *Extended linear chain compounds* 3, edited by J. S. Miller, New York Plenum Press (1983).

- B. R. Coles, B. V. B. Sarkissian and R. H. Taylor, *Phil. Mag.* **37**, 489 (1978).

- J. des Cloizeaux and J. J. Pearson, *Phys. Rev.* **128**, 2131 (1962).

- M. Greven, R. J. Birgeneau, Y. Endoh, M. A. Kastner, B. Keimer, M. Matsuda, G. Shirane and T. R. Thurston, *Phys. Rev. Lett.* **72**, 1096 (1994).

- K. Kanoda, *Hyp. Int.*, **104**, 235 (1997).

- M. A. Kastner, R. J. Birgeneau, G. Shirane and Y. Endoh, *Rev. Mod. Phys.* **70**, 897 (1998).

- B. W. Lovett, S. J. Blundell, F. L. Pratt, Th. Jestädt, W. Hayes, S. Tagaki and M. Kurmoo, *Phys. Rev. B*, **61**, 12241, (2000).

- W. Lubczynski, S. V. Demishev, J. Singleton, J. M. Caulfield, L. du Croo de Jongh, C. J. Kepert, S. J. Blundell, W. Hayes, M. Kurmoo and P. Day, *J. Phys.: Condens. Matter*, **8**, 6005, (1996)

- N. D. Mathur, F. M. Grosche, S. R. Julian, I. R. Walker, D. M. Freye, R. K. W. Haselwinner and G. G. Lonzarich, *Nature*, **394**, 39 (1998).

- S. Nagata, P. H. Keesom and H. R. Harrison, *Phys. Rev. B* **19**, 1633 (1979).

- S. S. Saxena, P. Agarwal, K. Ahilan, F. M. Grosche, R. K. W. Haselwimmer, M. J. Steiner, E. Pugh, I. R. Walker, S. R. Julian, P. Monthoux, G. G. Lonzarich, A. Huxley, I. Sheikin, D. Braithwaite and J. Flouquet, *Nature*, **406**, 587 (2000).

- D. A. Tennant, T. G. Perring, R. A. Cowley and S. E. Nagler, *Phys. Rev. Lett.* **70**, 4003 (1993).

- Y. Tokura, A. Urushibara, Y. Moritomo, T. Arima, A. Asamitsu, G. Kido and N. Furukawa, *J. Phys. Soc. Jpn.*, **63**, 3391 (1994).

# Appendix A

# Units in electromagnetism

Units in electromagnetism represent a potential minefield of complication. SI units are used throughout this book, but much of the twentieth century literature uses non-SI notation. Therefore this appendix attempts to unravel some of the mysteries of the old system and it may be used to interpret older literature for readers who have been brought up on the new system.

From the end of the nineteenth century until just after the second world war, the **cgs system** was the system of units preferred by the international committees who are concerned with such things. In the cgs system, distance is measured in centimetres, mass in grams, and time in seconds, hence the name cgs. In the **SI system**, we use metres (100 times as big as centimetres), kilograms (1000 times as big as grams) and seconds. This difference in choice of units of distance and mass appears to be rather innocent, but ends up causing great changes in other units. For example, force has the dimensions of mass $\times$ length / time$^2$, so in cgs the unit of force, the dyne, is $1000 \times 100 = 10^5$ times smaller than a Newton, the SI unit of force. The cgs unit of energy is the erg, a dyne cm, so this is $10^5 \times 10^2 = 10^7$ times smaller than the Joule, the SI unit of energy.

Unfortunately, that is not the end of it, and when it comes to the definitions of magnetic field things get decidedly worse. This arises because the choice of definitions of the fields **B**, **H** and **M** are different in the two systems:

$$\mathbf{B} = \mu_0(\mathbf{H} + \mathbf{M}) \quad \text{(SI)} \qquad \mathbf{B} = \mathbf{H} + 4\pi\mathbf{M} \quad \text{(cgs)} \qquad \text{(A.1)}$$

Thus B and H have the same dimensions in the cgs system but confusingly have different units (the former in Gauss (G), the latter in Oersted (Oe)). The difference in relation between $H$ and $M$ arises because the two systems make different choices in the equations defining the magnitude of the magnetic field, $H$, at a distance $r$ from a magnetic monopole $q$:

$$H = \frac{q}{4\pi r^2} \quad \text{(SI)} \qquad H = \frac{q}{r^2} \quad \text{(cgs)} \qquad \text{(A.2)}$$

The omission of the factor of $4\pi$ makes this equation, and similar ones relating the magnetic field from a magnetic dipole, rather simpler in the cgs system. However the factor of $4\pi$ is not removed entirely from the theory but pops up in a more unwanted place in eqn A.1. The cgs system defined $H$ according to non-existent monopoles from eqn A.2, so that a unit monopole produces a field of one Oersted a distance of one centimetre away (and if another unit monopole is lurking there it will experience a force of one dyne). In SI, $H$ is more helpfully defined in terms of current (since people normally use a current source in a laboratory to produce a magnetic field, rather than devising a suitable arrangement of unit magnetic monopoles). Thus a long solenoid with $n$ turns per metre and carrying a current of $I$ Amps produces a field given by $H = nI$ in the SI system.

The logic in the SI system is that the $4\pi$ appears in equations like eqn A.2 to remind us that the field from a point magnetic monopole has spherical symmetry (the $4\pi$ is related to the surface area of a sphere, $4\pi r^2$). The same is true of the electric field $E$ a distance $r$ from a point charge,

$$E = \frac{q}{4\pi\epsilon_0 r^2} \quad \text{(SI)} \quad E = \frac{q}{r^2} \quad \text{(cgs)}, \quad \text{(A.3)}$$

in which similar arguments apply. With SI, we have to live with a few extra factors, which fit from symmetry arguments, in equations defining the field from point charges and dipoles. The rest of the equations come out much cleaner and with fewer arbitrary factors.

The one thing to be said in favour of the cgs system is that electric and magnetic fields come out with the same dimensions, so that certain relativistic equations can look better. Also in free space $\mathbf{B} = \mathbf{H}$ which has a certain symmetry (although this can create an opportunity for confusion). But extra factors of $c$ get littered all over the place in the cgs system, so that for example the equation for the Lorentz force on a charge moving with velocity $\mathbf{v}$ in electric and magnetic fields becomes

$$\mathbf{F} = q(\mathcal{E} + \mathbf{v} \times \mathbf{B}) \quad \text{(SI)} \quad \mathbf{F} = q\left(\mathcal{E} + \frac{\mathbf{v} \times \mathbf{B}}{c}\right) \quad \text{(cgs).} \quad \text{(A.4)}$$

But there is much worse to come. It turns out that there are two ways to set up electromagnetism in cgs, with either the esu (electrostatic units) or the emu (electromagnetic units) with conversion factors needed to go between them. In magnetism the emu system is usually adopted. The emu unit of magnetic moment, usually abbreviated just to the 'emu', is a unit with dimensions of volume (which in cgs is $cm^3$). Magnetic susceptibility, $\chi = M/H$, is dimensionless in both systems, but in cgs is often written as emu $cm^{-3}$, which is still dimensionless even if it doesn't appear so at first glance. There is always the troublesome factor of $4\pi$ to remember when converting between cgs susceptibility and SI susceptibility which arises because $\mathbf{B} = \mathbf{H} + 4\pi\mathbf{M}$ in the cgs system (see above). For example, the demagnetization factor, N, is dimensionless and is $0 < N < 1$ in SI but $0 < N < 4\pi$ in cgs. Unfortunately the emu is still widely used in some parts of magnetism research. Table A summarizes the most useful results for converting SI to and from cgs.

Though not an SI unit, the Bohr magneton $\mu_B = 9.274 \times 10^{-24}$ J T$^{-1}$ is a useful measure of magnetic moment since it corresponds to the magnetic moment of a 1s electron in hydrogen. For a paramagnet, the molar susceptibility $\chi_m$ is given by Curie's law (adapting eqn 2.44) which is in SI units

$$\chi_m = \frac{\mu_0 N_A \mu_{eff}^2 \mu_B^2}{3k_B T} \quad \text{(A.5)}$$

where $N_A$ is Avogadro's number. Hence $\chi_m T$ is independent of temperature (see Fig. 2.10) and this can be related to the effective moment. Hence by rearranging eqn A.5, one has

$$\mu_{eff} = [3k_B/\mu_0 N_A \mu_B^2]^{1/2}\sqrt{\chi_m T}, \quad \text{(A.6)}$$

**Table A.1** Units in the SI system and the cgs system. The abbreviations are m = metre, g = gramme, N = Newton, J = Joule, T = Tesla, G = Gauss, A = Amp, Oe = Oersted, Wb = Weber, Mx = Maxwell. The term emu is short for electromagnetic unit. Note that magnetic susceptibility is dimensionless in SI units.

| Quantity | symbol | SI unit | | cgs unit |
|---|---|---|---|---|
| Length | $x$ | $10^{-2}$ | m | = 1 cm |
| Mass | $m$ | $10^{-3}$ | kg | = 1 g |
| Force | $F$ | $10^{-5}$ | N | = 1 dyne |
| Energy | $E$ | $10^{-7}$ | J | = 1 erg |
| Magnetic induction | **B** | $10^{-4}$ | T | = 1 G |
| Magnetic field strength | **H** | $10^3/4\pi$ | $A\,m^{-1}$ | = 1 Oe |
| Magnetic moment | $\mu$ | $10^{-3}$ | $J\,T^{-1}$ or $A\,m^2$ | = 1 erg $G^{-1}$ |
| Magnetization (= moment per volume) | **M** | $10^3$ | $A\,m^{-1}$ or $J\,T^{-1}\,m^{-3}$ | = 1 Oe |
| Magnetic susceptibility | $\chi$ | $4\pi$ | | = 1 emu $cm^{-3}$ |
| Molar susceptibility | $\chi_m$ | $4\pi \times 10^{-6}$ | $m^3\,mol^{-1}$ | = 1 emu $mol^{-1}$ |
| Mass susceptibility | $\chi_g$ | $4\pi \times 10^{-3}$ | $m^3\,kg^{-1}$ | = 1 emu $g^{-1}$ |
| Magnetic flux | $\phi$ | $10^{-8}$ | $T\,m^2$ or Wb | = 1 $G\,cm^{-2}$ or Mx |
| Demagnetization factor | N | | $0 < N < 1$ | $0 < N < 4\pi$ |

so that

$$\mu_{\text{eff}} = 797.8\sqrt{\chi_m^{\text{SI}}T} \quad \text{(SI)} \tag{A.7}$$

$$\mu_{\text{eff}} = 2.827\sqrt{\chi_m^{\text{cgs}}T} \quad \text{(cgs)} \tag{A.8}$$

where $\mu_{\text{eff}}$ is measured in Bohr magnetons per formula unit, $\chi_m^{\text{SI}}$ is measured in $m^3\,mol^{-1}$, and $\chi_m^{\text{cgs}}$ is measured in emu $mol^{-1}$. These numerical relationships can be useful for extracting effective moments from graphs of $\chi_m T$ against $T$.

# Further reading

- B. I. Bleaney and B. Bleaney, *Electricity and magnetism*, OUP 1989.
- J. D. Jackson, *Classical electrodynamics*, Wiley 1962.

# Electromagnetism

In this appendix several key results in electromagnetism are briefly reviewed. The further reading should be consulted for detailed derivations of these results.

## B.1 Magnetic moments

The Lorentz force $d\mathbf{F}$ on a straight segment of wire of length $dr$ carrying current $I$ in a magnetic field of flux density $\mathbf{B}$ is given by $d\mathbf{F} = I\,d\mathbf{r} \times \mathbf{B}$ (see Fig. B.1).

This result can now be applied to calculate the couple on the elementary loop of wire in Fig. B.2. The couple is given by

$$dG_x = -I\,dx\,dy\,B_y$$
$$dG_y = I\,dx\,dy\,B_x, \tag{B.1}$$

so that defining the magnetic moment of the loop by $d\mu_z = I\,dx\,dy$, and repeating the argument for arbitrary directions of the loop, the couple $d\mathbf{G}$ is given by

$$d\mathbf{G} = d\boldsymbol{\mu} \times \mathbf{B}. \tag{B.2}$$

For a loop of finite size, this yields by integration

$$\mathbf{G} = \boldsymbol{\mu} \times \mathbf{B}. \tag{B.3}$$

There will be a net force on the loop only if the magnetic field is non-uniform. Thus the force in the $x$ direction on the elementary loop in Fig. B.2 is

$$dF_x = I\,dy\left(B_z + \frac{\partial B_z}{\partial x}dx\right) - I\,dy\,B_z = d\mu_z \frac{\partial B_z}{\partial x}, \tag{B.4}$$

so that

$$d\mathbf{F} = (d\boldsymbol{\mu} \cdot \nabla)\mathbf{B} \tag{B.5}$$

and by integrating for a magnetic moment of finite size,

$$\mathbf{F} = (\boldsymbol{\mu} \cdot \nabla)\mathbf{B}. \tag{B.6}$$

It is therefore only a non-uniform magnetic field which can exert a force on a magnetic moment.

A magnetic moment $\boldsymbol{\mu}$ aligned with a magnetic field $\mathbf{B}$ has zero torque on it (by eqn B.3, $\mathbf{G} = 0$ if $\boldsymbol{\mu}$ and $\mathbf{B}$ are parallel). Rotating $\boldsymbol{\mu}$ with respect to $\mathbf{B}$

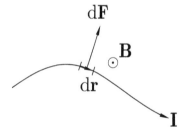

Fig. B.1 The force on a current element is $d\mathbf{F} = I\,d\mathbf{r} \times \mathbf{B}$.

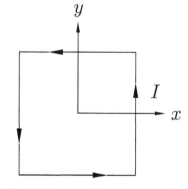

Fig. B.2 A square elementary current loop (with sides of length $dx$ and $dy$) oriented perpendicular to the $z$-direction.

therefore can only be accomplished by applying a torque. The work done $W$ by the torque to rotate $\boldsymbol{\mu}$ to an angle $\theta$ with respect to the magnetic field is

$$W = \int_0^\theta G \, d\theta' = \int_0^\theta \mu B \sin\theta' \, d\theta' = \mu B(1 - \cos\theta). \tag{B.7}$$

The potential energy $U$ of the magnetic moment can therefore be written as $-\mu B \cos\theta$ (ignoring the constant term). Hence we can write

$$U = -\boldsymbol{\mu} \cdot \mathbf{B}. \tag{B.8}$$

A magnetic moment $\boldsymbol{\mu}$ in a non-uniform field $\mathbf{B}$ has a force on it in the $x$ direction equal to $\mu \partial B / \partial x$ by eqn B.6. To stop it moving in the $x$ direction, one would have to apply a restoring force $-\mu \partial B / \partial x$. If it were allowed to move a distance $\delta x$ into a region of stronger field, the work done by this force would be

This argument can be generalized to show that
$$\delta W = -\boldsymbol{\mu} \cdot \delta\mathbf{B}.$$

$$dW = -\mu \frac{\partial B}{\partial x} \delta x = -\mu \delta B, \tag{B.9}$$

and this is an increment to the sample's free energy (in agreement with the results of Exercise 2.10).

The magnetic field due to a magnetic moment at the origin is

$$\mathbf{B}(\mathbf{r}) = -\mu_0 \nabla \left( \frac{\boldsymbol{\mu} \cdot \mathbf{r}}{4\pi r^3} \right) \tag{B.10}$$

and the magnetic vector potential (in the most convenient choice of gauge) is

$$\mathbf{A}(\mathbf{r}) = \frac{\mu_0}{4\pi} \frac{\boldsymbol{\mu} \times \mathbf{r}}{r^3}. \tag{B.11}$$

## B.2   Maxwell's equations in free space

James Clerk Maxwell (1831–1879)

Maxwell's equations in free space are

$$\nabla \cdot \boldsymbol{\mathcal{E}} = \rho / \epsilon_0 \tag{B.12}$$

$$\nabla \cdot \mathbf{B} = 0 \tag{B.13}$$

$$\nabla \times \boldsymbol{\mathcal{E}} = -\frac{\partial \mathbf{B}}{\partial t} \tag{B.14}$$

$$\nabla \times \mathbf{B} = \mu_0 \mathbf{J} + \epsilon_0 \mu_0 \frac{\partial \boldsymbol{\mathcal{E}}}{\partial t}, \tag{B.15}$$

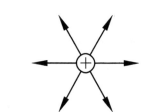

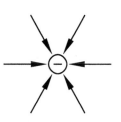

**Fig. B.3** Electric field diverges away from positive charges and converges into negative charges.

and describe the interrelationships between the electric field $\boldsymbol{\mathcal{E}}$, the magnetic induction $\mathbf{B}$, the charge density $\rho$ and the current density $\mathbf{J}$. Equation B.12 shows that electric field diverges away from positive charges and converges into negative charges; charge density therefore acts as a source or a sink of electric field (see Fig. B.3). Equation B.13 shows that magnetic fields have no such divergence; thus there are no magnetic charges (monopoles) and lines of $\mathbf{B}$ field must just exist in loops; they can never start or stop anywhere. Equation B.14 shows that you only get loops of electric field around regions in space in which there is a changing magnetic field. This leads to Faraday's law of electromagnetic induction. Equation B.15, in the absence of a changing electric field, shows that loops of magnetic induction are found around electric currents. Note that everything in this section has been for electromagnetic fields in free space.

# B.3 Free and bound currents

Once matter is brought into the picture, things get a little more complicated and it is worth trying to understand why. The treatment here will just consider the effects for magnetism, although analogous thinking lies behind the corresponding derivations for electric fields. Real matter contains a different type of electric current from the familiar currents we are used to thinking about flowing down wires. Atoms contain microscopic **bound currents** due to the electrons circulating around the nucleus. Thus the total current **J** in eqn B.15 can be separated into two components

$$\mathbf{J} = \mathbf{J}_{\text{free}} + \mathbf{J}_{\text{bound}}, \tag{B.16}$$

due to the free currents (10 amps flowing down a copper wire) and bound currents (electrons orbiting an atom).

If **M** is uniform in a sample, the only net bound current that flows is around the edge of the material. If **M** is non-uniform, bound currents will be generated inside the material. We need a general relation between the bound currents and the magnetization, and this is that

$$\mathbf{J}_{\text{bound}} = \nabla \times \mathbf{M}. \tag{B.17}$$

A derivation of eqn B.17 will now be sketched (for more details see the further reading). The vector potential **A** due to a single point dipole at the origin is given in eqn B.11. Now consider a magnetized specimen, contained in volume $V$, with magnetization $\mathbf{M}(\mathbf{r}')$ at position $\mathbf{r}'$. The vector potential at position $\mathbf{r}$ is

$$\mathbf{A}(\mathbf{r}) = \frac{\mu_0}{4\pi} \int_V d^3 r' \, \frac{\mathbf{M}(\mathbf{r}') \times (\mathbf{r} - \mathbf{r}')}{|\mathbf{r} - \mathbf{r}'|^3}. \tag{B.18}$$

This equation can be simplified using $\nabla_r(1/r) = -\mathbf{r}/r^3$, the vector identity

$$\nabla \times (f\mathbf{g}) = f\nabla \times \mathbf{g} - \mathbf{g} \times \nabla f, \tag{B.19}$$

where $f$ and $\mathbf{g}$ are scalar and vector functions respectively, and transforming the integral of $\mathbf{r}'$ to one over $\mathbf{r}$. The result is that the vector potential at $\mathbf{r}$ can be written as

$$\mathbf{A}(\mathbf{r}) = \frac{\mu_0}{4\pi} \int_V \frac{\nabla \times \mathbf{M} \, d\tau}{r} + \frac{\mu_0}{4\pi} \int_S \nabla \times \left(\frac{\mathbf{M}}{r}\right) d\tau \tag{B.20}$$

$$= \frac{\mu_0}{4\pi} \int_V \frac{\nabla \times \mathbf{M} \, d\tau}{r} + \frac{\mu_0}{4\pi} \int_V \frac{\mathbf{M} \times d\mathbf{S}}{r} \tag{B.21}$$

$$= \frac{\mu_0}{4\pi} \int_V \frac{\mathbf{J}_{\text{bound}} \, d\tau}{r} + \frac{\mu_0}{4\pi} \int_S \frac{\mathbf{K}_{\text{bound}} \, dS}{r} \tag{B.22}$$

where the first integral is over the volume of the sample, and the second integral is over the surface of the sample. The bound current in eqn B.22 agrees with eqn B.17. The surface current is given by $\mathbf{K}_{\text{bound}} = \mathbf{M} \times \hat{\mathbf{n}}$ and the vector $\hat{\mathbf{n}}$ is perpendicular to the surface.

The problem with eqn B.15 is that the curl of **B** is related not only to the free current density but also to the more inconvenient bound current density. Therefore we are motivated to define a new 'corrected' field which only 'sees' the free current density, although as shown below this will be only partially successful. Using the result for the bound volume currents, we define a new field $\mathbf{H}$ (called the **magnetic field strength**, or just the $H$ field) by

$$\mathbf{B} = \mu_0(\mathbf{H} + \mathbf{M}) \tag{B.23}$$

and hence we can write a new version of eqn B.15 which includes only $\mathbf{J}_{\text{free}}$. Thus in the absence of changing electric fields

$$\nabla \times \mathbf{H} = \mathbf{J}_{\text{free}}. \tag{B.24}$$

This is useful because it relates the $\mathbf{H}$-field to a quantity that can be read on an ammeter (which $\mathbf{J}_{\text{bound}}$ cannot be). Unfortunately this procedure, having simplified one of Maxwell's equation, complicates another. Equation B.13 changes from the simple $\nabla \cdot \mathbf{B} = 0$ to the slightly more tortuous

$$\nabla \cdot \mathbf{H} = -\nabla \cdot \mathbf{M}. \tag{B.25}$$

Thus $\mathbf{H}$, unlike $\mathbf{B}$, is not divergence-free (or **solenoidal**, to use the mathematical jargon). $\mathbf{H}$ behaves as if magnetic monopoles (i.e. the magnetic equivalent of charges) exist.

## B.4 Maxwell's equations in matter

Maxwell's equations in the presence of matter therefore become

$$\nabla \cdot \mathbf{D} = \rho_{\text{free}} \tag{B.26}$$
$$\nabla \cdot \mathbf{B} = 0 \tag{B.27}$$
$$\nabla \times \boldsymbol{\mathcal{E}} = -\frac{\partial \mathbf{B}}{\partial t} \tag{B.28}$$
$$\nabla \times \mathbf{H} = \mathbf{J}_{\text{free}} + \frac{\partial \mathbf{D}}{\partial t}, \tag{B.29}$$

where a similar distinction is made between free and bound charges and where the **electric displacement**, $\mathbf{D}$, is given by

$$\mathbf{D} = \epsilon_0 \mathbf{E} + \mathbf{P} \tag{B.30}$$

and $\mathbf{P}$ is the electric polarization.

## B.5 Boundary conditions

[1] These are traditionally called 'pill-boxes' in the physics literature, although pills have long since stopped being sold in such shaped boxes!

A significant difference between the $B$ and $H$ fields is revealed when they meet a boundary between two different regions. The boundary conditions between two regions 1 and 2 (see Fig. B.4) can be derived as follows. Consider the shoe-polish-tin-shaped box[1] straddling the surface. $\nabla \cdot \mathbf{B} = 0$ implies that $\int_S \mathbf{B} \cdot d\mathbf{S} = 0$ using the divergence theorem. Since the surface area of the box is mostly represented by the top and bottom surfaces of the box,[2] we have that the perpendicular component of $\mathbf{B}$ is continuous, i.e. that

[2] And we can make sure this is the case by making sure that we let the box thickness tend to zero before its radius tends to zero.

$$B_{\perp 1} = B_{\perp 2}. \tag{B.31}$$

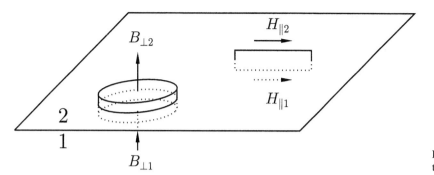

**Fig. B.4** Electromagnetic boundary conditions between regions 1 and 2.

The argument won't work for **H** because $\nabla \cdot \mathbf{H} = -\nabla \cdot \mathbf{M}$ which may be non-zero. The perpendicular component may be discontinuous if the magnetization diverges inside the box (which it may well do if region 1 is magnetized and region 2 is not). The argument for the parallel components comes from the equation $\nabla \times \mathbf{H} = \mathbf{J}_{\text{free}}$ which implies that $\oint \mathbf{H} \cdot d\mathbf{l} = \int \mathbf{J}_{\text{free}} \cdot d\mathbf{S}$. As the Amperean loop shown on the right-hand side of Fig. B.4 shrinks to zero, so does the free current enclosed by the loop and so the parallel component of **H** is continuous, i.e. that

$$H_{\|1} = H_{\|2}. \tag{B.32}$$

This argument won't work for **B** because $\nabla \times \mathbf{B} = \mu_0(\mathbf{J}_{\text{free}} + \mathbf{J}_{\text{bound}})$ and so the Amperean loop, although enclosing no free current as it shrinks to zero, can enclose some bound surface current in the limit that the loop shrinks to zero.

## Further reading

- B. I. Bleaney and B. Bleaney, *Electricity and magnetism*, OUP 1989.

- D. J. Griffiths, *Introduction to electrodynamics*, Prentice-Hall 1989.

- J. D. Jackson, *Classical electrodynamics*, John-Wiley 1962.

# Appendix C

# Quantum and atomic physics

[1] Here the symbol * denotes complex conjugation.

## C.1 Quantum mechanics

In this section we briefly review some basic features of quantum mechanics. The fundamental object in quantum mechanics is the **wave function** (also known as the **state function**) $\psi$ which contains all the knowledge one obtains about a system from observation. The wave function $\psi$ is a complex function and $|\psi|^2 = \psi\psi^*$ is the probability density.[1] To every observable quantity, there corresponds a Hermitian **operator**. When you make a measurement, the result you obtain is one of the eigenvalues of that operator. Hermitian operators have real eigenvalues and eigenfunctions which are orthogonal to each other. If the $i^{\text{th}}$ eigenfunction of the operator $\hat{A}$ (operators are given 'hats') is $\phi_i$ (assumed normalized) and has an eigenvalue $a_i$, then

$$\hat{A}\phi_i = a_i\phi_i. \tag{C.1}$$

The expected value that you obtain after a measurement of operator $\hat{A}$ is given by

$$\langle\hat{A}\rangle = \int d\tau\, \psi^*\hat{A}\psi, \tag{C.2}$$

where $d\tau$ is a volume element. The wave function $\psi$ can be expanded in terms of the eigenfunctions of $\hat{A}$, i.e.

$$\psi = \sum_i c_i\phi_i, \tag{C.3}$$

so that $\langle\hat{A}\rangle$ is then given by

$$\langle\hat{A}\rangle = \sum_i |c_i|^2 a_i, \tag{C.4}$$

using the orthonormality of $\phi_i$. The probability of obtaining the $i^{\text{th}}$ eigenvalue, $a_i$, is $|c_i|^2$. Measurement is a drastic process. If you take a wave function $\psi$, measure the physical quantity which is associated with operator $\hat{A}$ and obtain the result $a_n$ (the $n^{\text{th}}$ eigenvalue of $\hat{A}$), the system is forced into state function $\phi_n$. This, at least, is the Copenhagen interpretation.

The **commutator** of two operators $\hat{A}$ and $\hat{B}$ is defined by $[\hat{A}, \hat{B}] = \hat{A}\hat{B} - \hat{B}\hat{A}$. An expression defining a commutator is known as a **commutation relation**.

**Example C.1**

The commutator $[\hat{x}, \hat{p}] = i\hbar$ where $\hat{x}$ and $\hat{p}$ are the position and momentum operators respectively.

If the two operators $\hat{A}$ and $\hat{B}$ commute, $([\hat{A}, \hat{B}] = 0)$, they are said to be compatible and measurement of one does not affect the value of the other in any way. If they do not commute, there exists an uncertainty relation between them so that

$$\Delta A \, \Delta B \geq \tfrac{1}{2}|\langle i[\hat{A}, \hat{B}]\rangle|, \tag{C.5}$$

where $\Delta A \equiv \sqrt{\langle(\hat{A} - \langle\hat{A}\rangle)^2\rangle}$.

The time dependence of $\psi$ in the absence of a measurement is given by the Schrödinger equation

$$\hat{\mathcal{H}}\psi = i\hbar\frac{\mathrm{d}\psi}{\mathrm{d}t} \tag{C.6}$$

where $\hat{\mathcal{H}}$ is the Hamiltonian. The time dependence of the expected value of an operator is given by

$$\frac{\mathrm{d}\langle\hat{A}\rangle}{\mathrm{d}t} = \frac{1}{i\hbar}\langle[\hat{A}, \hat{\mathcal{H}}]\rangle, \tag{C.7}$$

if the operator $\hat{A}$ itself is not time dependent. An observable whose operator commutes with the Hamiltonian is a conserved quantity and is known as a constant of the motion or a good quantum number.

## C.2   Dirac bra and ket notation

In this section we review some notation, due to Paul Dirac, which is used sporadically in this text and often in the magnetism literature. The scalar product between two vectors $\mathbf{a}$ and $\mathbf{b}$ is given by $\mathbf{a} \cdot \mathbf{b} = \sum_i a_i b_i$. The length of $\mathbf{a}$ is given by $\sqrt{\mathbf{a} \cdot \mathbf{a}}$ which must be real and positive. The notation makes the inner product look completely symmetric between $\mathbf{a}$ and $\mathbf{b}$, but this hides some subtlety. If the vectors are complex, the scalar product should be written

Paul A. M. Dirac (1902–1984)

$$\mathbf{a}^\dagger \cdot \mathbf{b} = \sum_i a_i^* b_i = (a_1^* \quad a_2^* \quad a_3^*) \begin{pmatrix} b_1 \\ b_2 \\ b_3 \end{pmatrix}. \tag{C.8}$$

This definition ensures that $\mathbf{a}^\dagger \cdot \mathbf{a} = \sum_i |a_i|^2$ is real and positive, and so the length of the complex vector is well defined. Equation C.8 emphasizes that the two vectors in the scalar product should not really be considered on an equal footing: one is a row vector, the other a column vector; one is complex conjugated, the other is not. Mathematicians say that actually the two vectors in a scalar product 'live' in different spaces, so that here $\mathbf{b}$ lives in a vector space and $\mathbf{a}$ lives in its 'dual space'. To turn a vector into its dual, you need to take the adjoint, signified by a † sign (it means 'complex conjugate and transpose'), so that

$$\begin{pmatrix} a_1 \\ a_2 \\ a_3 \end{pmatrix}^{\dagger} = (a_1^* \quad a_2^* \quad a_3^*). \tag{C.9}$$

However in quantum mechanics, the state function is sometimes a vector, like $\mathbf{a} = \begin{pmatrix} a_1 \\ a_2 \end{pmatrix}$, and sometimes a continuous function $a(x)$. If a function, then the scalar product between $a(x)$ and $b(x)$ is

$$a^{\dagger} \cdot b = \int dx\, a^*(x) b(x). \tag{C.10}$$

Dirac circumvented this notational problem by writing the state function as a ket: $|a\rangle$, whether he was dealing with finite-component vectors or functions (which can be regarded as infinite dimensional vectors). This is *just* notation, so that $|a\rangle + |b\rangle = |c\rangle$ is simply another way of writing $\mathbf{a} + \mathbf{b} = \mathbf{c}$, or equivalently $\vec{a} + \vec{b} = \vec{c}$ or $a(x) + b(x) = c(x)$. One can turn the ket $|a\rangle$ into a bra, $\langle a|$, by using the adjoint

$$|a\rangle^{\dagger} = \langle a|. \tag{C.11}$$

The scalar product between $\langle a|$ and $|b\rangle$ is then written as

$$\langle a|b\rangle = \sum_i a_i^* b_i = \int dx\, a^*(x) b(x), \tag{C.12}$$

a bra-(c)-ket (Dirac here demonstrating evidence of humour). The expression for an expectation value of an operator can be written

$$\langle \hat{A} \rangle = \langle \psi | \hat{A} | \psi \rangle. \tag{C.13}$$

The notation conveys the asymmetry of the scalar product, is widely used and greatly simplifies many computations. It is particularly useful in labelling particular states, as in the following:

$$|\text{Schrödinger's cat}\rangle = \frac{|\text{alive}\rangle + |\text{dead}\rangle}{\sqrt{2}}. \tag{C.14}$$

## C.3   The Bohr model

The Bohr model is a semiclassical model describing the motion of an electron in a hydrogen atom. It gives the right results, but for the wrong reason. Nevertheless, it is simple and therefore useful for quickly deriving results such as the dependence of the radius of an atomic orbit on nuclear charge $Z$ and effective mass.

Consider an electron orbiting a nucleus of charge $+Ze$ (see Fig. C.1). Equating electrostatic and centripetal forces yields

$$\frac{Ze^2}{4\pi\epsilon_0 r^2} = \frac{m_e v^2}{r}. \tag{C.15}$$

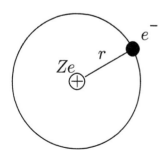

**Fig. C.1** The Bohr model.

The angular momentum of the electron $m_e v r$ is quantized in units of $\hbar$ so we can write

$$m_e v r = n\hbar, \qquad \text{(C.16)}$$

where $n$ is an integer (the principal quantum number). Hence the radius of the orbit can be written

$$r = \frac{4\pi \epsilon_0 \hbar^2 n^2}{Z e^2 m_e} = \left( \frac{4\pi \epsilon_0 \hbar^2}{m_e e^2} \right) \frac{n^2}{Z} = \frac{a_0 n^2}{Z}, \qquad \text{(C.17)}$$

These results can be written as $a_0 = \frac{h}{m\alpha c}$ and $E = -\frac{1}{2}m(\alpha c)^2 \frac{Z^2}{n^2}$ where $\alpha$ is the fine structure constant.

where $a_0$ is the **Bohr radius** defined by

$$a_0 = \frac{4\pi \epsilon_0 \hbar^2}{m_e e^2}. \qquad \text{(C.18)}$$

The energy $E$ is then given by

$$E = \frac{1}{2}mv^2 - \frac{Ze^2}{4\pi \epsilon_0 r} \qquad \text{(C.19)}$$

$$= -\frac{1}{2}\left( \frac{Ze^2}{4\pi \epsilon_0 r} \right) \qquad \text{(C.20)}$$

$$= -\frac{\hbar^2}{2m_e} \frac{1}{a_0^2} \frac{Z^2}{n^2} \qquad \text{(C.21)}$$

$$= -\frac{m_e e^4}{(4\pi \epsilon_0 \hbar)^2} \frac{Z^2}{n^2}. \qquad \text{(C.22)}$$

## C.4 Orbital angular momentum

The angular momentum operator is given by

$$\hbar \hat{\mathbf{L}} = \hat{\mathbf{r}} \times \hat{\mathbf{p}} = -i\hbar \hat{\mathbf{r}} \times \nabla. \qquad \text{(C.23)}$$

The $z$ component of the angular momentum thus has an operator

$$\hbar \hat{L}_z = i\hbar \left[ y \frac{\partial}{\partial x} - x \frac{\partial}{\partial y} \right] \qquad \text{(C.24)}$$

$$= -i\hbar \frac{\partial}{\partial \phi} \qquad \text{(C.25)}$$

which has eigenfunctions given by $e^{im_l \phi}$ and eigenvalues $m_l \hbar$, where $m_l$ is the magnetic quantum number. The following commutation relations can be easily proved:

$$[\hat{L}_i, \hat{L}^2] = 0 \qquad \text{(C.26)}$$

and

$$[\hat{L}_i, \hat{L}_j] = i\epsilon_{ijk} \hat{L}_k \qquad \text{(C.27)}$$

where the **alternating tensor** $\epsilon_{ijk}$ is defined by

$$\epsilon_{ijk} = \begin{cases} 1 & \text{if } ijk \text{ is an even permutation of 123} \\ -1 & \text{if } ijk \text{ is an odd permutation of 123} \\ 0 & \text{if any two of } i, j \text{ or } k \text{ are equal.} \end{cases} \qquad \text{(C.28)}$$

**Table C.1** $P_l^{|m_l|}(\cos\theta)$.

|  | $m_l = 0$ | $m_l = 1$ | $m_l = 2$ | $m_l = 3$ |
|---|---|---|---|---|
| $l = 0$ | 1 | – | – | – |
| $l = 1$ | $\cos\theta$ | $\sin\theta$ | – | – |
| $l = 2$ | $\frac{1}{2}(3\cos^2\theta - 1)$ | $\frac{1}{3}\sin\theta\cos\theta$ | $\frac{1}{3}\sin^2\theta$ | – |
| $l = 3$ | $\frac{1}{2}(5\cos^3\theta - 3\cos\theta)$ | $\frac{2}{3}\sin\theta(5\cos^2\theta - 1)$ | $\frac{1}{15}\sin^2\theta\cos\theta$ | $\frac{1}{15}\sin^3\theta$ |

Equation C.27 is shorthand for

$$[\hat{L}_x, \hat{L}_y] = i\hat{L}_z$$

and cyclic permutations. The equation uses Einstein's summation convention in which any twice-repeated index is assumed summed. For example $a_i b_i$ is short for $\sum_i a_i b_i$. The alternating tensor is useful in expressions such as

$$(\mathbf{a} \times \mathbf{b})_i = \epsilon_{ijk} a_j b_k.$$

In this equation $j$ and $k$ are twice-repeated and thus assumed summed. Hence for example

$$(\mathbf{a} \times \mathbf{b})_1 = \epsilon_{123} a_2 b_3 + \epsilon_{132} a_3 b_2$$
$$= a_2 b_3 - a_3 b_2.$$

The operator $\hat{L}^2$ has eigenfunctions

$$|l, m_l\rangle = Y_{lm_l}(\theta, \phi) \propto P_l^{m_l}(\cos\theta)e^{im_l\phi}, \tag{C.29}$$

known as spherical harmonics, with eigenvalues $l(l+1)$ where $l$ is the angular momentum quantum number and $P_l^{m_l}(\cos\theta)$ is an associated Legendre polynomial (see Table C.1).

Thus

$$\hat{L}^2|l, m_l\rangle = l(l+1)|l, m_l\rangle. \tag{C.30}$$

Similarly

$$\hat{L}_z|l, m_l\rangle = m_l|l, m_l\rangle. \tag{C.31}$$

The raising and lowering operators $\hat{L}_\pm$ are defined by

$$\hat{L}_\pm = \hat{L}_x \pm i\hat{L}_y. \tag{C.32}$$

Hence one can show that

$$\hat{L}_\pm|l, m_l\rangle = \sqrt{l(l+1) - m_l(m_l \pm 1)}|l, m_l \pm 1\rangle. \tag{C.33}$$

Certain states have no orbital angular momentum. Singlet states ($S = 0$) are a good example, and this arises because they have real wave functions, as can be proved straightforwardly. In the absence of a magnetic field the Hamiltonian is real so that if $\psi$ is an eigenfunction of the Hamiltonian $\hat{\mathcal{H}}$ with energy $E$, then so is $\psi^*$. ($\hat{\mathcal{H}}\psi = E\psi$ so $\hat{\mathcal{H}}\psi^* = E\psi^*$.) But since the state is by definition a singlet, $\psi = \psi^*$ and $\psi$ is then real. The operators for all components of $\hat{\mathbf{L}}$ contain $i$ so that the expectation values $\langle\psi|\hat{L}_\alpha|\psi\rangle$ for $\alpha = x, y, z$ are all purely imaginary if $\psi$ is real. But expectation values are measurable quantities and therefore must be pure real; hence they must all be zero and the state has no orbital angular momentum.

## C.5   The hydrogen atom

Most of the atoms considered in this book are not hydrogen, but the Schrödinger equation can only be solved exactly for the hydrogen atom. In this section, some results concerning the hydrogen atom are tabulated for easy reference.

The Schrödinger equation for a spherically symmetrical potential can be solved by separation of variables and the eigenfunction $\Psi$ can be written as a

**Table C.2** Radial wave functions.

| $n$ | $l$ | $L_{n+l}^{2l+1}(x)$ | $R_{nl}(\rho)$ |
|---|---|---|---|
| 1 | 0 | $L_1^1(x) = -1$ | $2(Z/a_0)^{3/2}e^{-\rho/2}$ |
| 2 | 0 | $L_2^1(x) = 2x - 4$ | $(1/2\sqrt{2})(Z/a_0)^{3/2}(2-\rho)e^{-\rho/2}$ |
| 2 | 1 | $L_3^3(x) = -6$ | $(1/2\sqrt{6})(Z/a_0)^{3/2}\rho e^{-\rho/2}$ |
| 3 | 0 | $L_3^1(x) = -3x^2 + 18x - 18$ | $(1/9\sqrt{3})(Z/a_0)^{3/2}(6-6\rho+\rho^2)e^{-\rho/2}$ |
| 3 | 1 | $L_4^3(x) = 24x - 96$ | $(1/9\sqrt{6})(Z/a_0)^{3/2}(4-\rho)\rho e^{-\rho/2}$ |
| 3 | 2 | $L_5^5(x) = -120$ | $(1/9\sqrt{30})(Z/a_0)^{3/2}\rho^2 e^{-\rho/2}$ |

product of radial and angular parts:

$$\Psi = R(r)\Theta(\theta)\Phi(\phi) \tag{C.34}$$

The equation for $\Phi(\phi)$ is

$$\frac{d^2\Phi}{d\phi^2} + m_l^2\Phi = 0 \tag{C.35}$$

which can be readily solved to yield solutions of the form

$$\Phi = Ae^{im_l\phi}. \tag{C.36}$$

Solutions to the radial part of the wave function can be found and are of the form

$$R_{nl}(r) = -2\left(\frac{Z}{na_0}\right)^{\frac{3}{2}}\left[\frac{(n-l-1)!}{n[(n+l)!]^3}\right]^{\frac{1}{2}}\rho^l e^{-\rho/2}L_{n+l}^{2l+1}(\rho) \tag{C.37}$$

where $n$ is the principal quantum number,

$$\rho = \frac{2Zr}{na_0}, \tag{C.38}$$

and the associated Laguerre polynomials are given by

$$L_\alpha^\beta(x) = \frac{d^\beta}{dx^\beta}\left(e^x\frac{d^\alpha}{dx^\alpha}[e^{-x}x^\alpha]\right) = \sum_{j=0}^{\alpha-\beta}\frac{\alpha!^2(-1)^{j+\beta}}{(j+\beta)!j!(\alpha-\beta-j)!}x^j. \tag{C.39}$$

For atomic states the quantum numbers are in the range $n \geq 1, 0 \leq l \leq n - 1$ and $-l \leq m_l \leq l$. Table C.2 lists some radial wave functions for $n = 1, 2, 3$. Some radial wave functions $R(r)$ for the hydrogen atom are plotted in Fig. C.2. Also shown is the radial probability density $r^2R(r)^2$.

## C.6   The g-factor

The energy of the electron in a magnetic field $B$ is

$$E = g\mu_B m_s B \tag{C.40}$$

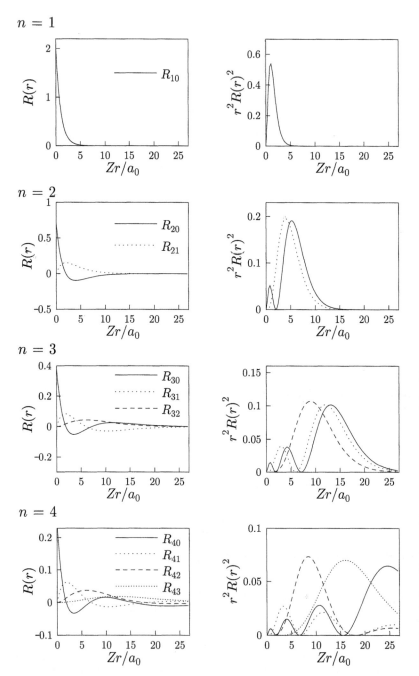

**Fig. C.2** The radial wave functions $R(r)$ (left) and the corresponding probability density $r^2 R(r)^2$ (right) for $n = 1, 2, 3$ and $4$.

where $g$ is known as the g-factor. The energy levels therefore split by an amount $g\mu_B B$. A natural consequence of Dirac's theory of the electron (outside the scope of this book) is that $g$ is precisely equal to 2. Actually the g-factor is not quite 2 but takes the value

$$g = 2\left(1 + \frac{\alpha}{2\pi} + \cdots\right) = 2.0023\ldots \tag{C.41}$$

where $\alpha$ is the **fine structure constant**, a dimensionless quantity given by

$$\alpha = \frac{e^2}{4\pi\epsilon_0\hbar c} = \frac{1}{137.04}. \tag{C.42}$$

This theoretical value obtained from quantum electrodynamics (QED) agrees with experiment to an astonishing degree of precision. The discrepancy from $g = 2$ can be explained as follows: electromagnetic interactions are due to the action of virtual photons. An electron can in fact emit a virtual photon, only to later reabsorb it. If you measure the magnetic moment of the electron during the life of one of these virtual photons, you will actually be measuring the magnetic moment of the electron–photon pair which will include an extra orbital component. You may also measure the system during a time in which more than one virtual photon has been created, though this sort of process becomes less probable as the number of virtual photons increases. On average then, the system has a magnetic moment slightly higher than what you might first expect, and this accounts for the non-zero value of $g - 2$. The expression is a power series in $\alpha$ with each successive term reflecting the contributions of progressively more convoluted creations and absorptions of virtual photons.

## C.7   d orbitals

The angular parts of the d wave functions ($l = 2$), in the form most often considered in condensed matter physics, can be constructed as linear combinations of functions of the form $Y_{2m}e^{im_l\phi}$ where the functions $Y_{2m}$ are those listed in the $l = 2$ row of Table C.1:

$$Y_{xy} = \frac{Y_{22} - Y_{2-2}}{i\sqrt{2}} \tag{C.43}$$

$$Y_{x^2-y^2} = \frac{Y_{22} + Y_{2-2}}{\sqrt{2}} \tag{C.44}$$

$$Y_{yz} = \frac{-Y_{21} - Y_{2-1}}{i\sqrt{2}} \tag{C.45}$$

$$Y_{zx} = \frac{-Y_{21} + Y_{2-1}}{\sqrt{2}} \tag{C.46}$$

$$Y_{z^2} = Y_{20} \tag{C.47}$$

which yields the following results (with the prefactors included for normalization)

$$Y_{xy} = \sqrt{\frac{15}{16\pi}} \sin^2\theta \sin 2\phi = \sqrt{\frac{15}{4\pi}} \frac{xy}{r^2} \tag{C.48}$$

$$Y_{x^2-y^2} = \sqrt{\frac{15}{16\pi}} \sin^2\theta \cos 2\phi = \sqrt{\frac{15}{16\pi}} \frac{x^2 - y^2}{r^2} \tag{C.49}$$

$$Y_{yz} = \sqrt{\frac{15}{4\pi}} \sin\theta \cos\theta \sin\phi = \sqrt{\frac{15}{4\pi}} \frac{yz}{r^2} \tag{C.50}$$

$$Y_{zx} = \sqrt{\frac{15}{4\pi}} \sin\theta \cos\theta \cos\phi = \sqrt{\frac{15}{4\pi}} \frac{zx}{r^2} \tag{C.51}$$

$$Y_{z^2} = \sqrt{\frac{15}{16\pi}} (3\cos^2\theta - 1) = \sqrt{\frac{15}{16\pi}} \frac{2z^2 - x^2 - y^2}{r^2}. \tag{C.52}$$

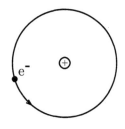

nucleus frame

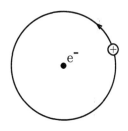

electron frame

**Fig. C.3** The intrinsic spin–orbit interaction. In the frame of the nucleus the electron orbits the nucleus, but an inertial frame comoving with the electron it is the nucleus which orbits.

[2]There are an infinite set of Lorentz transformations corresponding to the various instantaneous rest frames of the electron around its orbit. The Lorentz transformations do not commute when the direction of velocity changes and this is something our derivation has ignored. The net Lorentz transformation contains a rotation and it turns out that this Thomas precession yields a factor of precisely one-half.

L. H. Thomas (1903–1992)

These orbitals, together with the s and p orbitals, are shown in Fig. 3.1.

## C.8  The spin–orbit interaction

The spin–orbit interaction in an atom arises as follows. Consider an electron orbiting an atom. This is depicted at the top of Fig. C.3 in the nucleus rest frame. In the lower diagram, the atom is shown in an inertial frame comoving with the electron in which the nucleus appears to be orbiting the electron. The orbiting nucleus constitutes a current which gives rise to a magnetic field at the origin equal to

$$\mathbf{B} = \frac{\boldsymbol{\mathcal{E}} \times \mathbf{v}}{c^2}, \tag{C.53}$$

where

$$\boldsymbol{\mathcal{E}} = -\nabla V(r) = -\frac{\mathbf{r}}{r}\frac{dV(r)}{dr} \tag{C.54}$$

is the electric field at the electron due to the nucleus and $V(r)$ is the corresponding potential energy. Equation C.53 comes from the transformation of electric and magnetic fields in special relativity. This magnetic field interacts with the spin of the electron to give a term in the Hamiltonian

$$H_{\mathrm{so}} = -\frac{1}{2}\mathbf{m} \cdot \mathbf{B} \tag{C.55}$$

$$= \frac{e\hbar^2}{2m_ec^2r}\frac{dV(r)}{dr}\mathbf{S} \cdot \mathbf{L} \tag{C.56}$$

where the orbital angular momentum is given by $\hbar\mathbf{L} = m_e\mathbf{r} \times \mathbf{v}$ and the magnetic moment $\mathbf{m} = (ge\hbar/2m)\mathbf{S}$, and where the factor of $\frac{1}{2}$ in eqn C.55 is the relativistic Thomas factor. This result can be obtained extremely elegantly with the relativistic correction automatically included using the Dirac equation.[2] This effect is known as the intrinsic spin–orbit interaction and is an interaction between the spin and the orbital part of an electron's wave function in an atom. For the Coulomb field in a hydrogen-like atom

$$\frac{1}{r}\frac{dV(r)}{dr} = \frac{Ze}{4\pi\epsilon_0 r^3} \tag{C.57}$$

and for electronic states with quantum numbers $l$ and $n$ one has

$$\langle r^{-3}\rangle = \frac{Z^3}{a_0^3 n^3 l(l + \frac{1}{2})(l + 1)} \tag{C.58}$$

so that the spin–orbit splitting is given by

$$\frac{Z^4 e^2 \hbar^2 \langle \mathbf{S} \cdot \mathbf{L}\rangle}{4\pi\epsilon_0 a_0^3 n^3 l(l + \frac{1}{2})(l + 1)}. \tag{C.59}$$

## C.9  Landé g-factor

The Landé g-factor $g_J$ is given by

$$g_J = \frac{3}{2} + \frac{S(S + 1) - L(L + 1)}{2J(J + 1)} \tag{C.60}$$

The expression for $g_J$ in eqn C.60 is obtained by starting with an expression for the magnetic moment operator:

$$\hat{\mu} = \mu_B(g_L\hat{\mathbf{L}} + g_S\hat{\mathbf{S}}). \tag{C.61}$$

In this expression $g_L = 1$ and $g_S = 2$ are the g-factors for orbital and spin angular momentum respectively. Now $\mathbf{S}$ and $\mathbf{L}$ may not be good quantum numbers in many atoms, but $\mathbf{J}$ is. Therefore the component of the magnetic moment which is parallel to $\mathbf{J}$ will be a conserved quantity but the component perpendicular to $\mathbf{J}$ will not be. We therefore write

$$\hat{\mu} = g_J\mu_B\hat{\mathbf{J}}, \tag{C.62}$$

where $g_J$ is a constant to be determined. Thus eqn C.62 implies that $g_J$ is the projection of $\hat{\mathbf{L}} + 2\hat{\mathbf{S}}$ on to $\hat{\mathbf{J}}$. Multiplying both sides of eqn C.61 by $\hat{\mathbf{J}}$ yields

$$\hat{\mu} \cdot \mathbf{J} = \mu_B(g_L\hat{\mathbf{L}} \cdot \hat{\mathbf{J}} + g_S\hat{\mathbf{S}} \cdot \hat{\mathbf{J}}). \tag{C.63}$$

Multiplying both sides of eqn C.62 by $\hat{\mathbf{J}}$ yields[3]

$$\hat{\mu} \cdot \mathbf{J} = g_J\mu_B\hat{\mathbf{J}}^2 = g_J\mu_B J(J + 1). \tag{C.64}$$

Also, $\mathbf{L}^2 = (\mathbf{J} - \mathbf{S})^2 = \mathbf{J}^2 + \mathbf{S}^2 - 2\mathbf{S} \cdot \mathbf{J}$ so that

$$\mathbf{S} \cdot \mathbf{J} = \frac{1}{2}(\mathbf{J}^2 - \mathbf{L}^2 + \mathbf{S}^2) \tag{C.65}$$

and $\mathbf{S}^2 = (\mathbf{J} - \mathbf{L})^2 = \mathbf{J}^2 + \mathbf{L}^2 - 2\mathbf{L} \cdot \mathbf{J}$ so that

$$\mathbf{L} \cdot \mathbf{J} = \frac{1}{2}(\mathbf{J}^2 + \mathbf{L}^2 - \mathbf{S}^2). \tag{C.66}$$

Equating eqns C.63 and C.64, and inserting the results from eqns C.65 and C.66 yields

$$g_J = g_L\left(\frac{J(J + 1) + L(L + 1) - S(S + 1)}{2J(J + 1)}\right)$$
$$+ g_S\left(\frac{J(J + 1) - L(L + 1) + S(S + 1)}{2J(J + 1)}\right) \tag{C.67}$$

which reduces to eqn C.60 if $g_L = 1$ and $g_S = 2$.

[3] In this, and subsequent steps, we use the fact that the eigenvalue of $\mathbf{S}^2$ is $S(S + 1)$, of $\mathbf{L}^2$ is $L(L + 1)$ and of $\mathbf{J}^2$ is $J(J + 1)$.

## C.10   Perturbation theory

Consider a Hamiltonian $\hat{\mathcal{H}}_0$ which has known eigenfunctions $|\phi_i\rangle$ with known eigenvalues $E_i$, so that

$$\hat{\mathcal{H}}_0|\phi_i\rangle = E_i|\phi_i\rangle. \tag{C.68}$$

The eigenfunctions $|\phi_i\rangle$ are all orthogonal so that

$$\langle\phi_i|\phi_j\rangle = \int d\tau\, \phi_i^*\phi_j = \delta_{ij} \tag{C.69}$$

212 *Quantum and atomic physics*

where d$\tau$ is a volume element. Now suppose that a perturbation $\hat{V}$ is added to $\hat{\mathcal{H}}_0$ so that the new Hamiltonian is $\hat{\mathcal{H}}$ where

$$\hat{\mathcal{H}} = \hat{\mathcal{H}}_0 + \hat{V}. \tag{C.70}$$

The system starts in a state $|\phi_k\rangle$ with energy $E_k$ before the addition of the perturbation. The new eigenfunction of the system will be $|\psi\rangle$ and the eigenvalue $E$ where

$$\hat{\mathcal{H}}|\psi\rangle = E|\psi\rangle, \tag{C.71}$$

but both $|\psi\rangle$ and $E$ are unknown so far. $|\psi\rangle$ can be expanded in terms of the old eigenfunctions so that

$$|\psi\rangle = \sum_j a_j |\phi_j\rangle. \tag{C.72}$$

Inserting this into eqn C.71, premultiplying by $\langle\phi_i|$ and integrating over the appropriate volume gives

$$\sum_j a_j \int d\tau \, \phi_i^*(\hat{\mathcal{H}}_0 + \hat{V})\phi_j = E \sum_j a_j \int d\tau \, \phi_i^*\phi_j, \tag{C.73}$$

so that using the orthogonality of the original eigenfunctions (eqn C.69) yields

$$\sum_j V_{ij}a_j = (E - E_i)a_i, \tag{C.74}$$

which is known as the **secular equation**, where

$$V_{ij} = \langle\phi_i|\hat{V}|\phi_j\rangle = \int d\tau \, \phi_i^*\hat{V}\phi_j \tag{C.75}$$

is the **matrix element** of the perturbation.

From now on, we shall assume the non-degenerate case, i.e. we will assume that none of the states are degenerate.

If the perturbation is small, $|\psi\rangle$ will be very close to the starting state $|\phi_k\rangle$ so that $a_k \approx 1$ and $|a_i| \ll 1$ for $i \neq k$. Since the wave function has changed only marginally, one might imagine that the new energy is given by

$$E \approx E_k, \tag{C.76}$$

in other words that it remains completely unchanged. We can improve on this zeroth order approximation by using the results $a_k \approx 1$ and $a_i \ll 1$ for $i \neq k$. Substituting these into the secular equation can give some approximate expressions for the energy and for $a_i$. For $i = k$, the sum in the secular equation is dominated by the term for which $j = k$ giving

$$E \approx E_k + V_{kk}, \tag{C.77}$$

the shift of the energy of the state is given simply by $V_{kk}$. This result, which contains the 'first-order perturbation theory' correction to the energy, is used very often in the text. The secular equation can also be evaluated for the case $i \neq k$, yielding

$$a_i = \frac{V_{ik}}{E_k - E_i} \tag{C.78}$$

(where $E$ has been replaced by $E_k$). Putting eqn C.78 back into eqn C.74 yields the next level of approximation for the energy, namely

$$E \approx E_k + V_{kk} + \sum_{i \neq k} \frac{|V_{ik}|^2}{E_k - E_i} \tag{C.79}$$

which is the 'second-order perturbation theory' result. This process can be continued and will generate a power series expression for $E$, with each successive term in the series containing a higher power of the matrix elements of the perturbation.

# Further reading

- A. I. Rae, *Introduction to quantum mechanics*, 3rd edition, IOP Publishing 1992.
- J. J. Sakurai, *Modern quantum mechanics*, 2nd edition, Addison-Wesley 1994.
- P. A. M. Dirac, *The principles of quantum mechanics*, 4th edition, OUP 1958.
- B. H. Bransden and C. J. Joachain, *Physics of atoms and molecules*, Longman 1983.
- P. W. Atkins, *Molecular quantum mechanics*, OUP 1983.
- G. K. Woodgate, *Elementary atomic structure*, OUP 1980.

# Appendix D

# Energy in magnetism and demagnetizing fields

## D.1 Energy

Calculating the energy of a magnetized medium in a magnetic field is a surprisingly subtle business. Various expressions can be obtained but they can refer to different things. In magnetism it depends on whether one is talking about the energy of the magnetic moment alone, or the magnetic moment plus whatever it is that is providing the magnetic field, and exactly how this energy is partitioned up.

For example, if you place a screwdriver close to the pole pieces of a large magnet, the screwdriver is strongly attracted into the region of maximum field. Replacing a bit of the air in the gap between the pole pieces with the magnetized screwdriver therefore seems to be an energetically favourable thing, suggesting that the presence of the magnetizable medium (the screwdriver) is lowering the energy. However, it takes more energy to establish a current in a magnet which contains an iron core than in one without an iron core, suggesting that the presence of the magnetizable medium (in this case the iron core) is increasing the energy. So does the energy decrease or increase? It depends on which energy you are considering. In the first example we are only measuring the energy saved by the force we feel on our arm holding the screwdriver, and we are ignoring the extra work needed to be done by the magnet power supply in maintaining the current at the level it had before we started waving the screwdriver around. This emphasizes that great care is needed in treating the energetics of magnetic materials (for a detailed treatment, see the article by Heine in the further reading).

In Appendix B, we found that $\delta W = -\boldsymbol{\mu}\cdot\delta\mathbf{B}$ is the appropriate free energy to consider for magnetized media. This expression will be useful in formulating magnetism in statistical mechanics (see Appendix E). In this appendix, we concentrate on the way in which the presence of a ferromagnetic material can change both $\mathbf{H}$ and $\mathbf{B}$ because of the introduction of demagnetizing fields which themselves cost energy.

## D.2 Demagnetizing factors

When the magnetization $\mathbf{M}$ inside a ferromagnetic body meets the surface, it has to suddenly stop. Hence there is a divergence of $\mathbf{M}$. Using the equation

$$\nabla \cdot \mathbf{H} = -\nabla \cdot \mathbf{M}, \tag{D.1}$$

(a)                              (b)                              (c)

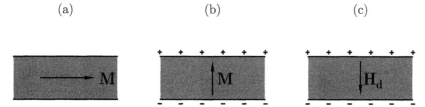

**Fig. D.1** Demagnetization in an infinite flat plate, viewed in cross-section. (a) If the magnetization lies in the plane of the plate, no magnetic poles are created on the surface of the plate (except for tiny ones at the ends). (b) If the magnetization is perpendicular to the plane of the plate, as shown, there is a negative (positive) divergence of **M** on the top (bottom) surface which produces a positive (negative) divergence of **H**. This results in positive (negative) magnetic poles on the top (bottom) surface as shown. (c) This results in a demagnetization field which runs from positive magnetic poles to negative magnetic poles.

(which follows from $\nabla \cdot \mathbf{B} = 0$, see appendix B) we find that there is an equal and opposite divergence of **H**. The situation is as if magnetic monopoles have been left on the surface of the ferromagnet, and these monopoles act as sources of **H**. The resulting **H** field is known as a **demagnetizing field**. This situation is illustrated for a simple case in Figure D.1 which describes the magnetization in an infinite flat plate or thin film of ferromagnetic material. If the magnetization lies in the plane of the plate, the only divergence of **M** is at the ends which, we suppose, are an infinite distance away. Therefore there is no demagnetizing field. If the magnetization is perpendicular to the plane of the plate, magnetic poles are created on top and bottom surfaces and give rise to a demagnetizing field $\mathbf{H_d} = -\mathbf{M}$ inside the plate.

The demagnetizing field can be an extremely complicated function of position for a ferromagnet of arbitrary shape. However, it takes a relatively simple form in the case of an ellipsoidal ferromagnet. In this case it is uniform inside the ferromagnet with a value of $\mathbf{H_d}$ equal to

$$\mathbf{H_d} = -\mathsf{N}\mathbf{M} \tag{D.2}$$

where $\mathsf{N}$ is the demagnetizing tensor. Thus in general we can write

$$(H_d)_i = -\sum_j \mathsf{N}_{ij} M_j. \tag{D.3}$$

If **M** is along one of the principal axes of the ellipse, $\mathsf{N}_{ij}$ can be diagonalized so that

$$\mathsf{N} = \begin{pmatrix} N_x & 0 & 0 \\ 0 & N_y & 0 \\ 0 & 0 & N_z \end{pmatrix}. \tag{D.4}$$

The demagnetizing tensor satisfies

$$\mathrm{Tr}\mathsf{N} = N_x + N_y + N_z = 1. \tag{D.5}$$

## Example D.1

We now examine some special cases.

(1) For a sphere, $N_x = N_y = N_z = \frac{1}{3}$, so that

$$\mathbf{H_d} = -\frac{\mathbf{M}}{3} \qquad (D.6)$$

(2) For a very long cylindrical rod parallel to $z$, $N_x = N_y = \frac{1}{2}$, $N_z = 0$. This is because if the magnetization lies exactly along the rod, the magnetic poles created will be at either end of the rod which can be assumed to be too far away to matter.

(3) For a flat plate perpendicular to $z$, $N_x = N_y = 0$, $N_z = 1$. This is the case considered above.

## D.3   A ferromagnet of arbitrary shape

In this section we will show how to compute the energy due to the demagnetizing energy for a ferromagnet of arbitrary shape in an applied magnetic field. For this problem, the magnetic fields $\mathbf{H(r)}$ and $\mathbf{B(r)}$ can be broken up into two components:

$$\mathbf{H(r)} = \mathbf{H_a(r)} + \mathbf{H_d(r)} \qquad (D.7)$$
$$\mathbf{B(r)} = \mathbf{B_a(r)} + \mathbf{B_d(r)}, \qquad (D.8)$$

where the applied field is $\mathbf{H_a(r)} = \mathbf{B_a(r)}/\mu_0$ and the demagnetizing field is $\mathbf{H_d(r)}$ and we write $\mathbf{B_d(r)}$ as a sum of the flux density due to the demagnetizing field $\mu_0\mathbf{H_d(r)}$ and $\mu_0\mathbf{M(r)}$ inside the material, i.e. $\mathbf{B_d(r)} = \mu_0(\mathbf{H_d(r)}+\mathbf{M(r)})$. The demagnetizing field arises from the magnetic poles which are produced on the surface of the magnetic body wherever $\nabla \cdot \mathbf{M(r)} \neq 0$.

We now state an important result. The demagnetizing fields $\mathbf{B_d(r)}$ and $\mathbf{H_d(r)}$ satisfy

$$\int_{\text{all space}} \mathbf{B_d} \cdot \mathbf{H_d} \, d\tau = 0. \qquad (D.9)$$

The proof is as follows: since $\nabla \times \mathbf{H_d} = 0$, the demagnetizing field $\mathbf{H_d}$ can be written in terms of the gradient of a scalar function $\phi$ (e.g. $\mathbf{H_d} = -\nabla\phi$). Now

$$\nabla \cdot (\phi \, \mathbf{B_d}) \equiv \phi \nabla \cdot \mathbf{B_d} + (\nabla\phi) \cdot \mathbf{B_d} = (\nabla\phi) \cdot \mathbf{B_d} \qquad (D.10)$$

where the last equality follows from $\nabla \cdot \mathbf{B_d} = 0$. Then

$$\int_{\text{all space}} \mathbf{B_d} \cdot \mathbf{H_d} \, d\tau = -\int_{\text{all space}} \nabla \cdot (\phi \, \mathbf{B_d}) \, d\tau = \lim_{R \to \infty} \int_{\text{surface}} \phi \, \mathbf{B_d} \cdot \mathbf{dS} = 0,$$
$$(D.11)$$

where the surface integral is taken over a sphere whose radius $R$ tends to infinity. The integral is zero if the ferromagnetic body is of finite extent, so that as $R \to \infty$, $\mathbf{B_d} \sim R^{-2}$ and $\phi \sim R^{-1}$ at worst.

If the applied field is not present, then the energy $E$ of the demagnetizing field is simply the integral over all space of the energy density $\frac{1}{2}\mu_0\mathbf{H_d^2}$. This can be converted to an integral inside the volume $V$ of the ferromagnetic body by using eqn D.9, so that the energy can be expressed as

$$
\begin{aligned}
E = \frac{\mu_0}{2} \int_{\text{all space}} \mathbf{H}_{\mathrm{d}}^2 \, d\tau &= -\frac{1}{2} \int_{\text{all space}} (\mathbf{B}_{\mathrm{d}} - \mu_0 \mathbf{M}) \cdot \mathbf{H}_{\mathrm{d}} \, d\tau \\
&= -\frac{\mu_0}{2} \int_{\text{all space}} \mathbf{M} \cdot \mathbf{H}_{\mathrm{d}} \, d\tau \\
&= -\frac{\mu_0}{2} \int_{V} \mathbf{M} \cdot \mathbf{H}_{\mathrm{d}} \, d\tau,
\end{aligned}
\tag{D.12}
$$

where the last line follows from the fact that $\mathbf{M} = 0$ outside the body. It should be noted that even though this final integral is only over the body, it expresses the energy of all stray fields, including those *outside* the body.

If there is an applied field, then the energy of the ferromagnetic body may be written as the difference between the energy density in the total field $\frac{1}{2}\mu_0 \mathbf{H}^2$ and the energy density in the applied field $\frac{1}{2}\mu_0 \mathbf{H}_{\mathrm{a}}^2$. This removes the infinite energy contribution from a uniform applied field integrated over all space. Thus the energy can be usefully written as

$$
E = \frac{\mu_0}{2} \int_{\text{all space}} (\mathbf{H}^2 - \mathbf{H}_{\mathrm{a}}^2) \, d\tau.
\tag{D.13}
$$

Using the fact that $\nabla \times \mathbf{H}_{\mathrm{a}} = 0$, we have

$$
\int_{\text{all space}} \mathbf{H}_{\mathrm{a}} \cdot \mathbf{B}_{\mathrm{d}} \, d\tau = 0.
\tag{D.14}
$$

Hence, by also using the following equations:

$$
\mathbf{H}^2 = \mathbf{H}_{\mathrm{a}}^2 + 2\mathbf{H}_{\mathrm{d}} \cdot \mathbf{H}_{\mathrm{a}} + \mathbf{H}_{\mathrm{d}}^2,
\tag{D.15}
$$

$$
\mathbf{H}_{\mathrm{d}} \cdot \mathbf{H}_{\mathrm{a}} = \mathbf{H}_{\mathrm{a}} \cdot \left( \frac{\mathbf{B}_{\mathrm{d}}}{\mu_0} - \mathbf{M} \right),
\tag{D.16}
$$

and by also using eqn D.9, the energy $E$ can be written in the familiar form

$$
E = -\mu_0 \int_{V} \mathbf{M} \cdot \mathbf{H}_{\mathrm{a}} \, d\tau - \frac{\mu_0}{2} \int_{V} \mathbf{M} \cdot \mathbf{H}_{\mathrm{d}} \, d\tau,
\tag{D.17}
$$

the sum of a Zeeman term and a demagnetizing term, with each integral being taken only over the volume $V$ of the ferromagnetic body. The energy in eqn D.17 nevertheless implicitly includes the energy associated with the demagnetizing fields outside the body.

The factor of one-half in the demagnetizing term can be understood in terms of an avoidance of double-counting. The Zeeman term is the interaction of the magnetization with the applied field. The demagnetizing term is a self-energy so the moments which cause the field interact with their own field. The factor of one-half ensures that you don't count this energy twice.

## Example D.2

A ferromagnetic sphere of radius $a$ with uniform magnetization $\mathbf{M}$ is used here as a detailed example of the ideas leading to eqn D.12. It is sketched in Fig. D.2. The demagnetizing fields are given by

$$
\mathbf{H}_{\mathrm{d}} = -\frac{\mathbf{M}}{3}
\tag{D.18}
$$

$$
\mathbf{B}_{\mathrm{d}} = \frac{2\mu_0 \mathbf{M}}{3},
\tag{D.19}
$$

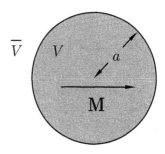

**Fig. D.2** A ferromagnetic sphere with uniform magnetization $\mathbf{M}$. The region inside the sphere of volume $V_{\text{sphere}} = \frac{4}{3}\pi a^3$ is denoted $V$, the region outside the sphere is denoted $\bar{V}$.

so that inside the sphere (a space of volume $V_{\text{sphere}} = \frac{4}{3}\pi a^3$, denoted by $V$) simple integration yields the following results:

$$-\frac{\mu_0}{2}\int_V \mathbf{M}\cdot\mathbf{H_d}\,d\tau = \frac{1}{6}\mu_0 M^2 V_{\text{sphere}} \tag{D.20}$$

$$\frac{\mu_0}{2}\int_V \mathbf{H_d^2}\,d\tau = \frac{1}{18}\mu_0 M^2 V_{\text{sphere}} \tag{D.21}$$

$$\frac{1}{2}\int_V \mathbf{B_d}\cdot\mathbf{H_d}\,d\tau = -\frac{1}{9}\mu_0 M^2 V_{\text{sphere}} \tag{D.22}$$

Outside the sphere (a space denoted by $\bar{V}$), $\mathbf{B_d} = \mu_0\mathbf{H_d}$ (since $\mathbf{M} = 0$ outside the sphere), and the demagnetizing fields result from the magnetic dipole moment of the sphere. These have radial and polar components given by

$$(H_d)_r = 2Ma^3\cos\theta/3r^3$$
$$(H_d)_\theta = Ma^3\sin\theta/3r^3, \tag{D.23}$$

so that integration over this region outside the sphere gives:

$$\frac{\mu_0}{2}\int_{\bar{V}} \mathbf{H_d^2}\,d\tau = \frac{1}{2}\int_{\bar{V}} \mathbf{B_d}\cdot\mathbf{H_d}\,d\tau = \frac{1}{2\mu_0}\int_{\bar{V}} \mathbf{B_d^2}\,d\tau = \frac{1}{9}\mu_0 M^2 V_{\text{sphere}}. \tag{D.24}$$

Then integration over all space gives

$$-\frac{\mu_0}{2}\int_{\text{all space}} \mathbf{M}\cdot\mathbf{H_d}\,d\tau = \frac{\mu_0}{2}\int_{\text{all space}} \mathbf{H_d^2}\,d\tau$$
$$= \frac{1}{6}\mu_0 M^2 V_{\text{sphere}} \tag{D.25}$$

in agreement with eqn D.12. By combining eqns D.22 and D.24 one has

$$\frac{1}{2}\int_{\text{all space}} \mathbf{B_d}\cdot\mathbf{H_d}\,d\tau = 0 \tag{D.26}$$

in agreement with eqn D.9.

# Further reading

- V. Heine, *Proc. Camb. Phil. Soc.* **52**, 546 (1956) contains a profound treatment of the thermodynamics of bodies in electromagnetic fields.

- Treatments are also found in C. J. Adkins, *Equilibrium thermodynamics*, CUP 1983, and J. R. Waldram, *The theory of thermodynamics*, CUP 1985.

- Magnetostatic energy and demagnetizing factors are cov-ered in depth in A. Aharoni, *Introduction to the theory of ferromagnetism*, OUP 1996.

- I. Brevik, *J. Phys.: Condensed Matter* **7**, 8065 (1995) contains a helpful analysis of the problem of calculating the energy of a ferromagnetic body immersed in a ferromagnetic liquid.

# Statistical mechanics

**Appendix**

**E**

This appendix briefly sketches some results in statistical mechanics which will be useful in the main text.

## E.1 The partition function and thermodynamic functions

The probability $p_i$ of occupying a state $i$ with energy $E_i$ at temperature $T$ is

$$p_i = \frac{e^{-E_i/k_\mathrm{B}T}}{Z} \tag{E.1}$$

where $Z$ is the single-particle **partition function** which is used to normalize the probability such that

$$\sum_i p_i = 1. \tag{E.2}$$

This implies that $Z$ is given by

$$Z = \sum_i e^{-E_i/k_\mathrm{B}T}. \tag{E.3}$$

Defining $\beta = 1/k_\mathrm{B}T$, $\partial Z/\partial \beta$ can be evaluated, giving

$$\frac{\partial Z}{\partial \beta} = -\sum_i E_i e^{-\beta E_i}. \tag{E.4}$$

Now since we have found the probability distribution, we can use it to calculate expected values of quantities. The expected value of the energy $\langle E \rangle$ is given by

$$\langle E \rangle = \sum_i E_i p_i = \frac{\sum_i E_i e^{-\beta E_i}}{Z} = -\frac{1}{Z}\frac{\partial Z}{\partial \beta} = -\frac{\partial \ln Z}{\partial \beta}. \tag{E.5}$$

The entropy per particle is given by

$$S = -k_\mathrm{B} \sum_i p_i \ln p_i \tag{E.6}$$

and hence

$$S = -k_\mathrm{B} \sum_i \frac{e^{-\beta E_i}}{Z}[-\beta E_i - \ln Z] = k_\mathrm{B} \ln Z + \frac{\langle E \rangle}{T}. \tag{E.7}$$

In eqn E.8 we have used the result from Appendix B that the available free energy is $-\mu \cdot \delta \mathbf{B}$.

For $N$ independent distinguishable particles, the appropriate partition function becomes $Z^N$ and so the total energy is $E = N\langle E \rangle$. The first law of thermodynamics can be expressed as

$$dE = T\,dS - p\,dV - M\,dB. \tag{E.8}$$

The Helmholtz free energy $F$, also known as the Helmholtz function, is defined by

$$F = E - TS, \tag{E.9}$$

and is related to the partition function $Z$ by

$$F = -Nk_BT \ln Z. \tag{E.10}$$

Using eqns E.8 and E.9, $dF$ can be written

$$dF = -S\,dT - p\,dV - M\,dB \tag{E.11}$$

and so

$$M = -\left(\frac{\partial F}{\partial B}\right)_{T,V} = Nk_BT\left(\frac{\partial \ln Z}{\partial B}\right)_{T,V}. \tag{E.12}$$

The entropy is given by

$$S = -\left(\frac{\partial F}{\partial T}\right)_{V,B}. \tag{E.13}$$

## E.2　The equipartition theorem

Very often the energy of a system can be written as a sum of quadratic terms. For example, a mass $m$ on a spring of spring constant $K$ has an energy $E = \frac{1}{2}mv^2 + \frac{1}{2}Kx^2$, which contains two such quadratic terms. At low temperatures, where quantum effects dominate, we expect harmonic oscillator solutions and a ladder of energy levels. But at high temperatures, where $k_BT$ is much larger than the energy scale set by the spacing between the levels, the system can be considered classically.

The equipartition theorem states that in the classical limit, when $k_BT$ is much larger than the energy spacing between quantum levels, and the energy contains a number of independent quadratic terms, the expected value of the energy is equal to $\frac{1}{2}k_BT$ multiplied by the number of independent quadratic terms in the energy.

The proof is straightforward: if $E(x_1, x_2, \ldots x_n) = \sum_{i=1}^{n} \alpha_i x_i^2$ where $\alpha_i$ are constants, then

$$\langle E \rangle = \frac{\displaystyle\int_{-\infty}^{\infty}\int_{-\infty}^{\infty}\cdots\int_{-\infty}^{\infty} dx_1\, dx_2 \ldots dx_n\, e^{-\beta E(x_1, x_2, \ldots x_n)} E}{\displaystyle\int_{-\infty}^{\infty}\int_{-\infty}^{\infty}\cdots\int_{-\infty}^{\infty} dx_1 dx_2 \ldots dx_n\, e^{-\beta E(x_1, x_2, \ldots x_n)}}. \tag{E.14}$$

This looks formidable but the integrals for each variable can be easily separated. Hence

$$
\langle E \rangle = \sum_{i=1}^{n} \alpha_i \left( \frac{\int_{-\infty}^{\infty} \mathrm{d}x_i \, x_i^2 \mathrm{e}^{-\beta \alpha_i x_i^2} \prod_{j \neq i} \int_{-\infty}^{\infty} \mathrm{d}x_j \, \mathrm{e}^{-\beta \alpha_i x_j^2}}{\int_{-\infty}^{\infty} \mathrm{d}x_i \, \mathrm{e}^{-\beta \alpha_i x_i^2} \prod_{j \neq i} \int_{-\infty}^{\infty} \mathrm{d}x_j \, \mathrm{e}^{-\beta \alpha_i x_j^2}} \right)
$$

$$
= \sum_{i=1}^{n} \left( \frac{\frac{\alpha_i}{2} \sqrt{\frac{\pi}{\alpha_i^3 \beta^3}}}{\sqrt{\frac{\pi}{\alpha_i \beta}}} \right)
$$

$$
= \sum_{i=1}^{n} \frac{1}{2\beta}
$$

$$
= \frac{n}{2} k_B T, \tag{E.15}
$$

where use has been made of the standard integrals

$$
\int_{-\infty}^{\infty} \mathrm{d}x \, \mathrm{e}^{-\gamma x^2} = \sqrt{\frac{\pi}{\gamma}} \quad \text{and} \quad \int_{-\infty}^{\infty} \mathrm{d}x \, x^2 \mathrm{e}^{-\gamma x^2} = \frac{1}{2} \sqrt{\frac{\pi}{\gamma^3}}. \tag{E.16}
$$

**Example E.1**

Three examples of this result now follow:

(1) For a mass $m$ on a spring of spring constant $K$, $E = \frac{1}{2}mv^2 + \frac{1}{2}Kx^2$, (two quadratic terms), so that $\langle E \rangle = k_B T$.
(2) For a molecule of mass $m$ in a gas, $E = \frac{1}{2}mv_x^2 + \frac{1}{2}mv_y^2 + \frac{1}{2}mv_z^2$, (three quadratic terms), so that $\langle E \rangle = \frac{3}{2}k_B T$.
(3) For a solid containing $N$ atoms (and therefore $3N$ springs, and hence the energy contains $6N$ quadratic terms), $\langle E \rangle = 3N k_B T$.

In addition to adding $\frac{1}{2}k_B T$ to the energy, each quadratic degree of freedom adds $\frac{1}{2}k_B$ to the heat capacity.

# Further reading

- F. Reif, *Fundamentals of statistical and thermal physics*, Mc-Graw–Hill, 1965.

- J. R. Waldram, *The theory of thermodynamics*, CUP 1985.

- P. M. Chaikin and T. C. Lubensky, *Principles of condensed matter physics*, CUP 1995.

- A. M. Glazer and J. S. Wark, *Statistical mechanics*, OUP 2001.

# Appendix F

# Answers and hints to selected problems

(1.1) $9.27 \times 10^{-24}$ Am$^2$; 8.40 GHz; 34.7 $\mu$eV; 8.40 GHz. Here we are only considering *spin* angular momentum.

(1.4) This is the commutator of a scalar operator $(\hat{\mathbf{S}} \cdot \mathbf{X})$ and a vector operator $(\hat{\mathbf{S}})$. Work out one component of this, and so evaluate

$$[\hat{\mathbf{S}} \cdot \mathbf{X}, \hat{\mathbf{S}} \cdot \mathbf{Y}] = \frac{1}{4}((\boldsymbol{\sigma} \cdot \mathbf{X})(\boldsymbol{\sigma} \cdot \mathbf{Y}) - (\boldsymbol{\sigma} \cdot \mathbf{y})(\boldsymbol{\sigma} \cdot \mathbf{X}))$$
$$= \frac{\mathrm{i}}{4}(\boldsymbol{\sigma} \cdot (\mathbf{X} \times \mathbf{Y}) - \boldsymbol{\sigma} \cdot (\mathbf{Y} \times \mathbf{X}))$$
$$= \frac{\mathrm{i}}{2}\boldsymbol{\sigma} \cdot (\mathbf{X} \times \mathbf{Y})$$
$$= \mathrm{i}\mathbf{S} \times \mathbf{X} \cdot \mathbf{Y}, \qquad \text{(F.1)}$$

and the result follows.

(1.5) Using eqn 1.59 one can show that

$$\hat{S}_z(\hat{S}_\pm|S, S_z)) = (\hat{S}_\pm \hat{S}_z \pm \hat{S}_\pm)|S, S_z)$$
$$= (S_z \pm 1)(\hat{S}_\pm|S, S_z)), \qquad \text{(F.2)}$$

so that the z-component of the spin is raised or lowered. Equations 1.58 and 1.61 can be written as

$$\hat{S}_+\hat{S}_- - \hat{S}_-\hat{S}_+ = 2\hat{S}_z \qquad \text{(F.3)}$$
$$\hat{S}_+\hat{S}_- + \hat{S}_-\hat{S}_+ = 2(\hat{S}_x^2 + \hat{S}_y^2)$$
$$= \hat{\mathbf{S}}^2 - \hat{S}_z^2, \qquad \text{(F.4)}$$

so that adding or subtracting leads to

$$\hat{S}_+\hat{S}_- = \hat{\mathbf{S}}^2 - \hat{S}_z^2 + \hat{S}_z \qquad \text{(F.5)}$$
$$\hat{S}_-\hat{S}_+ = \hat{\mathbf{S}}^2 - \hat{S}_z^2 - \hat{S}_z. \qquad \text{(F.6)}$$

Hence

$$\langle S, S_z|\hat{S}_-\hat{S}_+|S, S_z\rangle = S(S+1) - S_z^2 - S_z$$
$$= S(S+1) - S_z(S_z+1) \qquad \text{(F.7)}$$

and

$$\langle S, S_z|\hat{S}_+\hat{S}_-|S, S_z\rangle = S(S+1) - S_z^2 + S_z$$
$$= S(S+1) - S_z(S_z-1) \qquad \text{(F.8)}$$

and the required normalization is therefore proved.

(1.6) Without loss of generality, put $\mathbf{B} = (0, 0, B)$ so that $\mathbf{B} \times \mathbf{r} = B(-y, -x, 0)$ and $\frac{1}{2}\nabla \times (\mathbf{B} \times \mathbf{r}) = (0, 0, B)$. Other choices are possible, e.g. $\mathbf{A} = -By(1, 0, 0)$ or $\mathbf{A} = Bx(0, 1, 0)$. In fact, $\mathbf{A} = B((\alpha - 1)y, \alpha x, 0)$ will work where $\alpha$ is a real number.

(1.7) Hint: evaluate $[\boldsymbol{\sigma} \cdot (\mathbf{p} + e\mathbf{A})]^2/2m_e$.

(1.8) Without loss of generality, the new axis can be at an angle $\theta$ to the $z$ axis in the $xz$ plane. Thus the operator is

$$\hat{S}_\theta = \cos\theta \hat{S}_z + \sin\theta \hat{S}_x. \qquad \text{(F.9)}$$

The eigenvector of this operator with eigenvalue 1 is found to be

$$(2(1 - \cos\theta))^{-1/2} \begin{pmatrix} \sin\theta \\ 1 - \cos\theta \end{pmatrix}, \qquad \text{(F.10)}$$

so that the required probability is

$$\frac{\sin^2\theta}{2(1 - \cos\theta)} = \cos^2\frac{\theta}{2}. \qquad \text{(F.11)}$$

(1.10) $H_d/H_a$: (a) $3 \times 10^{-5}$ (b) $-8.8 \times 10^{-4}$
$B_d/B_a$: (a) $-6 \times 10^{-5}$ (b) $1.76 \times 10^{-3}$

(1.12) Hint: $\sigma_m^2 = I$ so that

$$\sigma^n = \begin{cases} I & n \text{ even} \\ \sigma_m & n \text{ odd}. \end{cases} \qquad \text{(F.12)}$$

so that

$$\mathrm{e}^{\mathrm{i}\alpha\sigma_m} = I\left(1 - \frac{\alpha^2}{2!} + \frac{\alpha^4}{4!} - \cdots\right)$$
$$+ \mathrm{i}\sigma_m\left(\alpha - \frac{\alpha^3}{3!} + \frac{\alpha^5}{5!} - \cdots\right)$$
$$= I\cos\alpha + \mathrm{i}\sigma_m\sin\alpha. \qquad \text{(F.13)}$$

(1.14) (a) Torque$= I\ddot{\theta} = -p\mathcal{E}\sin\theta$ so that $\ddot{\theta} + p\mathcal{E}\sin\theta/I = 0$ which leads to simple harmonic motion when $\theta \ll 1$ with $\omega^2 = p\mathcal{E}/I$.
(b) The Hamiltonian is $\mathcal{H} = \hat{L}^2/2I - p\mathcal{E}\cos\theta$. In this calculation, several commutators need to be evaluated. You should find that

$$\left[\hat{\theta}, \hat{L}\right] = \mathrm{i}\hbar \qquad \text{(F.14)}$$
$$\left[\hat{\theta}, \hat{L}^2\right] = 2\mathrm{i}\hbar\hat{L} \qquad \text{(F.15)}$$
$$\left[\hat{L}, \cos\theta\right] = \mathrm{i}\hbar\sin\theta, \qquad \text{(F.16)}$$

and the results follow. The electric dipole is different because, unlike the magnetic moment, it is not associated with angular momentum. (An electric dipole is just two, opposite charges, which are spatially separated. A magnetic dipole is two, opposite charges, which are orbiting each other and so is associated with angular momentum.) Note that the Hamiltonian contains no dissipation, so the electric dipole oscillates forever, and the magnetic dipole precesses forever. Both dipoles will only line up with the external field if we allow them to change their energy, by exchanging energy with a heat bath. This is a dissipative process.

(2.1)  $-1.65 \times 10^{-15}$; $7.8 \times 10^{-12}$

(2.2)  A duck is mainly water and has a mass of 2–3 kg. An iron filing might be 0.1 mm$^3$=10$^{-10}$ m$^3$ (as a guess) and each iron atom carries 2.2 $\mu_B$. The density of iron is 7873 kg m$^{-3}$ and the relative atomic mass is 55.847 g. This can be used to show that the required field is 1 mT (the Earth's field is about 0.05 mT). A cow might be 400 kg, and assuming that it too is made of water means that with about 200 times more mass, and hence more volume, a much smaller field ($\sim 2 \mu T$) is needed to produce the same level of magnetization. Of course we have ignored the effect of haemoglobin.

(2.3)  $2.74 \times 10^{-5}$ JT$^{-1}$ if you assume Cu$^{2+}$ is spin $\frac{1}{2}$, g = 2, and use density and RMM. The answer is $3.35 \times 10^{-5}$ JT$^{-1}$ if you use measured susceptibility from table 2.1.

(2.4)  Hint: use $Z = 2\cosh(\mu_B B/k_B T)$ and $F = -nk_B T \log Z$. Then $E = -n(\mathrm{d}\log Z/\mathrm{d}\beta)$ and $C = \partial E/\partial T$. The entropy can be obtained either using $S = -(\partial F/\partial T)_B$ or $S = (E - F)/T$.

(2.5)  Hint: a useful intermediate result to prove is that

$$\frac{\partial \log Z_y}{\partial y} = B_J(y). \qquad \text{(F.17)}$$

(2.7)  (a) $^5I_8$ (b) $^4I_{15/2}$ (c) $^3H_6$ (d) $^1S_0$

(2.8)  Hints: (1) You will need to evaluate a term containing $\mathbf{p}\cdot\mathbf{A}+\mathbf{A}\cdot\mathbf{p}$. This is proportional to

$$\mathbf{p}\cdot\mu\times\mathbf{r}+\mu\times\mathbf{r}\cdot\mathbf{p}=2\mathbf{r}\times\mathbf{p}\cdot\mu, \qquad \text{(F.18)}$$

where the final equality works because $\mu$ does not depend on position, so commutes with $\mathbf{r}$ and $\mathbf{p}$, and you can commute $\mathbf{r}$ and $\mathbf{p}$ because they appear in a scalar triple product so you are never multiplying the same component of $\mathbf{r}$ and $\mathbf{p}$. (2) $\nabla^2(\frac{1}{r}) = \nabla \cdot \nabla(\frac{1}{r}) = 4\pi\delta(\mathbf{r})$. An analogous result is $(\mathbf{S} \cdot \nabla)\nabla(\frac{1}{r}) = \mathbf{S}\frac{4\pi}{3}\delta(\mathbf{r})$, where the factor of 3 is due to the fact that $\mathbf{S}$ picks out one of out of the three components of the singularity.

(2.9)  $1.89\times10^{-4}$ emu mol$^{-1}$; $9.68\times10^{-7}$ emu g$^{-1}$

(2.10)  At fixed $T$,

$$T\,\delta S = -\mathbf{B}\cdot\delta\mathbf{M} - \mathbf{M}\cdot\delta\mathbf{B} + \frac{\partial(nk_B T \log Z)}{\partial\mathbf{B}}\cdot\delta\mathbf{B}, \qquad \text{(F.19)}$$

but using $M = nk_B T(\partial \log Z/\partial B)$, the result follows.

(3.1)  Sc$^{2+}$ has one 3d electron ($l = 2$, $s = \frac{1}{2}$) and has 5 orbital states characterized by $l_z = -2, -1, 0, 1, 2$. The energy levels are thus given by $0, A, 4A$ (degeneracies 2,4 and 4 respectively, including spin) with the ground state being twofold degenerate ($E = 0$) if $A > 0$, or fourfold degenerate ($E = 4A$) if $A < 0$. Let us write states using the notation $|l_z, s_z\rangle$. If $A > 0$, the ground state levels are $|0, \frac{1}{2}\rangle$ and $|0, -\frac{1}{2}\rangle$ and are not split by the spin–orbit interaction (because $\lambda l_z s_z = 0$). The levels are split by a magnetic field in any direction, and lead to a Curie-like susceptibility providing $\mu_B B \ll k_B T \ll A$. This condition ensures that you do not populate excited states ($k_B T \ll A$) and you keep on the linear part of the Brillouin function ($\mu_B B \ll k_B T$).

If $A < 0$, the ground state levels are $|2, \frac{1}{2}\rangle, |2, -\frac{1}{2}\rangle |-2, \frac{1}{2}\rangle$ and $|-2, -\frac{1}{2}\rangle$ and the spin–orbit interaction acts so that $|2, -\frac{1}{2}\rangle$ and $|-2, \frac{1}{2}\rangle$ are lowered in energy by $\lambda$ while $|2, \frac{1}{2}\rangle$ and $|-2, -\frac{1}{2}\rangle$ are raised in energy by $\lambda$. The lower levels will only now be split by a magnetic field parallel to $z$ and so the susceptibility will be Curie-like if $\mu_B B \ll k_B T \ll \lambda$. The susceptibility will be temperature independent if the field is applied perpendicular to $z$.

(3.2)  Use

$$(1+x)^{-1/2} = 1 - \frac{x}{2} + \frac{(-\frac{1}{2})(-\frac{3}{2})}{2!}x^2 + \frac{(-\frac{1}{2})(-\frac{3}{2})(-\frac{5}{2})}{3!}x^3$$
$$+ \frac{(-\frac{1}{2})(-\frac{3}{2})(-\frac{5}{2})(-\frac{7}{2})}{4!}x^4 + \cdots$$
$$= 1 - \frac{x}{2} + \frac{3x^2}{8} - \frac{5x^3}{16} + \frac{35x^4}{128} + \cdots. \qquad \text{(F.20)}$$

(3.3)  Converting $V$ into spherical polars produces

$$V = \frac{Dr^4}{8}\left[\sin^4\theta(e^{4i\phi} + e^{-4i\phi} + 6) + 8\cos^2\theta - \frac{24}{5}\right]. \qquad \text{(F.21)}$$

For the matrix elements, all off-diagonal terms vanish except $V_{2,-2}$ and $V_{-2,2}$. The evaluation of the non-zero matrix elements are straightforward but tedious. Here is one example:

$$V_{1,1} = \int_0^\infty r^6 R^2(r)\,\mathrm{d}r \frac{D}{8}\int_0^\pi \left(6\sin^2\theta + 8\cos^4\theta - \frac{24}{5}\right)$$
$$\times 4\sin^3\theta\cos^2\theta\,\mathrm{d}\theta\int_0^{2\pi}\mathrm{d}\phi$$
$$= -4A. \qquad \text{(F.22)}$$

The eigenvalues and eigenvectors are

| Eigenvalue | Eigenvector |
|---|---|
| $A + 5A\left(1 + \dfrac{4\mu_B^2 B^2}{25A^2}\right)^{1/2}$ | $\dfrac{(|2\rangle + x_+|-2\rangle)}{(1+x_+^2)^{1/2}}$ |
| $-4A + \mu_B B$ | $|1\rangle$ |
| $6A$ | $|0\rangle$ |
| $-4A - \mu_B B$ | $|-1\rangle$ |
| $A - 5A\left(1 + \dfrac{4\mu_B^2 B^2}{25A^2}\right)^{1/2}$ | $\dfrac{(|2\rangle + x_-|-2\rangle)}{(1+x_-^2)^{1/2}}$ |

where

$$x_\pm = \pm\left(1 + \left(\frac{2\mu_B B}{5A}\right)^2\right)^{1/2} - \frac{2\mu_B B}{5A}. \qquad \text{(F.23)}$$

Hence in the limit of low field, the eigenvalues and eigenvectors are

| Eigenvalue | Eigenvector |
|---|---|
| $6A + \dfrac{2\mu_B^2 B^2}{5A}$ | $\dfrac{(|2\rangle + |-2\rangle)}{\sqrt{2}}$ |
| $-4A + \mu_B B$ | $|1\rangle$ |
| $6A$ | $|0\rangle$ |
| $-4A - \mu_B B$ | $|-1\rangle$ |
| $-4A - \dfrac{2\mu_B^2 B^2}{5A}$ | $\dfrac{(|2\rangle - |-2\rangle)}{\sqrt{2}}$ |

and so if $k_B T \ll A$, then you only populate the lowest levels which are a triplet if $A > 0$ or a doublet if $A < 0$. Considering these lowest levels only, the partition function can be written

$$Z = \begin{cases} 1 + e^{-\mu_B B/k_B T} + e^{\mu_B B/k_B T} & A > 0 \\ 1 + e^{-2\mu_B^2 B^2/5Ak_B T} & A < 0 \end{cases}, \qquad \text{(F.24)}$$

and the results follow using $F = -Nk_B T \log Z$ and $M = -\partial F/\partial B$.

(3.4)  $I = \frac{3}{2}$, $A = 2 \times 10^{-6}$ eV. Not much can be said about the nuclear moment. Line broadening due to phonons is large at high temperature. Low temperatures are often needed to resolve fine structure.

(3.5)  Measure $g_J J = 7$ (for $Gd^{3+}$), 5 (for $Fe^{3+}$) and 3 (for $Cr^{3+}$). This agrees with the predictions of Hund's rules for all of them except $Cr^{3+}$ for which one would expect $J = \frac{3}{2}$ and $g_J = \frac{2}{5}$. This is because orbital quenching occurs and means that $g_J = 2$. (Orbital quenching also occurs for $Fe^{3+}$ but you do not notice because $L = 0$ is predicted from Hund's rules anyway.)

(3.6)  The fields for 60 MHz are

| | |
|---|---|
| $^1$H | 1.41 T |
| $^2$H | 9.18 T |
| $^{13}$C | 5.60 T |
| $^{19}$F | 1.50 T |

(3.7)  0.322 T

(3.8)  First derive matrix representations for $\hat{S}_z$ and $\hat{S}_x$ for an $S = 1$ particle:

$$\hat{S}_z = \begin{pmatrix} 1 & 0 & 0 \\ 0 & 0 & 0 \\ 0 & 0 & -1 \end{pmatrix} \quad \text{(F.25)}$$

$$\hat{S}_x = \frac{1}{\sqrt{2}} \begin{pmatrix} 0 & 1 & 0 \\ 1 & 0 & 1 \\ 0 & 1 & 0 \end{pmatrix}. \quad \text{(F.26)}$$

Hence, the Hamiltonian is

$$\hat{\mathcal{H}} = \begin{pmatrix} D + g_\parallel \mu_B B \cos\theta & \frac{g_\perp \mu_B B \sin\theta}{\sqrt{2}} & 0 \\ \frac{g_\perp \mu_B B \sin\theta}{\sqrt{2}} & 0 & \frac{g_\perp \mu_B B \sin\theta}{\sqrt{2}} \\ 0 & \frac{g_\perp \mu_B B \sin\theta}{\sqrt{2}} & D - g_\parallel \mu_B B \cos\theta \end{pmatrix} \quad \text{(F.27)}$$

and the eigenvalues can be found to give the required result. When $\theta = 0$ the eigenvalues simplify to

$$E = 0, \, D \pm g_\parallel \mu_B B. \quad \text{(F.28)}$$

When $\theta = \pi/2$ the eigenvalues simplify to

$$E = D, \, \frac{D}{2} \pm \sqrt{\left(\frac{D}{2}\right)^2 + (g_\perp \mu_B B)^2}. \quad \text{(F.29)}$$

(3.9)  82.4 ms$^{-1}$; $7.7 \times 10^{-19}$ ms$^{-1}$; $9.4 \times 10^{14}$ Hz; $9.1 \times 10^{-9}$ Hz; $2.3 \times 10^7$ Hz; 0.1 $\mu$eV; $7.7 \times 10^{-4}$ cm$^{-1}$.

(3.10)  If the lowest frequency you can easily measure has a period of 20 $\mu$s, it has a frequency of 50 kHz, corresponding to a field of about $4 \times 10^{-4}$ T. The fraction of muons living for 20 $\mu$s or longer is $e^{-20/2.2} = 1.1 \times 10^{-4}$ corresponding to just over $10^3$ of the $10^7$ muons implanted.

This result follows from the exponential decay of muons with mean lifetime $\tau_\mu$. The fraction living longer than $T$ is given by

$$\frac{\int_T^\infty e^{-t/\tau_\mu} \, dt}{\int_0^\infty e^{-t/\tau_\mu} \, dt} = e^{-T/\tau_\mu}. \quad \text{(F.30)}$$

If the pulse-width is 50 ns, destructive interference can occur if the spin–precession period is 100 ns, corresponding to a frequency of 10 MHz, giving an upper limit on the field of $\sim$0.07 T.

If $B = 0.4$ T, $f = 54.2$ MHz.

(3.11)  For $Fe^{3+}$, $J = \frac{5}{2}$ and $g_J = 2$. The saturation moment is $g_J J \mu_B = 5\mu_B$. Susceptibility measures $\mu_{\text{eff}}^2 = g_J^2 J(J+1)\mu_B^2$ and hence

$$g_J \sqrt{J(J+1)}\mu_B = 2 \times \sqrt{\frac{5}{2} \times \frac{7}{2}}\mu_B = \sqrt{35}\,\mu_B = 5.92\mu_B. \quad \text{(F.31)}$$

The saturation moment involves $\hat{J}_z$ because it is a measurement of the moment saturated along a particular direction. In contrast, susceptibility involves $\hat{J}^2$.

(4.1)  There are several ways to derive this result. One way is to start with the vector potential from dipole 1 which is

$$\mathbf{A}_1 = \frac{\mu_0}{4\pi} \nabla \times \frac{\mu_1}{r}, \quad \text{(F.32)}$$

and this produces a magnetic field

$$\mathbf{B}_1 = \nabla \times \mathbf{A}_1 = \frac{\mu_0}{4\pi}\left[\nabla\left(\nabla \cdot \frac{\mu_1}{r}\right) - \nabla^2\left(\frac{\mu_1}{r}\right)\right], \quad \text{(F.33)}$$

and gives rise to an energy

$$\begin{aligned} E &= -\mu_2 \cdot \mathbf{B}_1 \\ &= -\frac{\mu_0}{4\pi}\left[(\mu_1 \cdot \nabla)(\mu_1 \cdot \nabla)\frac{1}{r} - (\mu_1 \cdot \mu_2)\nabla^2 \frac{1}{r}\right], \\ &= \frac{\mu_0}{4\pi r^3}\left[\mu_1 \cdot \mu_2 - \frac{3(\mu_1 \cdot \mathbf{r})(\mu_1 \cdot \mathbf{r})}{r^2}\right]. \end{aligned} \quad \text{(F.34)}$$

It is also possible to derive the result by using the magnetic scalar potential $\phi_1 = \mu_1 \cdot \mathbf{r}/4\pi r^3$ and then $\mathbf{B}_1 = -\mu_0 \nabla \phi_1$.

(4.2)  1 Å: (a) 1.855 T (b) 1.855 mT; 10 Å: (a) 0.927 T (b) 0.927 mT.

(4.3)  The lattice constant in Fe (which is bcc) is $a = 2.87$ Å. The nearest neighbour distance is therefore $a/\sqrt{2}$ and so the dipole-dipole energy is $\sim (\mu_0 \times (2.2\mu_B)^2/4\pi r^3) = 30$ $\mu$eV. J $\approx k_B T_C \sim 0.09$ eV, which is about 3000 times larger.

(4.4)  The bandwidth$\sim 0.05$ eV and the Coulomb energy $\sim 1$ eV. $J \sim -\frac{t^2}{U} = -2.5$ meV. Hence if $|J| = k_B T_N$, then $T_N = 29$ K.

(4.5)  The partition function is

$$Z = 1 + e^{-\beta\Delta}(e^{\beta g\mu_B B} + 1 + e^{-\beta g\mu_B B}). \quad \text{(F.35)}$$

This can be used to derive $F$ and hence $M$ given by

$$M = \frac{2n\beta(g\mu_B)^2 B}{3 + e^{\beta\Delta}}, \quad \text{(F.36)}$$

and the result for $\chi \ll 1$ then follows. Note that at high temperatures $T \gg \Delta/k_B$, $\chi \to n\mu_0 g^2\mu_B^2/2k_B T$ which is a Curie law. It corresponds to all levels occupied, so that the susceptibility is $\frac{3}{4}$ of an $S = 1$ system (for which $S(S+1) = 2$) and $\frac{1}{4}$ of an $S = 1$ system (for which $S(S+1) = 0$) so is $\frac{3}{2}$ of a classical Curie law.

At low temperatures, $T \ll \Delta/k_B$, then if $\Delta > 0$ one has $\chi \propto e^{-\Delta/k_B T} \to 0$ (non-magnetic because only the singlet is occupied), but if $\Delta < 0$ one has $\chi \to 2n\mu_0 g^2\mu_B^2/3k_B T$ (corresponding to the occupied triplet, $S(S+1) = 2$, and so twice a classical Curie law).

(4.6)  Hint: First show that

$$\mu_{I1} \cdot \mu_{I2} \propto \hat{I}_{1z}\hat{I}_{2z} + \frac{1}{2}(\hat{I}_{1+}\hat{I}_{2-} + \hat{I}_{1-}\hat{I}_{2+}) \quad \text{(F.37)}$$

and using $\mathbf{r} = r(\sin\theta\cos\phi, \sin\theta\sin\phi, \cos\theta)$, show that

$$\mu_I \cdot \hat{\mathbf{r}} \propto \cos\theta\hat{I}_z + \frac{1}{2}\sin\theta[e^{-i\phi}\hat{I}_+ + e^{i\phi}\hat{I}_-]. \quad \text{(F.38)}$$

The results then follow fairly quickly.

(4.8) Note that if $\mathbf{S} = \mathbf{S}_1 + \mathbf{S}_2 + \mathbf{S}_3$ then

$$\mathbf{S}^2 = \mathbf{S}_1^2 + \mathbf{S}_2^2 + \mathbf{S}_3^2 + 2(\mathbf{S}_1 \cdot \mathbf{S}_2 + \mathbf{S}_2 \cdot \mathbf{S}_3 + \mathbf{S}_3 \cdot \mathbf{S}_1). \quad \text{(F.39)}$$

Hence the Hamiltonian can be rewritten in terms of $\mathbf{S}^2$, $\mathbf{S}_1^2$, $\mathbf{S}_2^2$ and $\mathbf{S}_3^2$ giving $E = J(6 - S(S+1))$. Hence for the possible values of $S$ the required energies are obtained.

(5.1) $B_{mf} = 2.1 \times 10^3$ T. The number of atoms per unit volume in Fe are $n = 8.49 \times 10^{28}$ m$^{-3}$, and using $M = n \cdot 2.2\mu_B$ then $\mu_0 M = 2.2$ T. Hence $B_{mf}$ is $10^3$ times bigger than $\mu_0 M$.

(5.2) The magnetization is given by

$$\frac{M}{M_s} = B_S(y) = \alpha y - \beta y^3 + \cdots, \quad \text{(F.40)}$$

where $\alpha$ and $\beta$ are constants and

$$y = \frac{\gamma M}{T M_s}, \quad \text{(F.41)}$$

where $\gamma$ is a positive constant given by $T_C/\alpha$. Near $T_C$, these equations lead to

$$\left(\frac{M}{M_s}\right)^2 \approx \frac{\alpha^3}{\beta} \frac{(T_C - T)}{T_C} \quad \text{(F.42)}$$

and this leads to the required result using the values of $\alpha$ and $\beta$ that can be obtained from eqn 5.38.
The energy is given by $E = -\frac{1}{2}\lambda M^2$ and using eqn F.42 one can show that at $T_C$

$$\frac{d(M^2)}{dT} = -\frac{M_s^2 \alpha^3}{\beta T_C}, \quad \text{(F.43)}$$

and hence the result for the specific heat can be found.
In general one can now easily show that

$$C = \begin{cases} \frac{5}{2}nk_B \left[\frac{(2S+1)^2-1}{(2S+1)^2+1}\right]\left(3\left(\frac{T}{T_C}\right)^2 - 2\left(\frac{T}{T_C}\right)\right) & T \le T_C \\ 0 & T > T_C. \end{cases} \quad \text{(F.44)}$$

(5.4) The energy is given by

$$E = -2NS^2(J_1 \cos\theta + J_2 \cos 2\theta), \quad \text{(F.45)}$$

and energy minimization $\partial E/\partial\theta = 0$ implies that either $\sin\theta = 0$ (and hence $\theta = 0$ or $\pi$) or $\cos\theta = -J_1/4J_2$. These solutions are ferromagnetic ($\theta = 0$), antiferromagnetic ($\theta = \pi$) and helimagnetic ($\cos\theta = -J_1/4J_2$) and substitution back into the equation for the energy reproduces eqn 5.45. The helimagnetic solution is only possible if $|J_1| < 4|J_2|$.
Which solution gives the lowest energy can be evaluated by comparing the energies of each solution. Alternatively one can look at the stability of each solution by evaluating $\partial^2 E/\partial\theta^2$.

(5.5) When $y \gg 1$,

$$\tanh y = \frac{e^y - e^{-y}}{e^y + e^{-y}} = \frac{1 - e^{-2y}}{1 + e^{-2y}} \approx 1 - 2e^{-2y}. \quad \text{(F.46)}$$

This can be used together with

$$\frac{M}{M_s} = \tanh y, \quad \text{(F.47)}$$

and also with the fact that near $T_C$ one has

$$\frac{M}{M_s} = \frac{T}{T_C} y, \quad \text{(F.48)}$$

to derive the final result.

(5.6) Recall the mean-field theory for a ferromagnet. In that case the magnetization is given by

$$M = \frac{C}{\mu_0 T}(\lambda M + \mu_0 H), \quad \text{(F.49)}$$

where $C$ is the Curie constant. This rearranges to

$$M\left(1 - \frac{\lambda C}{\mu_0 T}\right) = \frac{HC}{T}, \quad \text{(F.50)}$$

and so the susceptibility is

$$\chi = \frac{M}{H} = \frac{C}{T - \theta} \quad \text{(F.51)}$$

where the Weiss temperature is $\theta = \lambda C/\mu_0$.
For this ferrimagnet, the magnetization in each sublattice is

$$M_1 = \frac{C_1}{\mu_0 T}(\mu_0 H - \lambda M_2) \quad \text{(F.52)}$$

$$M_2 = \frac{C_2}{\mu_0 T}(\mu_0 H - \lambda M_1) \quad \text{(F.53)}$$

which becomes

$$\begin{pmatrix} T & \lambda C_1/\mu_0 \\ \lambda C_2/\mu_0 & T \end{pmatrix}\begin{pmatrix} M_1 \\ M_2 \end{pmatrix} = H\begin{pmatrix} C_1 \\ C_2 \end{pmatrix}. \quad \text{(F.54)}$$

The determinant of the matrix in this equation is therefore zero if $T^2 - \lambda^2 C_1 C_2/\mu_0^2 = 0$ and hence one can deduce that the paramagnetic regime is when $T > \theta$ where $\theta = \lambda(C_1 C_2)^{1/2}/\mu_0$. The susceptibility is given by $\chi = (M_1 + M_2)/H$ which can be found by inverting eqn F.54 to give

$$\begin{pmatrix} M_1 \\ M_2 \end{pmatrix} = \frac{H}{T^2 - \theta^2}\begin{pmatrix} T & -\lambda C_1/\mu_0 \\ -\lambda C_2/\mu_0 & T \end{pmatrix}\begin{pmatrix} C_1 \\ C_2 \end{pmatrix} \quad \text{(F.55)}$$

and hence the final answer.

(5.7) The first part follows straightforwardly from an application of $\hat{S}^\pm = \hat{S}^x \pm i\hat{S}^y$.
For the Heisenberg ferromagnet, the term $\hat{S}_i^z \hat{S}_j^z$ operating on the ground state produces $S^2$ and the ground state is an eigenstate of $\hat{S}_i^z \hat{S}_j^z$. There are $N$ such terms and hence $E = -NS^2$. The terms $\hat{S}_i^+ \hat{S}_j^-$ and $\hat{S}_i^- \hat{S}_j^+$ produce nothing because in each case the raising operator annihilates the ground state. Thus the ground state is an eigenstate of both $\hat{S}_i^+ \hat{S}_j^-$ and $\hat{S}_i^- \hat{S}_j^+$ with eigenvalue zero.
For the Heisenberg antiferromagnet the two terms $\hat{S}_i^+ \hat{S}_j^-$ and $\hat{S}_i^- \hat{S}_j^+$ produce states other than the 'obvious' ground state. Therefore this 'obvious' ground state is not an eigenstate of the system.

(5.8) The range of possible momentum transfer is 0 nm$^{-1}$ to $\sqrt{2}k_{in} = 2\sqrt{2}\pi/\lambda = 29.6$ nm$^{-1}$.
(a) At 100 K, just get the chemical Bragg reflections at

$$Q = \left[\left(\frac{2\pi}{a}\right)^2 (h^2 + k^2) + \left(\frac{2\pi}{c}\right)^2 l^2\right]^{1/2}. \quad \text{(F.56)}$$

The symmetry is bcc, so reflections are seen when $h + k + l$ is even.
(b) At 10 K, also see the magnetic Bragg reflections.
The observed reflections for both cases are listed below. The magnitude of the scattering vector $\mathbf{Q}$ are listed, together with $2\theta$, the angle through which the incident beam is scattered. The magnetic reflections are indicated by an asterisk and the multiplicity, $M$, of each reflection is given in the final column.

| h | k | l | $Q$ (nm$^{-1}$) | $2\theta$ (°) | M | |
|---|---|---|---|---|---|---|
| 0 | 1 | 0 | 12.566 | 34.915 | * | 4 |
| 1 | 0 | 0 | 12.566 | 34.915 | * | |
| 1 | 1 | 0 | 17.772 | 50.208 | | 4 |
| 0 | 0 | 1 | 20.944 | 60.000 | * | 2 |
| 0 | 1 | 1 | 24.425 | 71.337 | | 8 |
| 1 | 0 | 1 | 24.425 | 71.337 | | |
| 0 | 2 | 0 | 25.133 | 73.740 | | 4 |
| 2 | 0 | 0 | 25.133 | 73.740 | | |
| 1 | 1 | 1 | 27.468 | 81.952 | * | 8 |
| 1 | 2 | 0 | 28.099 | 84.261 | * | 8 |
| 2 | 1 | 0 | 28.099 | 84.261 | * | |

(5.9) (a) The first expression for $\tilde{m}$ follows from the definition of tanh $y$. This can be rearranged to give

$$x^2 = \frac{1+\tilde{m}}{1-\tilde{m}}, \tag{F.57}$$

and taking the log of this equation gives eqn 5.57. Using eqn F.48 leads to eqn 5.58.

(b) Rearranging eqn 5.60 to make $x$ the subject leads to

$$(1-\tilde{m})x^2 - \tilde{m}x - (1+\tilde{m}) = 0, \tag{F.58}$$

which is a quadratic. This has real solutions if $4 - 3\tilde{m}^2 \geq 0$ and are simply found to be eqn 5.62. Now $1 \leq 4 - 3\tilde{m}^2 \leq 2$ if $0 \leq \tilde{m} \leq 1$, so for $x > 0$ we need the positive solution in eqn 5.62. This then yields

$$y = \log(\tilde{m} + \sqrt{4 - 3\tilde{m}^2}) - \log(2(1-\tilde{m})), \tag{F.59}$$

and in this case one finds $y = 3\tilde{m}T_C/2T$ and the final result follows.

(6.1) The average energy is found to be

$$\langle E \rangle = -\frac{NJ}{4}\tanh\left(\frac{J}{2k_BT}\right), \tag{F.60}$$

and the heat capacity per spin is then

$$C = \frac{J^2}{4k_BT^2\cosh^2(J/2k_BT)}. \tag{F.61}$$

This behaves like $J^2/4k_BT^2$ at high temperature, and as $J^2e^{-J/k_BT}/4k_BT^2$ at low temperature. It is thus zero at both $T = 0$ and $T = \infty$ and exhibits a broad maximum in between. There is thus no cusp and this reinforces the point that there are no phase transitions in purely one-dimensional models.

(6.2) (b) $\hbar\omega(\mathbf{q}) = S[J(0) - J(\mathbf{q}) + K(0)]$.

(c) (i) $\hbar\omega(\mathbf{q}) = 2SJ_0(1 - \cos qa) + 2SK_0$.

(ii) $\hbar\omega(\mathbf{q}) = 2SJ_0(2 - \cos q_xa - \cos q_ya) + 4SK_0$.

(iii)

$$\hbar\omega(\mathbf{q}) = 8SJ_0\left(1 - \cos\frac{q_xa}{2}\cos\frac{q_ya}{2}\cos\frac{q_za}{2}\right) + 8SK_0.$$

(6.3) At small $q$ one can use $\cos qa \approx 1 - (qa)^2/2$ and obtain

(i) $\hbar\omega = S(J_0a^2q^2 + 2K_0)$

(ii) $\hbar\omega = S(J_0a^2q^2 + 4K_0)$

(iii) $\hbar\omega = S(J_0a^2q^2 + 8K_0)$

In the Ising case ($J_0 = 0$), one has $\hbar\omega = 2dSK_0 \equiv \Delta$ where $d = 1, 2, 4$ for cases (i), (ii) and (iii) respectively. In this model, spin waves 'cost' the same energy $\Delta$ regardless of wave vector. The number of spin waves is then proportional to $e^{-\Delta/k_BT}$, and so their energy is proportional to $\Delta e^{-\Delta/k_BT}$. The specific heat is then easily found to be proportional to

$$\frac{\Delta^2}{k_BT^2}e^{-\Delta/k_BT},$$

in agreement with Exercise 6.1 with $\Delta$ in this problem being equal to $J$ in that problem.

In the Heisenberg case ($K_0 = 0$), $\hbar\omega = Dq^2$ where the spin-stiffness $D = SJ_0a^2$. The number of spin waves $N_s$ is given by

$$N_s \propto \int \frac{d^n q}{e^{Dq^2/k_BT} - 1}, \tag{F.62}$$

and at small $q$

$$\frac{1}{e^{Dq^2/k_BT} - 1} \approx \frac{k_BT}{Dq^2}. \tag{F.63}$$

The integral therefore diverges in one or two dimensions (where $d^n q$ is equal to $dq$ or $2\pi q\,dq$ respectively), but converges for three dimensions (where $d^n q$ is equal to $4\pi q^2\,dq$). No long range order is therefore possible for the pure (isotropic) Heisenberg model in one or two dimensions.

(6.4) In the antiferromagnetic case, one can follow the derivation for the ferromagnetic case, but you have to remember which sublattice you are on. One finds that

$$\begin{aligned}\hbar\dot{S}_j^+ &= 2iJS[2S_j^+ + S_{j-1}^+ + S_{j+1}^+] \quad j \text{ odd}\\ \hbar\dot{S}_j^+ &= -2iJS[2S_j^+ + S_{j-1}^+ + S_{j+1}^+] \quad j \text{ even.}\end{aligned} \tag{F.64}$$

Hence substituting in

$$S_j^+ = \begin{cases} ue^{i(qja-\omega t)} & j \text{ odd}\\ ve^{i(qja-\omega t)} & j \text{ even,}\end{cases} \tag{F.65}$$

one finds that

$$i\hbar\omega\begin{pmatrix}u\\v\end{pmatrix} = 4iJS\begin{pmatrix}1 & \cos qa\\-\cos qa & -1\end{pmatrix}\begin{pmatrix}u\\v\end{pmatrix}, \tag{F.66}$$

and hence

$$\begin{vmatrix}1 - \frac{\hbar\omega}{4JS} & \cos qa\\\cos qa & 1 + \frac{\hbar\omega}{4JS}\end{vmatrix} = 0 \tag{F.67}$$

so that $\hbar\omega = 4JS\sin qa$.

(6.5) Using $\partial F/\partial M = 0$, some rearrangement quickly yields

$$M^2 = -\frac{2a_0(T - T_C)}{4b} + \frac{\mu_0 H}{4bM}, \tag{F.68}$$

which is of the required form. Plotting $M^2$ against $M/H$ gives a straight-line, with the intercept on the $M^2$ axis changing sign at $T = T_C$.

(6.6) The model in appropriate units is

$$\epsilon = \frac{1}{2}\sin^2(\theta - \phi) - h\cos\phi. \tag{F.69}$$

For $\theta = 0$, energy minimization gives

$$\sin\phi(\cos\phi + h) = 0 \tag{F.70}$$

for which the solutions are $\phi = 0, \pi, \cos^{-1}(-h)$. Evaluation of $\partial^2 E/\partial\theta^2$ can be used to examine the stability of these solutions. For $\theta = \pi/2$, energy minimization gives

$$\sin\phi(h - \cos\phi) = 0 \tag{F.71}$$

for which the solutions are $\phi = 0, \pi, \cos^{-1}(h)$.

(6.7) The components of magnetization are

$$M_x = |\mathbf{M}|\sin\frac{\pi r}{R} \tag{F.72}$$

$$M_y = |\mathbf{M}|\cos\frac{\pi r}{R}, \tag{F.73}$$

and using results such as

$$\frac{\mathrm{d}r}{\mathrm{d}x} = \frac{\mathrm{d}}{\mathrm{d}x}\sqrt{x^2 + y^2 + z^2} = \frac{x}{r} \tag{F.74}$$

one can evaluate all the partial differential of $M_x$ and $M_y$. Hence

$$(\nabla M_x)^2 + (\nabla M_y)^2 = |\mathbf{M}|^2\frac{\pi^2}{R^2}, \tag{F.75}$$

from which the final result follows.

(6.8) 25 Gbits in$^{-2}$ is $3.9\times10^{13}$ bits m$^{-2}$ in sensible units.

(a) Taking $\pi a_0^2$ to be the area of a hydrogen atom gives $3.4\times10^{-7}$ bits (hydrogen atom area)$^{-1}$.

(b) Most Shakespearian plays are 20000–40000 words long; with maybe an average 5 letters per word, and about 40 plays, one gets about 4 Mbytes or 32 Mbits (8 bits per byte). Taking a UK stamp, I calculated about 600 copies of the complete works per stamp, although that could be easily doubled with data-compression. "I am ill at all these numbers" (Hamlet Act 2 Scene 2)

DNA stores 1 bit per base pair, which is about 1 bit per cubic nanometre. If you made a sheet of this, this would be equivalent to about $6\times10^5$ Gbits in$^{-2}$, but the great advantage of DNA is that it can coil and is a volume storage technique.

(6.9) The best way to tackle this is to put the magnetic field parallel to $-z$ which you can do without loss of generality. (Why $-z$? So that the ground state has all it spins aligned parallel to $z$. Electrons have negative charge and so their spins are antiparallel to their magnetic moments and will therefore align antiparallel to the applied field.) You should find that each magnon *adds* an energy $+g\mu_{\mathrm{B}}B$.

(6.11) The equation

$$\langle q|i\rangle = \frac{1}{\sqrt{N}}\sum_j e^{-i\mathbf{q}\cdot\mathbf{R}_j}\langle j|i\rangle \tag{F.76}$$

can be simplified using $\langle j|i\rangle = \delta_{ij}$. The result then follows.

Hint for the next part: $S^+$ kills the state it is operating on *unless* the spin it is operating on is $< S$.

The exercise shows that each spin has a small transverse component which is perpendicular to the direction of magnetization. The transverse components vary in space in just the manner depicted in Fig. 6.13.

(6.12) It is necessary to solve

$$\Delta E = 0 = \int_{-\infty}^{\infty}\left[2A\frac{\partial\theta}{\partial z}\frac{\partial}{\partial z}(\delta\theta) + \frac{\partial f(\theta)}{\partial\theta}\delta\theta\right]\mathrm{d}z \tag{F.77}$$

The first part of this integral can be evaluated by parts, yielding

$$\frac{\partial f}{\partial\theta} - 2A\frac{\partial^2 f}{\partial z^2} = 0. \tag{F.78}$$

The question carefully guides the reader through the rest.

(6.13) The wall energy per unit volume due to the length of walls in the structure and is therefore approximately $\sigma_{\mathrm{W}}L/D$. The spins inside the small triangles cost an anisotropy energy. The area of one of these triangles is $D^2/4$ and there are two of them in every horizontal length $D$, so their total cost is $2\times K\times(D^2/4)\times(1/D) = KD/2$. Summing these terms and differentiating yields the required result. $D = 20\,\mu\mathrm{m}$.

(7.1) Using $g(E_{\mathrm{F}}) = 3n/2E_{\mathrm{F}}$, the first result follows quite quickly. To consider the temperature dependence, it is worthwhile to review some mathematical details concerning the Fermi function. Consider $h(E) = \int_0^E g(E')\,\mathrm{d}E'$, so that $g(E) = \mathrm{d}h/\mathrm{d}E$ and consider the integral

$$I = \int_0^{\infty}\frac{\mathrm{d}h}{\mathrm{d}E}f(E)\,\mathrm{d}E$$

$$= [f(E)g(E)]_0^{\infty} - \int_0^{\infty}h(E)\frac{\mathrm{d}f}{\mathrm{d}E}\,\mathrm{d}E$$

$$= -\int_0^{\infty}h(E)\frac{\mathrm{d}f}{\mathrm{d}E}\,\mathrm{d}E. \tag{F.79}$$

Now put $x = (E - \mu)/k_{\mathrm{B}}T$ and hence

$$\frac{\mathrm{d}f}{\mathrm{d}E} = -\frac{1}{k_{\mathrm{B}}T}\frac{e^x}{(e^x + 1)^2}. \tag{F.80}$$

Writing

$$h(E) = \sum_{s=0}^{\infty}\frac{x^s}{s!}\left(\frac{\mathrm{d}^s h}{\mathrm{d}x^s}\right)_{x=0}, \tag{F.81}$$

we have that

$$I = \sum_{s=0}^{\infty}\frac{1}{s!}\left(\frac{\mathrm{d}^s h}{\mathrm{d}x^s}\right)_{x=0}\int_{-E_{\mathrm{F}}/k_{\mathrm{B}}T}^{\infty}\frac{x^s e^x\,\mathrm{d}x}{(e^x + 1)^2}. \tag{F.82}$$

The integral part of this can be simplified by replacing the lower limit by $-\infty$. It vanishes for odd $s$, but for even $s$

$$\int_{-\infty}^{\infty}\frac{x^s e^x\,\mathrm{d}x}{(e^x + 1)^2} = 2\int_0^{\infty}\frac{x^s e^x\,\mathrm{d}x}{(e^x + 1)^2}$$

$$= 2\int_0^{\infty}\mathrm{d}x\sum_{n=0}^{\infty}e^x x^s$$

$$\times\left[(n+1)(-1)^{n+1}e^{-nx}\right]$$

$$= 2\sum_{n=1}^{\infty}(-1)^{n+1}n\int_0^{\infty}x^s e^{-nx}\,\mathrm{d}x$$

$$= 2(s!)\sum_{n=1}^{\infty}\frac{(-1)^{n+1}}{n^s}$$

$$= 2(s!)(1 - 2^{1-s})\zeta(s), \tag{F.83}$$

where $\zeta(s)$ is the Riemann zeta function. ($\zeta(0) = -\frac{1}{2}$, $\zeta(2) = \frac{\pi^2}{6}$, $\zeta(4) = \frac{\pi^4}{90}$).

Thus the integral is

$$I = \sum_{s=0, s \text{ even}}^{\infty} 2 \left( \frac{d^s h}{dx^s} \right)_{x=0} (1 - 2^{1-s}) \zeta(s)$$

$$= h + \frac{\pi^2}{6} \left( \frac{d^2 h}{dx^2} \right)_{x=0} + \frac{7\pi^4}{360} \left( \frac{d^4 h}{dx^4} \right)_{x=0} + \cdots$$

$$= \int_{-\infty}^{\mu} g(E) \, dE + \frac{\pi^2}{6} (k_B T)^2 \left( \frac{dg}{dE} \right)_{E=\mu}$$

$$+ \frac{7\pi^4}{360} (k_B T)^4 \left( \frac{d^3 g}{dE^3} \right)_{E=\mu} + \cdots . \quad \text{(F.84)}$$

This implies that the number density is

$$n = \int_0^{\mu} g(E) \, dE + \frac{\pi^2}{6} (k_B T)^2 \left( \frac{dg}{dE} \right)_{E=\mu} + \cdots . \quad \text{(F.85)}$$

Now to first order

$$\int_0^{\mu} g(E) \, dE = \int_0^{E_F} g(E) \, dE + (\mu - E_F) g(E_F), \quad \text{(F.86)}$$

so for a constant number of electrons, $n$ takes the same value at all temperatures and so

$$\mu = E_F - \frac{\pi^2}{6} (k_B T)^2 \frac{g'(E_F)}{g(E_F)}. \quad \text{(F.87)}$$

The rôle of a magnetic field is then to shift the energy levels by $\pm \mu_B B$. The number density for each spin state is given by

$$n_{\pm} = \frac{1}{2} \int_0^{E_F} g(E) \, dE + (\mu - E_F \mp \mu_B B) g(E_F \mp \mu_B B)$$

$$+ \frac{\pi^2}{6} (k_B T)^2 \left( \frac{dg}{dE} \right)_{E=\mu \mp \mu_B B} + \cdots \quad \text{(F.88)}$$

and using the approximate relations

$$g(E_F \mp \mu_B B) = g(E_F) \mp \mu_B B \left( \frac{dg}{dE} \right)_{E=\mu} \quad \text{(F.89)}$$

and

$$\left( \frac{dg}{dE} \right)_{E=\mu \mp \mu_B B} = \left( \frac{dg}{dE} \right)_{E=\mu} \mp \mu_B B \left( \frac{d^2 g}{dE^2} \right)_{E=\mu} \quad \text{(F.90)}$$

one then has

$$n_+ - n_- = -\frac{\pi^2}{6} (k_B T)^2 \mu_B B \left[ \left( \frac{g'}{g} \right)^2 - \frac{g''}{g} \right] + \mu_B B g(E_F). \quad \text{(F.91)}$$

Hence the susceptibility is

$$\chi = \mu_0 \mu_B g(E_F) \left[ 1 - \frac{\pi^2}{6} (k_B T)^2 \left[ \left( \frac{g'}{g} \right)^2 - \frac{g''}{g} \right] \right]. \quad \text{(F.92)}$$

For $g(E) \propto E^{1/2}$ it is simple to show that

$$\frac{g'}{g} = \frac{1}{2 E_F}$$

$$\left( \frac{g'}{g} \right)^2 - \frac{g''}{g} = \frac{1}{2 E_F^2} \quad \text{(F.93)}$$

evaluated at the Fermi energy. The final result then follows. For a metal with $E_F \sim 10$ eV (e.g. Sn), $T_F \sim 10^5$ K, the correction is about $10^{-5}$.

(7.2)  (a) This is shown in the derivation of eqn 7.23.

(b) The crossover temperature is $2T_F/3 = 27$ K. The Pauli paramagnetic susceptibility is then $1.3 \times 10^{-8}$. The Landau diamagnetic susceptibility is then

$$\chi_L = - \left( \frac{m_e}{m^*} \right)^2 \frac{\chi_P}{3} = -4.3 \times 10^{-7}. \quad \text{(F.94)}$$

(7.3)  Using $A_\phi = \frac{1}{2} Br$, the Schrödinger equation becomes

$$\left[ \frac{-\hbar^2}{2m_e} \left( \frac{\partial^2}{\partial r^2} + \frac{1}{r} \frac{\partial}{\partial r} + \frac{1}{r^2} \frac{\partial^2}{\partial \phi^2} + \frac{\partial^2}{\partial z^2} \right) - \frac{i\hbar e B}{2m_e} \frac{\partial}{\partial \phi} \right.$$

$$\left. + \frac{e^2 B^2 r^2}{8m_e} \right] \psi(r, \phi, z) = E \psi(r, \phi, z). \quad \text{(F.95)}$$

One can then look for solutions of the form

$$\psi(r, \phi, z) = R(r) e^{il\phi} e^{ik_z z} \quad \text{(F.96)}$$

and this leads to

$$\left( -\frac{\hbar^2}{2m_e} \left( \frac{\partial^2}{\partial r^2} + \frac{1}{r} \frac{\partial}{\partial r} \right) - \frac{\hbar^2 l^2}{2m_e} + \frac{1}{2} m_e \omega_c^2 r^2 \right.$$

$$\left. + \frac{\hbar k_z^2}{2m_e} \right) R(r) = \left( E - \frac{l}{2} \hbar \omega_c \right) R(r). \quad \text{(F.97)}$$

The solutions are

$$\psi(r, \theta, z) = \exp \left[ i(l\phi + k_z z) - \frac{e B r^2}{4\hbar} \right] r^{|l|} L_{n-1}^{|l|} \left( \frac{e B r^2}{2\hbar} \right), \quad \text{(F.98)}$$

where $L_\alpha^\beta(x)$ is an associated Laguerre polynomial (Appendix C.5), and the energy is

$$E = \left( n + \frac{l}{2} + \left| \frac{l}{2} \right| - \frac{1}{2} \right) \hbar \omega_c + \frac{\hbar k_z^2}{2m_e}. \quad \text{(F.99)}$$

In this so-called symmetric gauge, all states circulate about the same point in space, so that rotational symmetry is captured at the expense of translational symmetry.

(7.5)  Hints:

$$\frac{z+1}{z-1} = 1 + \frac{2}{z-1} \quad \text{(F.100)}$$

and hence one can show

$$1 + \frac{1-z^2}{2z} \log \left( \frac{z+1}{z-1} \right) = \frac{2}{3z^2} + O \left( \frac{1}{z^3} \right). \quad \text{(F.101)}$$

Also, $\log(-1) = i\pi$, so that

$$\log \left| \frac{x+1}{x-1} \right| - \log \left( \frac{x+1}{x-1} \right) = i\pi. \quad \text{(F.102)}$$

(7.6)  (b) It is useful to start with

$$\chi_q = \mu_0 \mu_B^2 \frac{2m_e}{\hbar^2} I$$

where $I$ is an integral given by

$$I = \int_{|\mathbf{k}| < k_F} g(\mathbf{k}) \, d^n k \left[ \frac{1}{(k+q)^2 - k^2} + \frac{1}{(k-q)^2 - k^2} \right].$$

(F.103)

In one-dimension

$$\int_{|\mathbf{k}| < k_F} \frac{dk}{(k+q)^2 - k^2} = \int_{-k_F}^{k_F} \frac{dk}{q(q+2k)}$$

$$= \frac{1}{2q} \int_{q-2k_F}^{q+2k_F} \frac{du}{u}$$

$$= \frac{1}{2q} \int_{|q-2k_F|}^{q+2k_F} \frac{du}{u}$$

$$= \frac{1}{2q} \log \left| \frac{q + 2k_F}{q - 2k_F} \right|.$$

(F.104)

The result then follows.

In two-dimensions, the relevant integral to consider is

$$I = \int_0^{k_F} k \, dk \int_0^{2\pi} \frac{d\theta}{q(q - 2k\cos\theta)} + \frac{d\theta}{q(q + 2k\cos\theta)}$$

$$= 2 \int_0^{k_F} k \, dk \int_0^{2\pi} \frac{d\theta}{q^2 - 4k^2 \cos\theta}$$

(F.105)

Now

$$\int_0^{2\pi} \frac{d\theta}{q^2 - 4k^2 \cos\theta} = \begin{cases} \dfrac{2\pi}{q(q^2 - 4k^2)^{1/2}} & q > 2k_F \\ 0 & q < 2k_F \end{cases}$$

(F.106)

so that the integral becomes

$$I = \begin{cases} \pi \left[ 1 - \sqrt{1 - \left( \dfrac{2k_F}{q} \right)} \right] & q > 2k_F \\ \\ \pi & q < 2k_F \end{cases}$$

(F.107)

and the final result follows.

# Appendix G

# Symbols, constants and useful equations

| | | | |
|---|---|---|---|
| $a$ | unit cell dimension | $\mathbf{H}$ | magnetic field strength (magnetic field) |
| $a_0$ | Bohr radius | $\mathbf{H}_a$ | applied field |
| $A$ | area | $\mathbf{H}_d$ | demagnetizing field |
| $A$ | hyperfine coupling constant | $\mathbf{H}_i$ | internal field |
| $A$ | continuum exchange constant | $\mathcal{H}$ | Hamiltonian |
| $A$ | mass number | i | $= \sqrt{-1}$ |
| $\mathbf{A}$ | magnetic vector potential | $I$ | current |
| $b$ | scattering length | $I$ | nuclear spin quantum number |
| $\mathbf{B}$ | magnetic flux density (magnetic field) | $J$ | total angular momentum quantum number |
| $\mathbf{B}_a$ | applied field | $\mathsf{J}$ | exchange constant/integral |
| $\mathbf{B}_d$ | demagnetizing field | $k$ | wave vector |
| $\mathbf{B}_i$ | internal field | $k_B$ | Boltzmann's constant ($1.3807 \times 10^{-23}\,\mathrm{J\,K^{-1}}$) |
| $\mathbf{B}_{mf}$ | molecular field | $k_F$ | Fermi wave vector |
| $\mathrm{B}_J$ | Brillouin function | $K$ | Coulomb integral |
| $c$ | velocity of light in free space ($2.9979 \times 10^8\,\mathrm{m\,s^{-1}}$) | $K$ | anisotropy constant |
| $D$ | electric displacement | $K$ | Knight shift |
| $D$ | axial anisotropy constant | $K$ | spring constant |
| $\mathbf{dS}$ | surface element | $K_s$ | surface anisotropy constant |
| $dS$ | entropy change | $K_v$ | volume anisotropy constant |
| e | $= \exp(1) = 2.718281828$ | $l$ | orbital angular momentum quantum number |
| $e$ | magnitude of the electron charge ($1.6022 \times 10^{-19}\,\mathrm{C}$) | $l$ | Landau level index |
| $E$ | energy | $m_e$ | electron mass ($9.109 \times 10^{-31}\,\mathrm{kg}$) |
| $E_F$ | Fermi energy | $m_l$ | orbital magnetic quantum number |
| $E_S$ | singlet energy | $m_I$ | nuclear magnetic quantum number |
| $E_T$ | triplet energy | $m_J$ | magnetic quantum number |
| $\mathcal{E}$ | electric field | $m_n$ | neutron mass |
| $F$ | Helmholtz free energy (Helmholtz function) | $m_s$ | spin magnetic quantum number |
| $g(E)$ | density of states in energy space | $\mathbf{M}$ | magnetization (magnetic moment per unit volume) |
| $g(k)$ | density of states in wave vector space | $\mathbf{M}_\pm$ | magnetization on a sublattice of an antiferromagnet |
| $g$ | g-factor | $\mathbf{M}_s$ | saturation magnetization |
| $g_e$ | electron g-factor | $n$ | principal quantum number |
| $g_I$ | nuclear g-factor | $n$ | number of atoms/moments per unit volume |
| $g_l$ | g-factor for orbital angular momentum | $\mathsf{N}$ | demagnetizing tensor |
| $g_s$ | g-factor for spin angular momentum | $N$ | demagnetizing factor |
| $\mathsf{g}$ | effective g-tensor | $N$ | number of neutrons in an atomic nucleus |
| $\mathbf{G}$ | torque | $N_A$ | Avogadro's number |
| $\mathbf{G}$ | reciprocal lattice vector | $p$ | dipole moment |
| $h$ | Planck's constant ($6.626 \times 10^{-34}\,\mathrm{J\,s}$) | $\mathbf{p}$ | canonical momentum |
| $\hbar$ | Planck's constant $/2\pi$ ($1.0546 \times 10^{-34}\,\mathrm{J\,s}$) | $\hat{\mathbf{p}}$ | momentum operator |
| | | $\mathbf{q}$ | magnon wave vector |

| | | | |
|---|---|---|---|
| $\mathbf{q}$ | helix wave vector | $\epsilon_0$ | electric permittivity of free space ($8.854 \times 10^{-12} \mathrm{Fm}^{-1}$) |
| $\mathbf{Q}$ | scattering vector | | |
| $\mathbf{r}$ | position vector | $\epsilon_r$ | relative dielectric constant |
| $\hat{\mathbf{r}}$ | position operator | $\theta$ | angle |
| $R$ | electrical resistance | $\theta$ | Weiss temperature |
| $R_\mathrm{e}$ | extraordinary Hall coefficient | $\lambda$ | wavelength |
| $R_\mathrm{o}$ | ordinary Hall coefficient | $\lambda$ | spin–orbit constant |
| $S$ | entropy | $\lambda_\mathrm{el-ph}$ | electron–phonon coupling constant |
| $s, S$ | spin quantum number | $\lambda_\mathrm{spin-ph}$ | spin–phonon coupling constant |
| $\hat{\mathbf{S}}$ | spin angular momentum operator | $\boldsymbol{\mu}$ | magnetic moment |
| $\hat{S}_+$ | raising operator | $\mu_r$ | relative magnetic permeability |
| $\hat{S}_-$ | lowering operator | $\mu_\mathrm{B}$ | Bohr magneton |
| $\tilde{\mathbf{S}}$ | effective spin | $\mu_N$ | nuclear magneton |
| $t$ | time | $\mu_\mathrm{eff}$ | effective magnetic moment in $\mu_B$ per formula unit |
| $t$ | hopping integral | | |
| $t_\mathrm{p}$ | pulse length | $\mu_0$ | magnetic permeability of free space ($4\pi \times 10^{-7} \mathrm{H\,m}^{-1}$) |
| $T$ | temperature | | |
| $T_\mathrm{c}$ | critical temperature | $\nu$ | frequency |
| $T_\mathrm{C}$ | Curie temperature | $\xi$ | correlation length |
| $T_\mathrm{N}$ | Néel temperature | $\rho$ | resistivity |
| $T_\mathrm{P}$ | Peierls transition temperature | $\rho_\mathrm{H}$ | Hall resistivity |
| $T_\mathrm{SP}$ | spin-Peierls transition temperature | $\sigma$ | conductivity |
| $T_1$ | spin–lattice relaxation time | $\boldsymbol{\sigma}$ | Pauli spin matrices |
| $T_2$ | spin–spin relaxation time | $\tau$ | period |
| $T_\mathrm{F}$ | Fermi temperature | $\tau$ | superparamagnetic relaxation time |
| $U$ | Coulomb energy | $\phi$ | angle |
| $v$ | velocity | $\Phi$ | magnetic flux |
| $\mathbf{v}$ | velocity vector | $\chi$ | magnetic susceptibility |
| $V$ | potential or potential energy | $\chi_\mathrm{g}$ | mass susceptibility |
| $W$ | transition rate | $\chi_\mathrm{L}$ | Landau diamagnetic susceptibility |
| $Y_{lm_l}(\theta, \phi)$ | spherical harmonic | $\chi_\mathrm{m}$ | molar magnetic susceptibility |
| $Z$ | partition function | $\chi_\mathrm{P}$ | Pauli spin susceptibility |
| $Z$ | atomic number | $\chi_\mathbf{q}$ | $\mathbf{q}$-dependent susceptibility |
| $\alpha$ | fine structure constant | $\chi$ | spin wave function |
| $\beta$ | $= 1/k_\mathrm{B}T$ | $\psi$ | wave function |
| $\gamma$ | gyromagnetic ratio | $\Psi_\mathrm{S}$ | singlet wave function |
| $\gamma_\mathrm{e}$ | electron gyromagnetic ratio | $\Psi_\mathrm{T}$ | triplet wave function |
| $\gamma_\mu$ | muon gyromagnetic ratio | $\omega$ | angular frequency |
| $\gamma_\mathrm{N}$ | nuclear gyromagnetic ratio | $\omega_\mathrm{c}$ | cyclotron frequency |
| $\Delta$ | exchange splitting | $\omega_\mathrm{L}$ | Larmor precession frequency |

## Fundamental constants

| | | |
|---|---|---|
| Bohr radius | $a_0$ | $5.292 \times 10^{-11}$ m |
| Speed of light in free space | $c$ | $2.9979 \times 10^8$ m s$^{-1}$ |
| Electronic charge | $e$ | $1.6022 \times 10^{-19}$ C |
| Planck's constant | $h$ | $6.626 \times 10^{-34}$ J s |
| $h/2\pi =$ $\hbar$ | $\hbar$ | $1.0546 \times 10^{-34}$ J s |
| Boltzmann's constant | $k_B$ | $1.3807 \times 10^{-23}$ J K$^{-1}$ |
| Electron rest mass | $m_e$ | $9.109 \times 10^{-31}$ kg |
| Proton rest mass | $m_p$ | $1.6726 \times 10^{-27}$ kg |
| Avogadro's number | $N_A$ | $6.022 \times 10^{23}$ mol$^{-1}$ |
| Standard molar volume | | $22.414 \times 10^{-3}$ m$^3$ mol$^{-1}$ |
| Molar gas constant | $R$ | $8.315$ J mol$^{-1}$ K$^{-1}$ |
| Fine structure constant $\dfrac{e^2}{4\pi\varepsilon_0\hbar c} =$ | $\alpha$ | $(137.04)^{-1}$ |
| Permittivity of free space | $\varepsilon_0$ | $8.854 \times 10^{-12}$ F m$^{-1}$ |
| Magnetic permeability of free space | $\mu_0$ | $4\pi \times 10^{-7}$ H m$^{-1}$ |
| Bohr magneton | $\mu_B$ | $9.274 \times 10^{-24}$ A m$^2$ *or* J T$^-$ |
| Nuclear magneton | $\mu_N$ | $5.051 \times 10^{-27}$ A m$^2$ *or* J T$^-$ |
| Neutron magnetic moment | $\mu_n$ | $-1.9130\mu_N$ |
| Proton magnetic moment | $\mu_p$ | $2.7928\mu_N$ |

## Useful equations

**(1) Identities**:

$$e^{i\theta} = \cos\theta + i\sin\theta$$

$$\sin\theta = \frac{e^{i\theta} - e^{-i\theta}}{2i}$$

$$\cos\theta = \frac{e^{i\theta} + e^{-i\theta}}{2}$$

$$\sin(\theta + \phi) = \sin\theta\cos\phi + \cos\theta\sin\phi$$
$$\cos(\theta + \phi) = \cos\theta\cos\phi - \sin\theta\sin\phi$$

$$\cos^2\theta + \sin^2\theta = 1$$

$$\cos 2\theta = \cos^2\theta - \sin^2\theta$$

$$\sin 2\theta = 2\cos\theta\sin\theta$$

$$\sinh x = \frac{e^x - e^{-x}}{2}$$

$$\cosh x = \frac{e^x + e^{-x}}{2}$$

**(2) Series expansions** (valid for $|x| < 1$):

$$(1 + x)^n = 1 + nx + \frac{n(n-1)}{2!}x^2$$
$$+ \frac{n(n-1)(n-2)}{3!}x^3 + \cdots$$

$$e^x = 1 + x + \frac{x^2}{2!} + \frac{x^3}{3!} + \frac{x^4}{4!} + \cdots$$

$$\sin x = x - \frac{x^3}{3!} + \frac{x^5}{5!} - \cdots$$

$$\cos x = 1 - \frac{x^2}{2!} + \frac{x^4}{4!} - \cdots$$

$$\log(1 + x) = x - \frac{x^2}{2} + \frac{x^3}{3} - \cdots$$

**(3) Integrals**:

$$\int_0^\infty x^n e^{-x}\, dx = n!$$

$$\int_{-\infty}^\infty e^{-\alpha x^2}\, dx = \sqrt{\frac{\pi}{\alpha}}$$

$$\int_{-\infty}^\infty x^{2n} e^{-\alpha^2 x^2}\, dx = \frac{(2n)!}{n!\, 2^{2n}\alpha^n}\sqrt{\frac{\pi}{\alpha}}$$

Indefinite (with $a > 0$):

$$\int \frac{dx}{x^2 + a^2} = \frac{1}{a}\tan^{-1}\frac{x}{a}$$

$$\int \frac{dx}{x^2 - a^2} = \frac{1}{2a}\ln\left|\frac{x - a}{x + a}\right|$$

$$\int \frac{dx}{\sqrt{x^2 + a^2}} = \sinh^{-1}\frac{x}{a} \qquad \text{(G.1)}$$

$$\int \frac{dx}{\sqrt{x^2 - a^2}} = \begin{cases} \cosh^{-1}\frac{x}{a} & \text{if } x > a \\ -\cosh^{-1}\frac{x}{a} & \text{if } x < -a \end{cases}$$

$$\int \frac{dx}{\sqrt{a^2 - x^2}} = \sin^{-1}\frac{x}{a}$$

(4) **Vector operators**:

- grad acts on a scalar field to produce a vector field:

$$\text{grad}\,\psi = \nabla\psi = \left(\frac{\partial\psi}{\partial x}, \frac{\partial\psi}{\partial y}, \frac{\partial\psi}{\partial z}\right)$$

- div acts on a vector field to produce a scalar field:

$$\text{div}\mathbf{A} = \nabla\cdot\mathbf{A} = \frac{\partial A_x}{\partial x} + \frac{\partial A_y}{\partial y} + \frac{\partial A_z}{\partial z}$$

- curl acts on a vector field to produce another vector field:

$$\text{curl}\,\mathbf{A} = \nabla\times\mathbf{A} = \begin{vmatrix} \mathbf{i} & \mathbf{j} & \mathbf{k} \\ \partial/\partial x & \partial/\partial y & \partial/\partial z \\ A_x & A_y & A_z \end{vmatrix}$$

where $\psi(\mathbf{r})$ and $\mathbf{A}(\mathbf{r})$ are any given scalar and vector field respectively.

(5) **Vector identities**:

$$\nabla\cdot(\nabla\psi) = \nabla^2\psi$$
$$\nabla\times(\nabla\psi) = 0$$
$$\nabla\cdot(\nabla\times\mathbf{A}) = 0$$
$$\nabla\cdot(\psi\mathbf{A}) = \mathbf{A}\cdot\nabla\psi + \psi\nabla\cdot\mathbf{A}$$
$$\nabla\times(\psi\mathbf{A}) = \psi\nabla\times\mathbf{A} - \mathbf{A}\times\nabla\psi$$
$$\nabla\times(\nabla\times\mathbf{A}) = \nabla(\nabla\cdot\mathbf{A}) - \nabla^2\mathbf{A}$$
$$\nabla\cdot(\mathbf{A}\times\mathbf{B}) = \mathbf{B}\cdot\nabla\times\mathbf{A} - \mathbf{A}\cdot\nabla\times\mathbf{B}$$
$$\nabla(\mathbf{A}\cdot\mathbf{B}) = (\mathbf{A}\cdot\nabla)\mathbf{B} + (\mathbf{B}\cdot\nabla)\mathbf{A}$$
$$\qquad + \mathbf{A}\times(\nabla\times\mathbf{B}) + \mathbf{B}\times(\nabla\times\mathbf{A})$$
$$\nabla\times(\mathbf{A}\times\mathbf{B}) = (\mathbf{B}\cdot\nabla)\mathbf{A} - (\mathbf{A}\cdot\nabla)\mathbf{B}$$
$$\qquad + \mathbf{A}(\nabla\cdot\mathbf{B}) - \mathbf{B}(\nabla\cdot\mathbf{A})$$

These identities can be easily proved by application of the alternating tensor and use of the summation convention. The alternating tensor $\epsilon_{ijk}$ is defined according to:

$$\epsilon_{ijk} = \begin{cases} 1 & \text{if } ijk \text{ is an even permutation of 123} \\ -1 & \text{if } ijk \text{ is an odd permutation of 123} \\ 0 & \text{if any two of } i, j \text{ or } k \text{ are equal} \end{cases}$$

so that the vector product can be written

$$(\mathbf{A}\times\mathbf{B})_i = \epsilon_{ijk}A_j B_k.$$

The summation convention is used here, so that twice repeated indices are assumed summed. The scalar product is then

$$\mathbf{A}\cdot\mathbf{B} = A_i B_i.$$

Use can be made of the identity

$$\epsilon_{ijk}\epsilon_{ilm} = \delta_{jl}\delta_{km} - \delta_{jm}\delta_{kl}$$

where $\delta_{ij}$ is the Kronecker delta given by

$$\delta_{ij} = \begin{cases} 1 & i = j \\ 0 & i \neq j \end{cases}$$

The vector triple product is given by

$$\mathbf{A}\times(\mathbf{B}\times\mathbf{C}) = (\mathbf{A}\cdot\mathbf{C})\mathbf{B} - (\mathbf{A}\cdot\mathbf{B})\mathbf{C}.$$

(6) **Cylindrical coordinates**:

$$\nabla^2\psi = \frac{1}{r}\frac{\partial}{\partial r}\left(r\frac{\partial\psi}{\partial r}\right) + \frac{1}{r^2}\frac{\partial^2\psi}{\partial\phi^2} + \frac{\partial^2\psi}{\partial z^2}$$

$$\nabla\psi = \left(\frac{\partial\psi}{\partial r}, \frac{1}{r}\frac{\partial\psi}{\partial\phi}, \frac{\partial\psi}{\partial z}\right)$$

(7) **Spherical polar coordinates**:

$$\nabla^2\psi = \frac{1}{r^2}\frac{\partial}{\partial r}\left(r^2\frac{\partial\psi}{\partial r}\right) + \frac{1}{r^2\sin\theta}\frac{\partial}{\partial\theta}\left(\sin\theta\frac{\partial\psi}{\partial\theta}\right)$$

$$+ \frac{1}{r^2\sin^2\theta}\frac{\partial^2\psi}{\partial\phi^2}$$

$$\nabla\psi = \left(\frac{\partial\psi}{\partial r}, \frac{1}{r}\frac{\partial\psi}{\partial\theta}, \frac{1}{r\sin\theta}\frac{\partial\psi}{\partial\phi}\right)$$

# Index

Note. The most important page numbers for any entry are indicated by **bold** type